Molecular Basis of Hematopoiesis

Amittha Wickrema • Barbara Kee

Editors

Molecular
Basis of
Hematopoiesis

 Springer

Editors

Amittha Wickrema
Department of Medicine, MC2115
University of Chicago
5841 S. Maryland Avenue
Chicago, IL 60637
awickrem@medicine.bsd.uchicago.edu

Barbara Kee
Department of Pathology, MC1089
University of Chicago
5841 S. Maryland Avenue
Chicago, IL 60637
bKee@bsd.uchicago.edu

ISBN: 978-0-387-85815-9   e-ISBN: 978-0-387-85816-6
DOI: 10.1007/978-0-387-85816-6

Library of Congress Control Number: 2008940970

springer.com

Foreword

Hematopoiesis is among the most thoroughly studied tissue systems investigated in humans and other organisms. This is true for many reasons, especially because of the accessibility of samples from finger prick, venipunctures, or leukophoresis to bone marrow aspiration. The availability of mouse models and genetic mutations, some of which mimic human diseases, have enhanced its attractions as an experimentally tractable system for study. Virtually, all of us have suffered from defects in hematopoiesis, tiredness due to anemia, bruising due to too few platelets, or infections due to too few neutrophils. Moreover, "benign" diseases of the blood, hemophilia, thalassemia, and sickle cell anemia are sufficiently common in some areas so that they have attracted medical attention for centuries. Although leukemia, lymphoma, and multiple myeloma are less common, their almost uniformly fatal consequences (untreated) have also captured the attention of physicians and scientists alike for a very long time. Thus, a new look at the components of the hematopoietic system based on current technologies is needed and timely.

The study of hematopoiesis has led to many of our most critical concepts: different lineages of related cells, usually morphologically unique, with distinct specialized functions; the parent or stem cell resulting in these different lineages; external substances, now called cytokines that regulate the growth and differentiation of the various lineages; the molecules within cells that respond to external cytokines transmitting the signal to the nucleus where the relevant genes respond to the signal; the subversion of these normal pathways in genetic diseases and in malignant diseases. The study of these diseases has individually and collectively enriched our understanding of the regulation of stem cell growth and differentiation.

In this new era of more sophisticated understanding and the incredible technology ungirding it, hematopoiesis continually pushes the biological frontiers. As we are enamored by one of our newest "toys," microRNA, we have to recognize our ignorance of much of cell biology. MicroRNA binds to the 3′ untranslated region (3′UTR) of many transcribed genes; yet the 3′UTR of the same gene can be the target for many microRNAs. However, if the gene is not transcribed in a particular cell or at a particular stage in differentiation, then the microRNA is nonfunctional, or may have a different target. Alternatively, the 3′UTR may be shortened so that the targeted site, present at an earlier stage, may be lost in differentiation, although the gene may still be transcribed. Thus, as in any stage of the maturation of our own insights into a complex biological system, we must recognize the incomplete nature of our knowledge. This recognition is a stimulus to further progress, but it should also make us humble as we interpret what must be partial information. Thus, the information presented in this compilation is a reaffirmation of our progress, of what we have learned, while at the same time, the focus for the future is to emphasize our ignorance of a very complex, dynamic process and the questions we still need to answer.

Our progress has been so remarkable and so successful on the therapeutic level because of our insights into the molecular bases of processes involved in hematopoiesis. For example, we have become more sophisticated about the molecular bases of hemoglobin synthesis, the precise genetic rearrangements that transform undifferentiated lymphocytes into functioning B and T cells, and the remarkable combination of genetic and therefore molecular changes that transform normal hematopoietic cells into malignant ones. Most of the challenges to further progress reside at the level of our incomplete penetration into the biochemical and biophysical mechanisms underlying the molecular processes of these transforming events.

Thus, while we celebrate our successes, there are plenty of challenges to keep all of us active in innovative research for years to come.

Blum-Riese Distinguished Service   Janet D. Rowley, M.D.
Professor of Medicine
Molecular Genetics & Cell Biology
and Human Genetics
University of Chicago
Chicago, IL

Contents

Chapter 1
Hematopoietic Stem Cells

Patricia Ernst

Abstract Hematopoietic stem cells (HSCs) represent one of the first amenable experimental models that enhanced our understanding of stem cell behavior. Features of the hematopoietic system facilitating the early development of this experimental model include the fact that the hematopoietic system is a naturally regenerative organ comprising cells that are migratory in nature. The fact that a few or even single cells could be demonstrated to regenerate the entire hematopoietic system in an irradiated recipient made it possible to perform retrospective identification of the engrafting cell and the fate of its progeny. In addition, clinical interests such as gene therapy and regenerative medicine have driven research to understand the pathways that maintain and modify stem cell behavior in vitro and in vivo. This chapter reviews a selection of cell surface receptors, intracellular signaling molecules, regulators of cell death and proliferation, and transcriptional regulators that are known to play an important role in maintaining HSCs in the adult. A large number of mouse knockout studies have been performed that provide a framework of pathways that are essential for the maintenance of steady-state hematopoiesis and for maintenance in conditions in which the hematopoietic system must regenerate. To understand how manipulating these pathways affects stem cell function, the phenotypic and functional characterization of murine and human HSCs has been refined over time to include different mouse strains and conditions, including specific knockout animals. In turn, the analysis of loss- and gain-of-function models has enriched our understanding of the stability of HSC identity.

P. Ernst
Department of Genetics
Norris Cotton Cancer Center, Dartmouth Medical School
725 Remsen, HB7400
Hanover, NH03755, USA
e-mail: patricia.ernst@dartmouth.edu

A. Wickrema and B. Kee (eds.), *Molecular Basis of Hematopoiesis*,
DOI: 10.1007/978-0-387-85816-6_1, © Springer Science+Business Media, LLC 2009

What Defines the Stem Cell of the Hematopoietic System?

Hematopoiesis is the ongoing process by which blood cells are produced from the hematopoietic stem cell (HSC) whose progeny can be either another HSC or a cell with more limited developmental potential (Fig. 1.1). Some cell types are first defined by their physical properties, with subsequent description of their developmental and functional attributes. In contrast, HSCs were first described by a clear set of properties (self-renewal, multipotency) that derived from experimental evidence, and then the cell type satisfying these properties was identified physically. Experiments done in the early 1950s illustrated that rats could be protected from X-ray-induced anemia by injection of bone marrow from a nonirradiated rat, and that mice who had their spleen protected by lead shielding could recover serum immunoglobulin and red blood cells (Jacobson et al. 1949; Urso and Congdon 1957; Wissler et al. 1953). Irradiation not only killed mature hematopoietic cells, but also ablated HSC activity in the recipient. These early studies paved the way for a series of experiments that would use the transfer of bone marrow to an irradiated recipient as the basis for a quantitative assay to define the physical identity of stem cells and their progeny (Till and McCulloch 1961). To physically identify such a cell, it was necessary to impose strict definitions on the properties expected of the HSC. First, this cell should be multipotent and therefore capable of yielding all of the major blood cell types in the recipient animal. Second, the production of these cell types should be stable throughout the lifetime of the recipient; in practical terms this is usually >24 weeks in the mouse. This requirement in the definition of a stem cell takes into account the fact that short-lived but multipotent progenitor types have been identified in fractionated bone marrow; thus the stringent definition of the HSC is often termed long-term engrafting HSC (LT-HSC), to distinguish it from the more finite-lived or ST-HSCs (Fig. 1.1). Third, the putative stem cell must be capable of self-renewal, producing additional stem cells that satisfy the first two criteria. This is typically demonstrated by serial transplantation from cells of the first recipient into secondary and tertiary recipients. Cell populations transferred to a second recipient are often diluted such that the experiment includes a number of animals in which no donor-type stem cells are detected, allowing the application of Poisson statistics to derive the number of "units of engraftment" (typically competitive repopulation units) from the overall experiment. The quantitative nature of this assay makes it possible to fractionate cells based on cell properties (dye efflux capacity, size, cell cycle state, cell surface markers, described below) with the purpose of determining whether particular fractionation schemes represent a numeric *enrichment* in stem cell activity on a per cell basis and thus bring the investigator closer to a phenotypic description of the HSC. To date, there are many fractionation schemes that enrich the stem cell population, but most investigators find an upper limit of about 30–40% of single purified cells that will engraft an irradiated adult recipient and satisfy the criteria outlined above using the mouse as a model. This upper limit may reflect (1) the technical inefficiency of injecting single cells, (2) heterogeneity in the "state" of the cell given a homogenous phenotype, or (3) actual

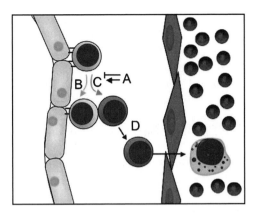

Fig. 1.1 Model of hematopoiesis with three tiers of lineage restriction. (**a**) Stem, progenitor, and developing lineage committed populations and their developmental relationships are shown. *LT-HSC* long-term engrafting hematopoietic stem cell, *ST-HSC* short-term engrafting hematopoietic stem cell, *MPP* multipotent progenitor, *CLP* common lymphocyte progenitor, *ETP* early thymic progenitor, *CMP* common myeloid progenitor, *MEP* megakaryocyte-erythroid progenitor, *GMP* granulocyte-macrophage progenitor, *DN* double negative thymocyte, *DP* double positive thymocyte, *NK* natural killer cell. (**b**) Idealized repopulation kinetics of LT- vs. ST-HSCs to illustrate their functional distinction (*see Color Plates*)

fluctuation in the correspondence between cell phenotype and functional capacity. It is significant that the location of the cell in the subject animal is not usually a criterion for HSC identity, although in many other systems (*D. Melanogaster* oogenesis and spermatogenesis for example) this is a very important characteristic used to identify the stem cell. The seemingly imprecise definition of the HSC may also reflect the flexibility built in to the hematopoietic system, since it must respond to environmental demands such as bleeding or infection.

Isolation of HSCs by Phenotype

Several purification strategies have been developed and refined over the years to enrich HSCs from bone marrow. Most strategies for enriching HSCs exploit cell-intrinsic dye efflux properties or the expression of specific cell surface proteins. As described above, the development of a quantitative assay to enumerate HSCs among total bone marrow cells first yielded an estimate of ~1/10,000 for the frequency of HSCs in unfractionated bone marrow (Szilvassy et al. 1990). The ability to purify this minority population was made practical by flow cytometry and cell sorting which allows the investigator to identify the presence or absence of a cell surface protein using fluorochrome-conjugated antibodies. As such antibodies were developed, a concept of "lineage-negative" (lin⁻) became an essential component of most HSC enrichment strategies, to exclude known progenitor and differentiated cell populations, such as developing B cells (Fig. 1.1) (Spangrude and Scollay 1990). Further enrichment utilized antibodies to key cell surface proteins such as c-Kit, Sca-1, and Thy1.1, although this strategy could not be used for all strains of mice (Christensen and Weissman 2001; Morrison and Weissman 1994). A more strain-independent strategy exploited the observation that the efflux and retention of particular dyes were correlated with HSC function. For example, bone marrow cells incubated with the cationic dye rhodamine-123 (Rho) could be enriched for HSC activity by selecting Rho$^{neg/low}$ cells (Bertoncello et al. 1985; Zijlmans et al. 1995). This property is thought to depend on two cellular characteristics: the ability to pump the dye out of the cell through multidrug resistance family proteins as well as the mitochondrial activation state of the cell (Kim et al. 1998). As with the cell surface markers, the Rho$^{neg/low}$ characteristic is most useful at enriching HSCs in combination with other markers. A related dye efflux characteristic is the basis for identifying the "side-population" (SP) phenotype. SP describes cells with the lowest retention of Hoechst 33342 and distinguishes them from the "major population" which does not as effectively efflux the dye. The SP phenotype is dependent on the *Bcrp1* gene product, a member of the multidrug resistance gene family (Zhou et al. 2001). SP cells are identified by incubation with the DNA intercalating Hoechst 33342 dye, followed by flow cytometric detection of two wavelengths of dye emission (Goodell et al. 1996). Like the Rho$^{neg/low}$ characteristic, this technique must be done on viable, metabolically active cells, and is most effective for enriching HSCs when used in conjunction with cell surface marker analysis.

The strategies used to purify HSCs described above were optimized in wild-type mice in the absence of any particular stress on the hematopoietic system. One of the main purposes of describing a phenotype that corresponds closely to the functional HSC was to apply this information to experimentally manipulated systems. As investigators attempted to use the cell surface and dye uptake properties to characterize stressed or genetically manipulated animals, it became clear that some of the phenotypes used to describe HSCs in normal animals would not be appropriate in these perturbed settings. Thus a deeper understanding of the molecular identity of HSCs was warranted. One approach was to carry out global expression analyses on HSCs enriched from wild-type mice to extend the list of genes whose expression was characteristic of the most highly purified HSCs. This approach identified genes that encoded additional cell surface proteins that could be strictly correlated with HSC function that were subsequently tested independently. In particular, the previously characterized signaling lymphocytic activation molecules (SLAM), CD244, CD150, and CD48, were capable of defining as pure a population of HSCs as many of the prior combinatorial strategies (Kiel et al. 2005). Two additional benefits of using SLAM markers were described. First, the smaller combination of cell surface proteins made it possible to examine HSCs in situ in histologic sections, which facilitates studies in which improper homing of HSCs is a potential outcome. Second, the stability of these proteins as markers of HSCs' diverse conditions such as development, aging, mobilization, and hematopoietic regeneration was demonstrated (Kim et al. 2006; Yilmaz et al. 2006a). Thus, the information gained from these studies facilitates the use of multiple corroborating approaches for the phenotypic identification/quantification of HSCs in a variety of experimental settings.

Identification of Human Hematopoietic Stem Cells

The first surrogate systems for the quantitative measure of human primitive hematopoietic progenitors were the long-term initiating culture (LTC-IC) assay or the cobblestone area forming colony assays (reviewed in Eaves et al. 1992). In these assays, human umbilical cord blood or bone marrow cells are plated on stromal support cell lines to yield colonies that produce myeloerythroid progeny for many weeks. Although this is a useful system measuring the frequency of stem/progenitor cells within particular populations, much effort went into developing systems in which human cells could be transplanted into the mouse, in order to provide a bone marrow microenvironment and presumably reflect properties of the "true" human HSC. The use of mice as recipients of human cells was made possible by efforts to develop particularly immunocompromised strains and by supplementing human cytokines that do not cross-react well between the two species. In addition to the above two systems, fetal ("preimmune") sheep have also been used as recipients (Zanjani et al. 1996).

The first xenotransplantation experiments using mice as recipients used sublethally irradiated beige/nude/XID mice (Nolta et al. 1994). This strain is impaired in its ability to eliminate human cells, since the beige mutation reduces NK cell function, the nude mutation blocks T-cell development, and the XID mutation blocks B-cell development. However, the leakiness of the immunocompromised phenotype in these animals prompted the use of other strains. For example, the NOD/SCID strain lacks B and T cells because of the SCID defect (a mutation in DNA-dependent protein kinase) required for antigen receptor gene rearrangement. These mice also have reduced NK cell function because of the NOD mutation. However, the SCID mutation renders these animals radiation-sensitive. For all of the xenograft systems, species-specific growth factor interactions render the engraftment and survival of human cells relatively inefficient when compared with mouse-to-mouse transplantation. One improvement to overall engraftment has been to perform direct intrafemoral injection to deliver the human cells (Mazurier et al. 2003). In addition, the use of different immunocompromised strains such as the $RAG^{-/-};\gamma C^{-/-}$ strain facilitates more efficient multilineage hematopoiesis (Legrand et al. 2008).

Using the NOD/SCID model and limiting dilution analysis, the repopulating cells (SCID repopulating cells or SRC) can be derived from experiments done as described above for the competitive repopulation unit in the mouse. Using these assays, an immunophenotype of $CD34^+lin^-CD38^-$ or Rho^{low} has been shown to harbor an SRC frequency of 1/30 cells and exhibit robust self-renewal using serial transplantation assays (McKenzie et al. 2007). Similar experiments have supported the concept that human $CD34^+lin^-CD38^-$ cells are quite heterogeneous (Guenechea et al. 2001). Collectively, these experiments begin to lay the groundwork for testing additional normal and malignant human populations in an experimentally tractable animal model.

Experimental Approaches for Defining Important Molecular Pathways in HSCs

Diverse experimental approaches have been used to identify cell interactions, signaling pathways, and transcriptional programs that are important for either the development or the maintenance of HSCs. Some of examples discussed in the following sections, such as the isolation of stem cell factor (SCF) and its receptor, were initiated with a classic combination of forward genetics and biochemistry. Subsequently, the development of gene targeting technology in the mouse facilitated the now commonplace "candidate approach" in which paradigms derived from model organisms such as *D. melanogaster* and *C. elegans* were applied and tested in the murine hematopoietic system. Studies of hedgehog and polycomb family members discussed below are examples of such paradigms. A related and fruitful approach to identifying novel regulators of hematopoiesis has been the targeted mutagenesis of genes frequently disrupted by chromosomal translocation in human leukemia (Jude et al. 2008). Many of the transcriptional regulators discussed below result from this approach. Traditional forward genetics as well as quantitative trait loci analyses have been used to discover novel genes that influence adult hematopoiesis. In the former strategy, animals are mutagenized and their progeny screened for a quantitative hematopoietic parameter (reviewed in Kile and Hilton 2005). Quantitative trait loci analysis employs existing mouse strains with known disparate phenotypes such as stem-cell frequency and uses recombinants between the strains to identify the genes responsible for HSC frequency (reviewed in Snoeck 2005). Both strategies have identified novel alleles of genes already known to play a role in hematopoiesis, such as c-Myb (Carpinelli et al. 2004; Sandberg et al. 2005), as well as novel genes of unknown function, such as latexin (Liang et al. 2007). Forward genetic screens in simpler model organisms have been carried out to identify developmental signals that control particular phases of cell cycle quiescence or proliferation (Hariharan and Haber 2003; Saito et al. 2004). These approaches will undoubtedly provide new information that helps us understand the developmental and steady-state control of HSC proliferation.

Signaling Molecules Critical for HSC Maintenance

One of the first growth factor–receptor interactions thought to be important for hematopoiesis was identified using a combination of genetics and biochemistry. Stem cell factor or steel factor (SCF or SF) is the ligand for the transmembrane tyrosine kinase, c-Kit, a signaling molecule that has a central and continuous role in late fetal and adult hematopoiesis. As early as the 1960s, the *White-spotting* (W) mutant mouse (spontaneous c-Kit mutant) and the *Sl* mouse (SF mutant) were proposed to represent mutants in a ligand–receptor pair essential for hematopoiesis (McCulloch et al. 1965). Nearly 30 years later, the molecular cloning of the receptor–ligand pair resulted in the development of antibody reagents that were instrumental in defining primitive hematopoietic populations (Katayama et al. 1993; Ogawa et al. 1991, 1993). Conversely, recombinant SCF is a critical component of most growth conditions designed to maintain or expand the most primitive hematopoietic cells in vitro (reviewed in Broudy 1997). The mechanisms by which c-Kit signaling maintains hematopoietic cells include the promotion of cell survival, proliferation, and migration (reviewed in Lyman and Jacobsen 1998).

Both a membrane-bound and a soluble form of SF are produced in murine and human cells through alternative splicing that includes or excludes a proteolytic cleavage site in the mature protein. Matrix metalloprotease 9 is a major participant in the liberation of soluble SF. Using a spontaneous mouse mutant (harboring the Sl^d allele which encodes only secreted SF) as well as mice engineered to lack the protease site, it was demonstrated that the soluble form of SF performs a unique function during hematopoietic regeneration after bone marrow damage. Animals in which the protease site is deleted (reduced systemic SF) exhibit delayed hematopoietic regeneration and increased mortality after sublethal irradiation (Tajima et al. 1998). In addition, matrix metalloprotease 9 knockout animals exhibit the same phenotype after myeloablation (Heissig et al. 2002). Steady-state hematopoiesis is normal in both models expressing only secreted SF. In contrast, Sl mutants lacking both forms of SF exhibit reduced steady-state hematopoiesis, illustrating that the membrane-bound form is sufficient to maintain HSC homeostasis. Studies using W mutant mice as recipients of normal bone marrow, as well as studies using blocking c-Kit antibodies demonstrate that c-Kit signaling is continuously needed for the maintenance of adult HSCs (Czechowicz et al. 2007; Miller et al. 1997; Ogawa et al. 1991). These studies have also supported the concept that the reduced HSC number in the W mouse results in an "empty niche" into which wild-type cells can engraft. These studies collectively suggest that c-Kit signaling (in contrast to Tie-2, see below) supports self-renewing proliferation without inducing differentiation, maintains tonic survival signaling, and supports regenerative proliferation subsequent to bleeding or myeloablation.

Flt-3 (or Flk-2) is a related receptor tyrosine kinase that exhibits some biological activities similar to those of c-Kit, particularly in vitro. The ligand for Flt-3 (Flt-3L), like SF, is a type I transmembrane protein, and a soluble form of the ligand is generated through alternative splicing (reviewed in Lyman and Jacobsen 1998). Also similar to c-Kit, the cell surface expression of Flt-3 (the tyrosine kinase receptor) is an important determinant for subdividing stem-cell-enriched bone marrow populations. The lin^-/Sca-1^+/c-Kit$^+$ (LSK) bone marrow population that is Flt-3^+ is more rapidly cycling and provides more short-term multipotent engraftment activity whereas the Flt-3-negative subpopulation is more quiescent and harbors most of the long-term multipotent HSCs (Adolfsson et al. 2001; Christensen and Weissman 2001; Passegue et al. 2005). The in vivo function of Flt-3 signaling is likely to act as HSCs differentiate to multipotent progenitors, due to the lack of Flt-3 on the most primitive LT-HSCs. In addition, Flt-3 signaling is essential for the differentiation of several cell types, including dendritic cells and lymphocytes (Lyman and Jacobsen 1998).

Thrombopoietin (THPO) is a cytokine that is essential to maintain megakary-opoiesis and platelet production, but work from several groups has made it clear that THPO has also a unique role in the development and maintenance of HSCs (Ku et al. 1996; Petit-Cocault et al. 2007; Sitnicka et al. 1996). THPO greatly supports the expansion of competitive repopulation units, colony forming units, and cell accumulation in vitro, when present with other cytokines. For example, highly purified single HSCs (lin$^-$/Rho^{low}/Hoechst^{low} or LSK) proliferate very poorly or not at all in the presence of THPO alone, but in conjunction with SF, yield very high cloning efficiency (no. of wells proliferating) and overall cell accumulation (Ku et al. 1996; Sitnicka et al. 1996). These studies suggested that THPO acts by increasing the number of cells entering the cell cycle and by reducing the time lag until cells enter the first cell cycle.

Complementary studies manipulating signaling through the receptor for THPO, c-Mpl, reveal more detail to the role of this signaling pathway in HSC maintenance. In the more complex setting of the bone marrow environment, c-Mpl stimulation increases β1-integrin levels and enhances the engraftment and quiescence of lin$^-$/Sca-1$^+$/c-Kit$^+$ (LSK) cells. Inhibition of c-Mpl signaling in vivo through the use of a neutralizing antibody results in a larger proportion of proliferating HSCs and a reduction in engraftment potential (Yoshihara et al. 2007). Although c-Mpl$^{-/-}$ bone marrow cells compete poorly with wild-type bone marrow cells for engraftment, c-Mpl$^{-/-}$ bone marrow does not exhibit the "empty niche" characteristics that the W mutant mice do, nor do the animals exhibit pan-cytopenia as might be expected for a rate-limiting effect on HSC maintenance (Abkowitz and Chen 2007). Some data suggest that THPO is not particularly effective at maintaining HSC potential in defined conditions in vitro (Uchida et al. 2003). These apparently conflicting results are likely to be the result of opposing THPO actions on the quiescent LT-HSC as compared to its differentiating progeny, and differences in comparing the behavior of single cells in defined conditions with cells in vivo integrating THPO signaling with other niche-specific signals.

Interestingly, gain-of-function (or transforming) alleles of c-Kit, Flt-3, and c-Mpl were identified as participants in leukemia or other cancers. For example, v-Kit is a constitutively active kinase that was identified in a feline sarcoma virus (Besmer et al. 1986). In addition, point mutations or amplification that results in constitutively active signaling have been found in several types of solid tumors and in acute myelogenous leukemia (AML) (Biermann et al. 2007; Hirota et al. 1998; Renneville et al. 2008). Flt-3-activating mutations are found in ~30% human AML cases, and v-Mpl is a constitutively active form of the THPO receptor that was identified in a mouse myeloproliferation-inducing retrovirus (Chabot et al. 1988; Souyri et al. 1990; Yokota et al. 1997).

Tie-2 is another tyrosine kinase receptor that also plays an important role in steady-state hematopoiesis. Both Tie-2 and the related Tie-1 are expressed on endothelial cells during embryogenesis and in the adult, as well as on subsets of hematopoietic cells. Tie-1 loss alone has little effect on embryo or adult hematopoiesis (Rodewald and Sato 1996). However, hematopoietic cells lacking both Tie-1 and Tie-2 develop normally in the fetal liver but fail to give rise to adult HSCs (Puri and Bernstein 2003). Tie-2 expression on adult HSCs subdivides partially pure HSC populations (i.e. LSK/SP cells) into a population that is mostly in G_0, is resistant to ablation with 5-fluorour-acil, and phenotypically corresponds to the LT-HSC (Arai et al. 2004). The ligand bound by Tie-2 is angiopoietin (Ang). Addition of this factor, together with THPO and SCF, to highly purified HSCs preserves their engraftment potential while it inhibits their proliferation in vitro. Furthermore, systemic delivery of Ang through viral expression or injection of recombinant protein enhances the number of SP cells and cells in G_0 within the LSK population (Arai et al. 2004). Thus, Tie-2/Ang signaling has been proposed to be most important for maintaining quiescence within the bone marrow niche, rather than promoting expansion and proliferation.

Intrinsic Regulators of HSCs: Role of Cell Death

Programmed cell death is a critical and actively regulated process throughout hematopoiesis and in mature hematopoietic cell types (Opferman 2007). Many of the cytokine receptors described above impart some of their biological effects through antiapoptotic signaling cascades (Kaushansky 2006). Experiments in which the antiapoptotic protein Bcl-2 was overexpressed in the hematopoietic system illustrated that protecting cells from death by raising Bcl-2 levels resulted in increased steady-state HSC pools and enhanced competitive engraftment by these cells. However, clonal analysis of purified HSCs in serum-free medium demonstrated that c-Kit or THPO signaling was still required for cell survival and that high Bcl-2 collaborates with these cytokines to increase the number of clones responding to the cytokine by deciding to proliferate, as well as the total cell accumulation per clone (Domen and Weissman 2000). Thus this experimental manipulation of cell death molecules suggests that a relatively low level of "paring down" of the HSC population occurs under steady-state conditions. Interestingly, the antiapoptotic Mcl-1 protein plays a uniquely critical role in maintaining the viability of HSCs, as conditional deletion of this gene is sufficient to result in the rapid death of these cells as well as multiple progenitor types (Opferman et al. 2005).

The interaction of the death receptor, Fas, with its cell-surface-bound ligand FasL is an important mechanism by which many hematopoietic populations are negatively regulated (Greil et al. 2003). Little to no Fas is detected on LSK cells under normal conditions; however, under inflammatory conditions (when cells are stimulated with tumor necrosis factor alpha (TNF-α)), these cells rapidly upregulate Fas and can be killed by ligation of this receptor (Bryder et al. 2001).

Intrinsic Regulators: Transmitting the Signals to Divide or Wait

Several pathways previously known to be essential for embryogenesis also play important roles during the maintenance of HSCs. Owing to the widespread use of these pathways, a common tool for the assessment of their role specifically during hematopoiesis is the Cre-loxP system for lineage-specific and inducible deletion of the gene of interest (Branda and Dymecki 2004). One such pathway investigated for its role in HSCs was the Wnt (Wingless/Int-1) pathway. Purified Wnt3a (together with SF and Bcl-2 expression) can act in vitro to promote cell proliferation that preserves the engraftment potential of HSCs (Willert et al. 2003). Canonical Wnt signaling ultimately results in the nuclear accumulation of β-catenin and activation of target genes in conjunction with LEF proteins (Cadigan and Nusse 1997). Constitutive activation of β-catenin by retroviral expression of a stable mutant expands HSCs in vitro (Reya et al. 2003). Acute activation of Wnt signaling, accomplished by excising exons encoding negative regulatory sequences of β-catenin (Harada et al. 1999), results in a transient increase in the number of LSK cells and an increase in the proportion that is in cycle (Kirstetter et al. 2006; Scheller et al. 2006). However, these cells failed to sustain hematopoiesis in a transplant setting or in the original animal, which succumbs to bone marrow failure within several weeks (Kirstetter et al. 2006). The rapid lethality may result from both the depletion of HSC function and a block in the development of multiple progenitors (Kirstetter et al. 2006; Scheller et al. 2006). This example illustrates how multiple experimental approaches are required to dissect the role of pathways that impinge on multiple aspects of HSC maintenance and differentiation (Fig. 1.2).

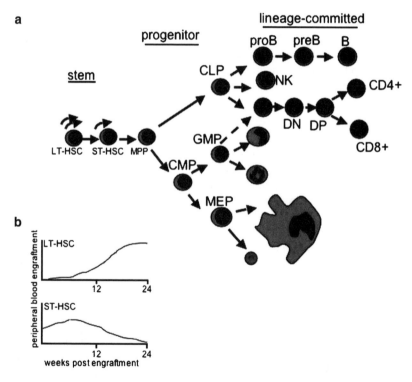

Fig. 1.2 Model of HSC cell division modes in the bone marrow environment. *Light purple* indicates a daughter cell that has preserved HSC identity and *darker purple* indicates a cell that has committed to a differentiative division. Signaling and intrinsic regulators described in the text act at the level of (**a**) regulating HSC proliferation in general, (**b**) enhancing the balance of self-renewal to generate HSC progeny, (**c**) enhancing differentiative divisions to generate ST-HSC or MPP progeny, or (**d**) enhancing proliferation as downstream of the HSC (*see Color Plates*)

The hedgehog (Hh) pathway was discovered in *D. Melanogaster* as a critical regulator of body plan organization and cell growth (McMahon et al. 2003). *Patched* (*Ptc*) encodes a transmembrane receptor that acts as a negative regulatory component of Hh signaling. Within the adult hematopoietic system, the effects of constitutive activation of the Hh signaling pathway can be studied using *Ptc* heterozygotes. These animals exhibit steady-state accumulation of phenotypically defined HSCs as well as an increase in the proportion of cycling cells within this population. Cells from these animals exhibit increased short-term engraftment in primary recipients. However, HSC activity upon secondary transplantation is reduced approximately threefold, indicating the functional exhaustion of the HSC pool in this mutant (Trowbridge et al. 2006).

The disruption of phosphoinositide-3 kinase signaling pathways through *PTEN* or *FoxO1/3/4* deletion using the *Mx1-Cre* transgene exhibits features in common with those models described above. In both the *PTEN* and *FoxO1/3/4* conditional knockout models, acute loss of gene function is associated with a transient increase in phenotypically defined HSCs, accompanied by an increase in the proportion of HSCs that are cycling (Miyamoto et al. 2007; Tothova et al. 2007; Yilmaz et al. 2006b; Zhang et al. 2006). The ability of these cells to engraft secondary recipients and to persist in the bone marrow of chimeras is both highly compromised. A slight increase in apoptosis appears to play a role in the homeostasis of HSCs in the *FoxO*1/3/4 knockout but not in the *PTEN* knockout (Tothova et al. 2007). Ultimately, both animals acquire a myeloproliferative syndrome or leukemia, but the defects in HSC proliferation and function are temporally and clonally separable from these diseases.

A class of small GTPases are important downstream mediators of chemokine signaling which influences HSC homing and mobilization (Cancelas et al. 2006). Cdc42 represents an example of such a protein that couples these signals to HSC proliferation. The disruption of the *Cdc42* gene (encoding a cytoplasmic Rho family GTPase) results in the transient increase in ST-HSCs (CD34$^+$/LSK immunophenotype), coupled with a rapid shift from quiescence to a proliferative state within the LSK pool overall (Yang et al. 2007). In this particular model, HSC mobilization (migration out of the bone marrow) to the blood and peripheral organs is substantial, although it is also observed to a lesser extent in the *FoxO*1/3/4 knockout (Tothova et al. 2007; Yang et al. 2007). Similar to the cases described above, *Cdc42*-deficient cells exhibit a severe reduction in their ability to engraft recipient animals.

This group of pathways collectively illustrates that it is common for gene disruptions (*FoxO1/3/4*, *cdc42*) or pathway-activating mutants (*Ptc*$^{+/-}$, *PTEN*$^{-/-}$, stabilized β-catenin) to allow an accumulation of HSCs at the expense of the long-term preservation of HSC functions (see Fig. 1.2). This relationship suggests that multiple independent pathways are actively restricting proliferation within the stem cell pool, and signals exist in the environment to promote HSC proliferation when the above pathways are disrupted. The frequent co-occurrence of enhanced proliferation and reduced function within the HSC pool is consistent with studies illustrating that these processes are mechanistically linked (Lambert et al. 2003; Passegue et al. 2005).

Role of the Core Cell Cycle Machinery in HSC Homeostasis

To understand the effects of the signaling pathways described above on the balance of proliferation and quiescence within HSCs, it is essential to determine the effects of disrupting components of the core cell cycle machinery. Ultimately, signals must be translated to modulate the cell cycle. In mammals, multiple homologues of the cyclins, cyclin-dependent kinases (CDKs), and Rb family members exist and their differential expression within the hematopoietic system has been documented (Passegue et al. 2005). Many knockouts and compound mutants of core cell cycle components are viable, allowing an assessment of loss-of-function mutations in adult bone marrow.

The cyclic activity of cyclin–CDK complexes drives the progression of the cell cycle in eukaryotic cells. The CDK inhibitors of the INK family (p15, p16, and p18) bind and inhibit CDK4/6 whereas those of the KIP family (p21, p27, and p57) inhibit CDK2 complexes. As the CDK4/6 complex is active at G_1/S boundary and the CDK2 complex is active during S phase, the loss of inhibition due to CDK inhibitor knockout would be expected to differentially affect cells that have a G_0 phase (such as HSCs) vs. those that are rapidly cycling (such as myeloid progenitors). In fact the hematopoietic phenotypes of the knockouts performed to date indicate even more specialized roles than such a simple model would predict. A comparison of the p21 and p27 knockouts illustrates this complexity. The steady-state pool of HSCs in p21 knockout animals is larger than that of wild-type animals, by both functional and phenotypic assessment (Cheng et al. 2000). In addition, the balance of proliferation within the HSC pool is tipped toward cycling (G_1) cells rather than G_0 cells. In contrast, p27 knockout animals exhibit normal HSC number and proportion of cycling cells, yet they compete better than wild-type cells in a transplantation setting. This is due to a proliferative advantage within the $p27^{-/-}$ progenitor pools, as revealed by colony assays and the disproportionate accumulation of $p27^{-/-}$ cells in the peripheral blood of chimeras engrafted with equal numbers of $p27^{-/-}$ and wild-type cells. The level of p57 is quite high in the most quiescent subpopulations of HSCs (Passegue et al. 2005; Umemoto et al. 2005), including THPO-sustained quiescence (Yoshihara et al. 2007). Owing to the fact that p57 knockout animals exhibit multiple developmental defects leading to perinatal lethality (Zhang et al. 1997), it is not yet clear what role p57 plays in HSCs.

Of the early G_1/S inhibitors, the p18 and p16 knockouts have been most extensively analyzed within the hematopoietic system. Both genes are expressed at relatively high levels in purified LT-HSCs (Passegue et al. 2005), and p16 has been implicated in HSC senescence as an important downstream target of Bmi-1 (Lessard and Sauvageau 2003; Park et al. 2003, see below). The loss of p18 alone results in the accumulation of HSCs, as observed in the p21 knockout, yet the accumulation is accompanied by a preservation of the function of these cells; upon serial transplantation, the $p18^{-/-}$ HSCs continue to outperform competing against wild-type cells (Yuan et al. 2004). In fact, the loss of p18 is epistatic to the loss of p21, since the double mutant bone marrow does not exhibit premature exhaustion upon serial transplantation, and HSCs proliferate in vitro better than $p21^{-/-}$ cells (Yu et al. 2006).

Upregulation of p16 has been observed during aging and upon in-vitro-induced senescence of several cell types (Zindy et al. 1997), including HSCs (Janzen et al. 2006). Accordingly, the steady-state pools and function of HSCs in p16$^{-/-}$ animals are similar to those of wild-type HSCs, but in older animals, the p16$^{-/-}$ HSCs perform better in functional assays (Janzen et al. 2006).

Intrinsic HSC Regulators: Transcriptional Pathways

Many transcriptional regulators have been found to regulate hematopoiesis at all levels of differentiation (discussed in the other chapters). One fruitful strategy to identify transcriptional regulators that are essential for some aspect of hematopoietic development is to investigate those genes involved in chromosomal translocation in human leukemia, or as common integration sites in murine retroviral-induced leukemia (Jude et al. 2008). Several of these key regulators are important for the emergence of HSCs during development (Dzierzak and Speck 2008). Since these mutations are often lethal, the study of transcriptional regulators in adult hematopoiesis has benefited from the use of developmentally regulated, hematopoietic-specific *Cre* recombinase-expressing transgenes or inducible *Cre*-expressing transgenes (reviewed in Branda and Dymecki 2004).

The *Gfi1* gene was identified as a proto-oncogene activated by proviral insertion in a mouse model of T-cell lymphoma (Schmidt et al. 1996). *Gfi1* knockout HSCs proliferate more than their wild-type counterparts, exhibit a decrease in p21 expression, and profoundly reduced HSC function in transplantation experiments (Hock et al. 2004; Zeng et al. 2004). Thus, under normal homeostasis, *Gfi1* is thought to suppress the proliferation of HSCs, thereby preventing their depletion.

The excision of the tumor suppressor gene, *JunB*, results in some phenotypes in common with the *Gfi1* knockout described above. Owing to the embryonic lethality, *JunB* excision was achieved using an inducible strategy (the Mx-Cre transgene) like the β-catenin experiments described above. Upon JunB loss, an expanded HSC pool was observed, specifically in the LT-HSC population. This expansion was accompanied by a greater percentage of LT-HSCs in S/G$_2$/M and reduced engraftment (Passegue et al. 2004). Ultimately, these animals succumb to a myeloproliferative disorder that is initiated within the HSC pool, demonstrating that the accumulated HSCs are susceptible to additional events that result in leukemia.

MYC was discovered as a proto-oncogene in lymphoma, and has over the intervening two decades been implicated in a wide variety of cancers (Grisendi and Pandolfi 2005). The acute loss of *Myc* in adult bone marrow results in the transient accumulation of HSCs as measured by phenotype, then a progressive loss of bone marrow cells. Perinatal deletion of *Myc* results in the accumulation of lineage-negative, Sca-1-positive, c-Kit-negative cells, proposed to reflect an aberrant, senescent primitive progenitor (Baena et al. 2007). *Myc*-deficient HSCs produced by either perinatal or adult excision fail to engraft in a competitive or noncompetitive setting, demonstrating that the accumulated cells are functionally defective (Baena et al. 2007; Wilson et al. 2004). The accumulation of phenotypic HSCs appears to result from a combination of a slight proliferative increase within the HSC pool (possibly due to N-Myc substitution at Myc targets (Baena et al. 2007; Malynn et al. 2000) and misexpression of cell surface molecules that may retain HSCs within the bone marrow microenvironment (discussed in Chapter 3).

The *RUNX1/AML1* gene is disrupted by chromosomal translocation in a majority of childhood leukemia, often producing a fusion protein that is thought to act by a dominant-interfering mechanism (Meyers and Hiebert 1995). Excision of *Runx1* in adult bone marrow, again using the *Mx1-Cre* transgene, caused the stem and early progenitor fraction of cells to expand by as much as threefold (Growney et al. 2005; Ichikawa et al. 2004). As with the examples above, *Runx1*-deficient HSCs exhibited reduced repopulating activity in both competitive and noncompetitive transplant settings. Furthermore, transplant recipients of *Runx1*-null bone marrow cells experienced a progressive loss of *Runx1*-deficient cells in peripheral blood (Growney et al. 2005). The decline of peripheral white blood cells in *Myc*-deficient mice was accompanied by a progressive decrease in bone marrow cellularity, whereas *Runx1*-deficient bone marrow cells actually increased by a factor of 2, likely because of the continuous myeloproliferation that follows conditional ablation of *Runx1*. In summary, the examples discussed above exhibit an early reduction of quiescent cells within highly purified HSC populations, demonstrating that these transcriptional regulators act within HSCs to affect the appropriate balance of cycling vs. quiescent cells.

The Ets-domain protein MEF/Elf-4 represents an additional gene that has been implicated in AML and plays a role in steady-state HSC homeostasis. Constitutive loss of *Elf4* results in an increase in phenotypically defined LT-HSCs (here defined as LSK/CD34$^-$/Flt-3$^-$) with a phenotype that resembles some aspects of the c-Myc knockout. *Elf4* knockout cells outcompete wild-type cells in transplantation assays, and knockout animals recover faster than wild-type mice from myeloablation (Lacorazza et al. 2006). Unlike some of the transcriptional regulators described above, steady-state or cytokine-stimulated *Elf4*-null HSCs exhibit a reduction in S/G$_2$/M cells, reduced BrdU uptake, and a resistance to enter the cell cycle upon cytokine stimulation. Despite this cell-cycle perturbation within the HSC pool, *Elf4* knockout animals apparently maintain normal bone marrow cellularity. These examples illustrate that the loss of transcriptional regulators can lead to coordinately enhanced HSC number and function.

Groups of genes that perform important functions within several stem cell types, including within the hematopoietic system, are the Polycomb group (PcG) and Trithorax group (TrxG) of genes. These gene families were initially identified genetically in *D. melanogaster* and were demonstrated to influence the expression of their target genes in a manner that is transmissible through daughter cell divisions (Cavalli 2002). Mammalian genes related to both groups have been implicated in leukemia and other cancers (Sparmann and van Lohuizen 2006).

Bmi-1 represents one of the best-characterized PcG members with respect to function within multiple stem cell types. *Bmi-1*-deficient mice exhibit a progressive reduction in phenotypically defined HSCs; these cells engraft poorly and exhaust prematurely in serial transplantation experiments. Both gain- and loss-of-function experiments have demonstrated that *Bmi-1* plays a central role in HSC self-renewal, with an initial focus on the role of *Bmi-1* as a negative regulator of the CDK inhibitor, p16 (Iwama et al. 2004; Lessard and Sauvageau 2003; Park et al. 2003).

Several other PcG members, *M33*, *Mel-18*, and *Rae28*, have also been shown to be important for maintaining HSC function (Core et al. 1997; Kajiume et al. 2004; Ohta et al. 2002). Proliferation defects are not apparent in most of these loss-of-function models, but some studies report observations consistent with the premature or ectopic activation of a senescence pathway and p16 induction.

The *mixed lineage leukemia (Mll)* gene is the best-characterized TrG gene with respect to hematopoietic phenotypes. Using distinct *Cre* transgenes and gene disruption strategies, two groups identified a role for *Mll* in maintaining quiescent vs. cycling HSCs (Jude et al. 2007; McMahon et al. 2007). Upon acute deletion of *Mll*, the LSK/ CD48⁻ subpopulation (enriched in quiescent cells) exhibited ectopic entry into G_1. Shortly thereafter, an increased number of cells in S-phase were observed in the total LSK population based on BrdU labeling studies (Jude et al. 2007). An increase in LSK cells in cycle was also observed in fetal liver HSCs (McMahon et al. 2007). Although these observations predict the type of transient accumulation in LSK cells as described above ($Ptc^{+/-}$ or β-catenin activation for example), no such increase was observed. This suggests that the increase in cycling HSCs in these loss-of-function mutants was tightly coupled to the differentiation of these cells (Fig. 1.2c, d).

A major class of genes that are regulated by both TrG and PcG members are the homeodomain-containing transcriptional regulators or *Hox* genes. *Hox* genes were originally characterized in *D. melanogaster* as regulators of segment identity, a role that is conserved in vertebrates. In addition to this developmental role, *Hox* genes have been demonstrated to influence proliferation and differentiation in the hematopoietic system and in human and murine leukemia (Buske and Humphries 2000). *Hoxb4* in particular has been the object of intensive study, since overexpression of this gene expands HSCs in vitro and in vivo using cells of diverse sources (Antonchuk et al. 2002; Krosl et al. 2003; Kyba et al. 2002; Sauvageau et al. 1995). Several other *Hox* genes can expand hematopoietic populations when overexpressed, but *Hoxb4* is unique in its ability to do so without inducing leukemia and while preserving other aspects of HSC behavior. Evidence is accumulating that many of the signaling molecules discussed above transmit information that modulates multiple *Hox* family members, leading to self-renewal vs. differentiation decisions. For example, Bmi-1 levels, THPO signaling, activation of β-catenin, and MLL levels have been shown to influence the expression or activity of *Hox* family members (Jude et al. 2007; Kirito et al. 2003, 2004; Reya et al. 2003; van der Lugt et al. 1996).

Concluding Remarks

The complementary experimental approaches used in the study of the development and maintenance of the hematopoietic system have yielded a wealth of information relating to the genes and pathways needed to maintain HSCs in the adult animal. The examples discussed here represent a selection of studies to illustrate key pathways;

however, this discussion is by no means comprehensive. The impetus for investigators to focus on these particular pathways comes from a variety of sources, including paradigms from model organisms, biochemical and genetic data implicating the genes discussed, as well as microarray data demonstrating selective enrichment of particular genes in HSCs. The examples discussed reinforce several principles regarding HSC behavior that are likely to be applicable to other stem cell types. First, an extended period of quiescence is a key control point for regulating the ability of the hematopoietic system to recover from depletion. The deregulation of the quiescent state can occur upon deletion of cell surface proteins, signaling molecules, or transcriptional regulators. This deregulation can be in the direction of promoting quiescence (*Elf4* knockout) or more commonly in the direction of allowing excess proliferation (c-Mpl or Tie-2 blockade, β-catenin activation, PTEN loss, Hh activation, Cdc42 loss). Interestingly, in most cases either perturbation results in the loss of HSC activity, underscoringthe importance of regulating the balance of quiescence vs. proliferation. Second, the influence of cell–cell contact or cell–extracellular matrix contact modulates the outcome of signaling through particular pathways. Although defined conditions in vitro facilitate the detailed molecular characterization of the effects of signaling, the outcome may be different in vivo or in conditions allowing adhesive interactions to synergize with the signaling pathway tested. This was discussed in the c-Mpl signaling example, where the outcome on cell proliferation differed depending on the setting in which HSC proliferation was examined. More complex in vitro culture systems that incorporate niche components may help in the definition of synergistic signaling pathways that maintain HSC potential. Such conditions, coupled with exciting new imaging techniques, are likely to provide new insight into the processes that are involved in self-renewal and differentiation (Dykstra et al. 2006; Schroeder 2005; Wu et al. 2007). Third, many of the pathways described as essential for maintaining HSCs also participate in leukemia. This underscores the delicate balance that must be achieved in a regenerating organ system such as the hematopoietic system. During differentiation, phases of rapid proliferation are interspersed with phases of quiescence, and achieving the appropriate balance between these states is essential not only for maintaining homeostasis, but also to prevent the deregulated growth that characterizes leukemia.

Acknowledgments   This work was supported by NIH grants 2P20RR016437, DK067119, and HL090036. I am grateful to Frédéric Barabé, Erika Artinger, Justin Gaudet, Craig Jude, and Kristin Zaffuto for comments and discussion, and I apologize to authors whose work was omitted in the interests of brevity.

References

Abkowitz, J. L., and Chen, J. (2007). Studies of c-Mpl function distinguish the replication of hematopoietic stem cells from the expansion of differentiating clones. Blood *109*, 5186–5190.

Adolfsson, J., Borge, O. J., Bryder, D., Theilgaard-Monch, K., Astrand-Grundstrom, I., Sitnicka, E., Sasaki, Y., and Jacobsen, S. E. (2001). Upregulation of Flt3 expression within the bone marrow

Lin−Sca1+c-kit+ stem cell compartment is accompanied by loss of self-renewal capacity. Immunity *15*, 659–669.

Antonchuk, J., Sauvageau, G., and Humphries, R. K. (2002). HOXB4-induced expansion of adult hematopoietic stem cells ex vivo. Cell *109*, 39–45.

Arai, F., Hirao, A., Ohmura, M., Sato, H., Matsuoka, S., Takubo, K., Ito, K., Koh, G. Y., and Suda, T. (2004). Tie2/angiopoietin-1 signaling regulates hematopoietic stem cell quiescence in the bone marrow niche. Cell *118*, 149–161.

Baena, E., Ortiz, M., Martinez, A. C., and de Alboran, I. M. (2007). c-Myc is essential for hematopoietic stem cell differentiation and regulates Lin−Sca-1+c-Kit− cell generation through p21. Exp Hematol *35*, 1333–1343.

Bertoncello, I., Hodgson, G. S., and Bradley, T. R. (1985). Multiparameter analysis of transplantable hemopoietic stem cells. I. The separation and enrichment of stem cells homing to marrow and spleen on the basis of rhodamine-123 fluorescence. Exp Hematol *13*, 999–1006.

Besmer, P., Murphy, J. E., George, P. C., Qiu, F. H., Bergold, P. J., Lederman, L., Snyder, H. W., Jr., Brodeur, D., Zuckerman, E. E., and Hardy, W. D. (1986). A new acute transforming feline retrovirus and relationship of its oncogene v-kit with the protein kinase gene family. Nature *320*, 415–421.

Biermann, K., Goke, F., Nettersheim, D., Eckert, D., Zhou, H., Kahl, P., Gashaw, I., Schorle, H., and Buttner, R. (2007). c-KIT is frequently mutated in bilateral germ cell tumours and down-regulated during progression from intratubular germ cell neoplasia to seminoma. J Pathol *213*, 311–318.

Branda, C. S., and Dymecki, S. M. (2004). Talking about a revolution: the impact of site-specific recombinases on genetic analyses in mice. Dev Cell *6*, 7–28.

Broudy, V. C. (1997). Stem cell factor and hematopoiesis. Blood *90*, 1345–1364.

Bryder, D., Ramsfjell, V., Dybedal, I., Theilgaard-Monch, K., Hogerkorp, C. M., Adolfsson, J., Borge, O. J., and Jacobsen, S. E. (2001). Self-renewal of multipotent long-term repopulating hematopoietic stem cells is negatively regulated by Fas and tumor necrosis factor receptor activation. J Exp Med *194*, 941–952.

Buske, C., and Humphries, R. K. (2000). Homeobox genes in leukemogenesis. Int J Hematol 71, 301–308.

Cadigan, K. M., and Nusse, R. (1997). Wnt signaling: a common theme in animal development. Genes Dev *11*, 3286–3305.

Cancelas, J. A., Jansen, M., and Williams, D. A. (2006). The role of chemokine activation of Rac GTPases in hematopoietic stem cell marrow homing, retention, and peripheral mobilization. Exp Hematol *34*, 976–985.

Carpinelli, M. R., Hilton, D. J., Metcalf, D., Antonchuk, J. L., Hyland, C. D., Mifsud, S. L., Di Rago, L., Hilton, A. A., Willson, T. A., Roberts, A. W., et-al. (2004). Suppressor screen in Mpl−/− mice: c-Myb mutation causes supraphysiological production of platelets in the absence of thrombopoietin signaling. Proc Natl Acad Sci USA *101*, 6553–6558.

Cavalli, G. (2002). Chromatin as a eukaryotic template of genetic information. Curr Opin Cell Biol 14, 269–278.

Chabot, B., Stephenson, D. A., Chapman, V. M., Besmer, P., and Bernstein, A. (1988). The proto-oncogene c-kit encoding a transmembrane tyrosine kinase receptor maps to the mouse W locus. Nature *335*, 88–89.

Cheng, T., Rodrigues, N., Shen, H., Yang, Y., Dombkowski, D., Sykes, M., and Scadden, D. T. (2000). Hematopoietic stem cell quiescence maintained by p21cip1/waf1. Science *287*, 1804–1808.

Christensen, J. L., and Weissman, I. L. (2001). Flk-2 is a marker in hematopoietic stem cell differentiation: a simple method to isolate long-term stem cells. Proc Natl Acad Sci USA *98*, 14541–14546.

Core, N., Bel, S., Gaunt, S. J., Aurrand-Lions, M., Pearce, J., Fisher, A., and Djabali, M. (1997). Altered cellular proliferation and mesoderm patterning in Polycomb-M33-deficient mice. Development *124*, 721–729.

Czechowicz, A., Kraft, D., Weissman, I. L., and Bhattacharya, D. (2007). Efficient transplantation via antibody-based clearance of hematopoietic stem cell niches. Science *318*, 1296–1299.

Domen, J., and Weissman, I. L. (2000). Hematopoietic stem cells need two signals to prevent apoptosis; BCL-2 can provide one of these, Kitl/c-Kit signaling the other. J Exp Med *192*, 1707–1718.

Dykstra, B., Ramunas, J., Kent, D., McCaffrey, L., Szumsky, E., Kelly, L., Farn, K., Blaylock, A., Eaves, C., and Jervis, E. (2006). High-resolution video monitoring of hematopoietic stem cells cultured in single-cell arrays identifies new features of self-renewal. Proc Natl Acad Sci USA *103*, 8185–8190.

Dzierzak, E., and Speck, N. A. (2008). Of lineage and legacy: the development of mammalian hematopoietic stem cells. Nat Immunol *9*, 129–136.

Eaves, C. J., Sutherland, H. J., Udomsakdi, C., Lansdorp, P. M., Szilvassy, S. J., Fraser, C. C., Humphries, R. K., Barnett, M. J., Phillips, G. L., and Eaves, A. C. (1992). The human hematopoietic stem cell in vitro and in vivo. Blood Cells *18*, 301–307.

Goodell, M. A., Brose, K., Paradis, G., Conner, A. S., and Mulligan, R. C. (1996). Isolation and functional properties of murine hematopoietic stem cells that are replicating in vivo. J Exp Med *183*, 1797–1806.

Greil, R., Anether, G., Johrer, K., and Tinhofer, I. (2003). Tuning the rheostat of the myelopoietic system via Fas and TRAIL. Crit Rev Immunol *23*, 301–322.

Grisendi, S., and Pandolfi, P. P. (2005). Two decades of cancer genetics: from specificity to pleiotropic networks. Cold Spring Harb Symp Quant Biol *70*, 83–91.

Growney, J. D., Shigematsu, H., Li, Z., Lee, B. H., Adelsperger, J., Rowan, R., Curley, D. P., Kutok, J. L., Akashi, K., Williams, I. R., et-al. (2005). Loss of Runx1 perturbs adult hematopoiesis and is associated with a myeloproliferative phenotype. Blood *106*, 494–504.

Guenechea, G., Gan, O. I., Dorrell, C., and Dick, J. E. (2001). Distinct classes of human stem cells that differ in proliferative and self-renewal potential. Nat Immunol *2*, 75–82.

Harada, N., Tamai, Y., Ishikawa, T., Sauer, B., Takaku, K., Oshima, M., and Taketo, M. M. (1999). Intestinal polyposis in mice with a dominant stable mutation of the beta-catenin gene. EMBO J *18*, 5931–5942.

Hariharan, I. K., and Haber, D. A. (2003). Yeast, flies, worms, and fish in the study of human disease. N Engl J Med *348*, 2457–2463.

Heissig, B., Hattori, K., Dias, S., Friedrich, M., Ferris, B., Hackett, N. R., Crystal, R. G., Besmer, P., Lyden, D., Moore, M. A., et-al. (2002). Recruitment of stem and progenitor cells from the bone marrow niche requires MMP-9 mediated release of kit-ligand. Cell *109*, 625–637.

Hirota, S., Isozaki, K., Moriyama, Y., Hashimoto, K., Nishida, T., Ishiguro, S., Kawano, K., Hanada, M., Kurata, A., Takeda, M., et-al. (1998). Gain-of-function mutations of c-kit in human gastrointestinal stromal tumors. Science *279*, 577–580.

Hock, H., Hamblen, M. J., Rooke, H. M., Schindler, J. W., Saleque, S., Fujiwara, Y., and Orkin, S. H. (2004). Gfi-1 restricts proliferation and preserves functional integrity of haematopoietic stem cells. Nature *431*, 1002–1007.

Ichikawa, M., Asai, T., Saito, T., Seo, S., Yamazaki, I., Yamagata, T., Mitani, K., Chiba, S., Ogawa, S., Kurokawa, M., and Hirai, H. (2004). AML-1 is required for megakaryocytic maturation and lymphocytic differentiation, but not for maintenance of hematopoietic stem cells in adult hematopoiesis. Nat Med *10*, 299–304.

Iwama, A., Oguro, H., Negishi, M., Kato, Y., Morita, Y., Tsukui, H., Ema, H., Kamijo, T., Katoh-Fukui, Y., Koseki, H., et-al. (2004). Enhanced self-renewal of hematopoietic stem cells mediated by the polycomb gene product Bmi-1. Immunity *21*, 843–851.

Jacobson, L. O., Marks, E. K., and Lorenz, E. (1949). The hematological effects of ionizing radiations. Radiology *52*, 371–395.

Janzen, V., Forkert, R., Fleming, H. E., Saito, Y., Waring, M. T., Dombkowski, D. M., Cheng, T., DePinho, R. A., Sharpless, N. E., and Scadden, D. T. (2006). Stem-cell ageing modified by the cyclin-dependent kinase inhibitor p16INK4a. Nature *443*, 421–426.

Jude, C. D., Climer, L., Xu, D., Artinger, E., Fisher, J. K., and Ernst, P. (2007). Unique and independent roles for MLL in adult hematopoietic stem cells and progenitors. Cell Stem Cell *1*, 324–337.

Jude, C. D., Gaudet, J., Speck, N., and Ernst, P. (2008). Leukemia and hematopoietic stem cells: balancing proliferation and quiescence. Cell Cycle *7*, 586–591.

Kajiume, T., Ninomiya, Y., Ishihara, H., Kanno, R., and Kanno, M. (2004). Polycomb group gene mel-18 modulates the self-renewal activity and cell cycle status of hematopoietic stem cells. Exp Hematol *32*, 571–578.

Katayama, N., Shih, J. P., Nishikawa, S., Kina, T., Clark, S. C., and Ogawa, M. (1993). Stage-specific expression of c-kit protein by murine hematopoietic progenitors. Blood *82*, 2353–2360.

Kaushansky, K. (2006). Lineage-specific hematopoietic growth factors. N Engl J Med *354*, 2034–2045.

Kiel, M. J., Yilmaz, O. H., Iwashita, T., Yilmaz, O. H., Terhorst, C., and Morrison, S. J. (2005). SLAM family receptors distinguish hematopoietic stem and progenitor cells and reveal endothelial niches for stem cells. Cell *121*, 1109–1121.

Kile, B. T., and Hilton, D. J. (2005). The art and design of genetic screens: mouse. Nat Rev Genet *6*, 557–567.

Kim, I., He, S., Yilmaz, O. H., Kiel, M. J., and Morrison, S. J. (2006). Enhanced purification of fetal liver hematopoietic stem cells using SLAM family receptors. Blood *108*, 737–744.

Kim, M., Cooper, D. D., Hayes, S. F., and Spangrude, G. J. (1998). Rhodamine-123 staining in hematopoietic stem cells of young mice indicates mitochondrial activation rather than dye efflux. Blood *91*, 4106–4117.

Kirito, K., Fox, N., and Kaushansky, K. (2003). Thrombopoietin stimulates Hoxb4 expression: an explanation for the favorable effects of TPO on hematopoietic stem cells. Blood *102*, 3172–3178.

Kirito, K., Fox, N., and Kaushansky, K. (2004). Thrombopoietin induces HOXA9 nuclear transport in immature hematopoietic cells: potential mechanism by which the hormone favorably affects hematopoietic stem cells. Mol Cell Biol *24*, 6751–6762.

Kirstetter, P., Anderson, K., Porse, B. T., Jacobsen, S. E., and Nerlov, C. (2006). Activation of the canonical Wnt pathway leads to loss of hematopoietic stem cell repopulation and multilineage differentiation block. Nat Immunol *7*, 1048–1056.

Krosl, J., Austin, P., Beslu, N., Kroon, E., Humphries, R. K., and Sauvageau, G. (2003). In vitro expansion of hematopoietic stem cells by recombinant TAT-HOXB4 protein. Nat Med *9*, 1428–1432.

Ku, H., Yonemura, Y., Kaushansky, K., and Ogawa, M. (1996). Thrombopoietin, the ligand for the Mpl receptor, synergizes with steel factor and other early acting cytokines in supporting proliferation of primitive hematopoietic progenitors of mice. Blood *87*, 4544–4551.

Kyba, M., Perlingeiro, R. C., and Daley, G. Q. (2002). HoxB4 confers definitive lymphoid-myeloid engraftment potential on embryonic stem cell and yolk sac hematopoietic progenitors. Cell *109*, 29–37.

Lacorazza, H. D., Yamada, T., Liu, Y., Miyata, Y., Sivina, M., Nunes, J., and Nimer, S. D. (2006). The transcription factor MEF/ELF4 regulates the quiescence of primitive hematopoietic cells. Cancer Cell *9*, 175–187.

Lambert, J. F., Liu, M., Colvin, G. A., Dooner, M., McAuliffe, C. I., Becker, P. S., Forget, B. G., Weissman, S. M., and Quesenberry, P. J. (2003). Marrow stem cells shift gene expression and engraftment phenotype with cell cycle transit. J Exp Med *197*, 1563–1572.

Legrand, N., Weijer, K., and Spits, H. (2008). Experimental model for the study of the human immune system: production and monitoring of "human immune system" Rag2–/–gc–/– mice. Methods Mol Biol *415*, 65–82.

Lessard, J., and Sauvageau, G. (2003). Bmi-1 determines the proliferative capacity of normal and leukaemic stem cells. Nature *423*, 255–260.

Liang, Y., Jansen, M., Aronow, B., Geiger, H., and Van Zant, G. (2007). The quantitative trait gene latexin influences the size of the hematopoietic stem cell population in mice. Nat Genet *39*, 178–188.

Lyman, S. D., and Jacobsen, S. E. (1998). c-kit ligand and Flt3 ligand: stem/progenitor cell factors with overlapping yet distinct activities. Blood *91*, 1101–1134.

Malynn, B. A., de Alboran, I. M., O'Hagan, R. C., Bronson, R., Davidson, L., DePinho, R. A., and Alt, F. W. (2000). N-myc can functionally replace c-myc in murine development, cellular growth, and differentiation. Genes Dev *14*, 1390–1399.

Mazurier, F., Doedens, M., Gan, O. I., and Dick, J. E. (2003). Rapid myeloerythroid repopulation after intrafemoral transplantation of NOD-SCID mice reveals a new class of human stem cells. Nat Med *9*, 959–963.

McCulloch, E. A., Siminovitch, L., Till, J. E., Russell, E. S., and Bernstein, S. E. (1965). The cellular basis of the genetically determined hemopoietic defect in anemic mice of genotype Sl-Sld. Blood *26*, 399–410.

McKenzie, J. L., Takenaka, K., Gan, O. I., Doedens, M., and Dick, J. E. (2007). Low rhodamine 123 retention identifies long-term human hematopoietic stem cells within the Lin−CD34+CD38− population. Blood *109*, 543–545.

McMahon, A. P., Ingham, P. W., and Tabin, C. J. (2003). Developmental roles and clinical significance of hedgehog signaling. Curr Top Dev Biol *53*, 1–114.

McMahon, K. A., Hiew, S. Y., Hadjur, S., Veiga-Fernandes, H., Menzel, U., Price, A. J., Kioussis, D., Williams, O., and Brady, H. J. (2007). Mll has a critical role in fetal and adult hematopoietic stem cell self-renewal. Cell Stem Cell *1*, 338–345.

Meyers, S., and Hiebert, S. W. (1995). Indirect and direct disruption of transcriptional regulation in cancer: E2F and AML-1. Crit Rev Eukaryot Gene Expr *5*, 365–383.

Miller, C. L., Rebel, V. I., Helgason, C. D., Lansdorp, P. M., and Eaves, C. J. (1997). Impaired steel factor responsiveness differentially affects the detection and long-term maintenance of fetal liver hematopoietic stem cells in vivo. Blood *89*, 1214–1223.

Miyamoto, K., Araki, K. Y., Naka, K., Arai, F., Takubo, K., Yamazaki, S., Matsuoka, S., Miyamoto, T., Ito, K., Ohmura, M., et-al. (2007). Foxo3a is essential for maintenance of the hematopoietic stem cell pool. Cell Stem Cell *1*, 101–112.

Morrison, S. J., and Weissman, I. L. (1994). The long-term repopulating subset of hematopoietic stem cells is deterministic and isolatable by phenotype. Immunity *1*, 661–673.

Nolta, J. A., Hanley, M. B., and Kohn, D. B. (1994). Sustained human hematopoiesis in immunodeficient mice by cotransplantation of marrow stroma expressing human interleukin-3: analysis of gene transduction of long-lived progenitors. Blood *83*, 3041–3051.

Ogawa, M., Matsuzaki, Y., Nishikawa, S., Hayashi, S., Kunisada, T., Sudo, T., Kina, T., Nakauchi, H., and Nishikawa, S. (1991). Expression and function of c-kit in hemopoietic progenitor cells. J Exp Med *174*, 63–71.

Ogawa, M., Nishikawa, S., Yoshinaga, K., Hayashi, S., Kunisada, T., Nakao, J., Kina, T., Sudo, T., Kodama, H., and Nishikawa, S. (1993). Expression and function of c-Kit in fetal hemopoietic progenitor cells: transition from the early c-Kit-independent to the late c-Kit-dependent wave of hemopoiesis in the murine embryo. Development *117*, 1089–1098.

Ohta, H., Sawada, A., Kim, J. Y., Tokimasa, S., Nishiguchi, S., Humphries, R. K., Hara, J., and Takihara, Y. (2002). Polycomb group gene rae28 is required for sustaining activity of hematopoietic stem cells. J Exp Med *195*, 759–770.

Opferman, J. T. (2007). Life and death during hematopoietic differentiation. Curr Opin Immunol *19*, 497–502.

Opferman, J. T., Iwasaki, H., Ong, C. C., Suh, H., Mizuno, S., Akashi, K., and Korsmeyer, S. J. (2005). Obligate role of anti-apoptotic MCL-1 in the survival of hematopoietic stem cells. Science *307*, 1101–1104.

Park, I. K., Qian, D., Kiel, M., Becker, M. W., Pihalja, M., Weissman, I. L., Morrison, S. J., and Clarke, M. F. (2003). Bmi-1 is required for maintenance of adult self-renewing haematopoietic stem cells. Nature *423*, 302–305.

Passegue, E., Wagner, E. F., and Weissman, I. L. (2004). JunB deficiency leads to a myeloproliferative disorder arising from hematopoietic stem cells. Cell *119*, 431–443.

Passegue, E., Wagers, A. J., Giuriato, S., Anderson, W. C., and Weissman, I. L. (2005). Global analysis of proliferation and cell cycle gene expression in the regulation of hematopoietic stem and progenitor cell fates. J Exp Med *202*, 1599–1611.

Petit-Cocault, L., Volle-Challier, C., Fleury, M., Peault, B., and Souyri, M. (2007). Dual role of Mpl receptor during the establishment of definitive hematopoiesis. Development *134*, 3031–3040.

Puri, M. C., and Bernstein, A. (2003). Requirement for the TIE family of receptor tyrosine kinases in adult but not fetal hematopoiesis. Proc Natl Acad Sci USA *100*, 12753–12758.

Renneville, A., Roumier, C., Biggio, V., Nibourel, O., Boissel, N., Fenaux, P., and Preudhomme, C. (2008). Cooperating gene mutations in acute myeloid leukemia: a review of the literature. Leukemia *22*, 915–931.

Reya, T., Duncan, A. W., Ailles, L., Domen, J., Scherer, D. C., Willert, K., Hintz, L., Nusse, R., and Weissman, I. L. (2003). A role for Wnt signalling in self-renewal of haematopoietic stem cells. Nature *423*, 409–414.

Rodewald, H. R., and Sato, T. N. (1996). Tie1, a receptor tyrosine kinase essential for vascular endothelial cell integrity, is not critical for the development of hematopoietic cells. Oncogene 12, 397–404.

Saito, R. M., Perreault, A., Peach, B., Satterlee, J. S., and van den Heuvel, S. (2004). The CDC-14 phosphatase controls developmental cell-cycle arrest in C. elegans. Nat Cell Biol 6, 777–783.

Sandberg, M. L., Sutton, S. E., Pletcher, M. T., Wiltshire, T., Tarantino, L. M., Hogenesch, J. B., and Cooke, M. P. (2005). c-Myb and p300 regulate hematopoietic stem cell proliferation and differentiation. Dev Cell 8, 153–166.

Sauvageau, G., Thorsteinsdottir, U., Eaves, C. J., Lawrence, H. J., Largman, C., Lansdorp, P. M., and Humphries, R. K. (1995). Overexpression of HOXB4 in hematopoietic cells causes the selective expansion of more primitive populations in vitro and in vivo. Genes Dev 9, 1753–1765.

Scheller, M., Huelsken, J., Rosenbauer, F., Taketo, M. M., Birchmeier, W., Tenen, D. G., and Leutz, A. (2006). Hematopoietic stem cell and multilineage defects generated by constitutive beta-catenin activation. Nat Immunol 7, 1037–1047.

Schmidt, T., Zornig, M., Beneke, R., and Moroy, T. (1996). MoMuLV proviral integrations identified by Sup-F selection in tumors from infected myc/pim bitransgenic mice correlate with activation of the gfi-1 gene. Nucleic Acids Res 24, 2528–2534.

Schroeder, T. (2005). Tracking hematopoiesis at the single cell level. Ann NY Acad Sci 1044, 201–209.

Sitnicka, E., Lin, N., Priestley, G. V., Fox, N., Broudy, V. C., Wolf, N. S., and Kaushansky, K. (1996). The effect of thrombopoietin on the proliferation and differentiation of murine hematopoietic stem cells. Blood 87, 4998–5005.

Snoeck, H. W. (2005). Quantitative trait analysis in the investigation of function and aging of hematopoietic stem cells. Methods Mol Med 105, 47–62.

Souyri, M., Vigon, I., Penciolelli, J. F., Heard, J. M., Tambourin, P., and Wendling, F. (1990). A putative truncated cytokine receptor gene transduced by the myeloproliferative leukemia virus immortalizes hematopoietic progenitors. Cell 63, 1137–1147.

Spangrude, G. J., and Scollay, R. (1990). A simplified method for enrichment of mouse hematopoietic stem cells. Exp Hematol 18, 920–926.

Sparmann, A., and van Lohuizen, M. (2006). Polycomb silencers control cell fate, development and cancer. Nat Rev Cancer 6, 846–856.

Szilvassy, S. J., Humphries, R. K., Lansdorp, P. M., Eaves, A. C., and Eaves, C. J. (1990). Quantitative assay for totipotent reconstituting hematopoietic stem cells by a competitive repopulation strategy. Proc Natl Acad Sci USA 87, 8736–8740.

Tajima, Y., Moore, M. A., Soares, V., Ono, M., Kissel, H., and Besmer, P. (1998). Consequences of exclusive expression in vivo of Kit-ligand lacking the major proteolytic cleavage site. Proc Natl Acad Sci USA 95, 11903–11908.

Till, J. E., and McCulloch, C. E. (1961). A direct measurement of the radiation sensitivity of normal mouse bone marrow cells. Radiat Res 14, 213–222.

Tothova, Z., Kollipara, R., Huntly, B. J., Lee, B. H., Castrillon, D. H., Cullen, D. E., McDowell, E. P., Lazo-Kallanian, S., Williams, I. R., Sears, C., et-al. (2007). FoxOs are critical mediators of hematopoietic stem cell resistance to physiologic oxidative stress. Cell 128, 325–339.

Trowbridge, J. J., Scott, M. P., and Bhatia, M. (2006). Hedgehog modulates cell cycle regulators in stem cells to control hematopoietic regeneration. Proc Natl Acad Sci USA 103, 14134–14139.

Uchida, N., Dykstra, B., Lyons, K. J., Leung, F. Y., and Eaves, C. J. (2003). Different in vivo repopulating activities of purified hematopoietic stem cells before and after being stimulated to divide in vitro with the same kinetics. Exp Hematol 31, 1338–1347.

Umemoto, T., Yamato, M., Nishida, K., Yang, J., Tano, Y., and Okano, T. (2005). p57Kip2 is expressed in quiescent mouse bone marrow side population cells. Biochem Biophys Res Commun 337, 14–21.

Urso, P., and Congdon, C. C. (1957). The effect of the amount of isologous bone marrow injected on the recovery of hematopoietic organs, survival and body weight after lethal irradiation injury in mice. Blood 12, 251–260.

van der Lugt, N. M., Alkema, M., Berns, A., and Deschamps, J. (1996). The Polycomb-group homolog Bmi-1 is a regulator of murine Hox gene expression. Mech Dev 58, 153–164.

Willert, K., Brown, J. D., Danenberg, E., Duncan, A. W., Weissman, I. L., Reya, T., Yates, J. R., III, and Nusse, R. (2003). Wnt proteins are lipid-modified and can act as stem cell growth factors. Nature *423*, 448–452.

Wilson, A., Murphy, M. J., Oskarsson, T., Kaloulis, K., Bettess, M. D., Oser, G. M., Pasche, A. C., Knabenhans, C., Macdonald, H. R., and Trumpp, A. (2004). c-Myc controls the balance between hematopoietic stem cell self-renewal and differentiation. Genes Dev *18*, 2747–2763.

Wissler, R. W., Robson, M. J., Fitch, F., Nelson, W., and Jacobson, L. O. (1953). The effects of spleen shielding and subsequent splenectomy upon antibody formation in rats receiving total-body X-irradiation. J Immunol *70*, 379–385.

Wu, M., Kwon, H. Y., Rattis, F., Blum, J., Zhao, C., Ashkenazi, R., Jackson, T. L., Gaiano, N., Oliver, T., and Reya, T. (2007). Imaging hematopoietic precursor division in real time. Cell Stem Cell *1*, 541–554.

Yang, L., Wang, L., Geiger, H., Cancelas, J. A., Mo, J., and Zheng, Y. (2007). Rho GTPase Cdc42 coordinates hematopoietic stem cell quiescence and niche interaction in the bone marrow. Proc Natl Acad Sci USA *104*, 5091–5096.

Yilmaz, O. H., Kiel, M. J., and Morrison, S. J. (2006a). SLAM family markers are conserved among hematopoietic stem cells from old and reconstituted mice and markedly increase their purity. Blood *107*, 924–930.

Yilmaz, O. H., Valdez, R., Theisen, B. K., Guo, W., Ferguson, D. O., Wu, H., and Morrison, S. J. (2006b). Pten dependence distinguishes haematopoietic stem cells from leukaemia-initiating cells. Nature *441*, 475–482.

Yokota, S., Kiyoi, H., Nakao, M., Iwai, T., Misawa, S., Okuda, T., Sonoda, Y., Abe, T., Kahsima, K., Matsuo, Y., and Naoe, T. (1997). Internal tandem duplication of the FLT3 gene is preferentially seen in acute myeloid leukemia and myelodysplastic syndrome among various hematological malignancies. A study on a large series of patients and cell lines. Leukemia *11*, 1605–1609.

Yoshihara, H., Arai, F., Hosokawa, K., Hagiwara, T., Keiyo, T., Nakamura, Y., Gomei, Y., Iwasaki, H., Matsuoka, S., Miyamoto, K., et-al. (2007). Thrombin/MPL signaling regulates hematopoietic stem cell quiescence and interaction with the osteoblastic niche. Cell Stem Cell *1*, 685–697.

Yu, H., Yuan, Y., Shen, H., and Cheng, T. (2006). Hematopoietic stem cell exhaustion impacted by p18 INK4C and p21 Cip1/Waf1 in opposite manners. Blood *107*, 1200–1206.

Yuan, Y., Shen, H., Franklin, D. S., Scadden, D. T., and Cheng, T. (2004). In vivo self-renewing divisions of haematopoietic stem cells are increased in the absence of the early G1-phase inhibitor, p18INK4C. Nat Cell Biol *6*, 436–442.

Zanjani, E. D., Almeida-Porada, G., and Flake, A. W. (1996). The human/sheep xenograft model: a large animal model of human hematopoiesis. Int J Hematol *63*, 179–192.

Zeng, H., Yucel, R., Kosan, C., Klein-Hitpass, L., and Moroy, T. (2004). Transcription factor Gfi1 regulates self-renewal and engraftment of hematopoietic stem cells. EMBO J 23, 4116–4125.

Zhang, J., Grindley, J. C., Yin, T., Jayasinghe, S., He, X. C., Ross, J. T., Haug, J. S., Rupp, D., Porter-Westpfahl, K. S., Wiedemann, L. M., et-al. (2006). PTEN maintains haematopoietic stem cells and acts in lineage choice and leukaemia prevention. Nature *441*, 518–522.

Zhang, P., Liegeois, N. J., Wong, C., Finegold, M., Hou, H., Thompson, J. C., Silverman, A., Harper, J. W., DePinho, R. A., and Elledge, S. J. (1997). Altered cell differentiation and proliferation in mice lacking p57KIP2 indicates a role in Beckwith-Wiedemann syndrome. Nature *387*, 151–158.

Zhou, S., Schuetz, J. D., Bunting, K. D., Colapietro, A. M., Sampath, J., Morris, J. J., Lagutina, I., Grosveld, G. C., Osawa, M., Nakauchi, H., and Sorrentino, B. P. (2001). The ABC transporter Bcrp1/ABCG2 is expressed in a wide variety of stem cells and is a molecular determinant of the side-population phenotype. Nat Med *7*, 1028–1034.

Zijlmans, J. M., Visser, J. W., Kleiverda, K., Kluin, P. M., Willemze, R., and Fibbe, W. E. (1995). Modification of rhodamine staining allows identification of hematopoietic stem cells with preferential short-term or long-term bone marrow-repopulating ability. Proc Natl Acad Sci USA *92*, 8901–8905.

Zindy, F., Quelle, D. E., Roussel, M. F., and Sherr, C. J. (1997). Expression of the p16INK4a tumor suppressor versus other INK4 family members during mouse development and aging. Oncogene *15*, 203–211.

Chapter 2
The Road to Commitment: Lineage Restriction Events in Hematopoiesis

Robert Mansson, Sasan Zandi, David Bryder, and Mikael Sigvardsson

Abstract All the mature blood cells in circulation as well as in other tissues can be generated by one and the same hematopoietic stem cell. Upon cell division, the multipotent stem cell generates one daughter cell with stem cell characteristics while the other progeny initiates a sequence of proliferation and differentiation events. These result in a dramatic expansion in the number of progenitor cells that upon further maturation gradually lose their multipotency and gain the specific features of mature blood cell types. This chapter focuses on the biology of the earliest events in hematopoietic cell maturation where the initial decisions of the cell to enter defined developmental pathways and lineage restriction events are achieved.

The Roles of Early Progenitors in Hematopoiesis

Transplantation experiments in the 1950s showed that bone marrow (BM) contains one or several components capable of producing the factors needed for survival after lethal radiation (Lorenz et al. 1951). Further transplantation experiments revealed that the BM contained progenitor cells capable of generating colonies composed of several types of blood cells in the spleen of recipient animals (colony forming unit spleen, CFU-S) (Becker et al. 1963; Till and McCulloch 1961), suggesting the existence of multipotent hematopoietic progenitor cells. This work established that cellular, rather than humoral, factors are responsible for the ability of BM to rescue lethally irradiated mice. Further evidence for a clonal multipotent progenitor capable of multilineage hematopoiesis was obtained through transplantation of retrovirus marked BM cells, since the same integration sites could be found in both myeloid and lymphoid cells (Dick et al. 1985; Keller et al. 1985; Lemischka et al. 1986). The experiment that undisputedly demonstrated the single-cell origin of multilineage

M. Sigvardsson (✉)
Department for Clinical and Experimental Research, Faculty for Health Sciences,
Linköping Sweden, Lund Stem Cell Center and the Department for Immunology,
Medical Faculty in Lund, Sweden
e-mail: miksi@ibk.liu.se

A. Wickrema and B. Kee (eds.), *Molecular Basis of Hematopoiesis*,
DOI: 10.1007/978-0-387-85816-6_1, © Springer Science+Business Media, LLC 2009

hematopoiesis was presented in 1991 in the work of Smith and his collegues (Smith et al. 1991), where prospectively isolated hematopoietic stem cells, HSCs were transferred as single cells into conditioned hosts and subsequently assayed for their differentiation potential. Thus, it has been clearly established that a single multipotent progenitor is able to give rise to all types of blood cells through a differentiation process where developmental potentials are gradually lost while lineage-specific features are gained. Considering the highly diverse functions and characteristics of the different blood cells, this process would have to be considered one of the most complex continually ongoing differentiation processes in adults. Hematopoietic cell development may demand the stepwise generation of specific substages of progenitor cells that could well be critical for the completion of these developmental pathways. This process must involve a complex integration of extracellular and internal signals to regulate the cellular composition of blood in homeostasis as well as upon challenge caused by trauma or disease. In addition, while multipotent progenitors are crucial intermediates in development of blood cells, they also contribute with the absolute major part of the expansion in cell numbers necessary to provide the 10^{12} cells/day needed to maintain steady state hematopoiesis in humans (Ogawa 1993). This is because, even though most HSCs progress through the cell cycle within a time span of 30–60 days (Bradford et al. 1997; Cheshier et al. 1999), in homeostasis, approximately 75% of the adult HSCs display a prolonged G1 phase similar to what can be observed in quiescence (Cheshier et al. 1999). The prolonged cell cycle time has been suggested to be an integrated part of the maintenance of HSCs activity (Cheng et al. 2000), protecting stem cells from damage to macromolecules including DNA (Cheng et al. 2000). However, the preserved ability to enter cell cycle also creates a functional buffer in case of disturbances of homeostatic conditions (Cheng 2004; Randall and Weissman 1997). Thus, it does not appear as if HSCs contribute to the expansion of cell populations to any significant degree, rather, this is achieved mainly through proliferation of progenitor cells at different stages of development. The high proliferative potential of progenitor cells also makes them primary targets for cellular transformation in leukemia. Even though genetic aberrations can be found in HSCs, leukemia mostly almost manifests itself by massive expansion of progenitor cells (Rossi et al. 2008; Tan et al. 2006). Thus, early progenitors compose a crucial part of the hema-topoietic system in health and disease.

Anatomical Niches for Hematopoietic Progenitor Cells

The dominant site for early hematopoiesis in the adult is the BM, although circulating HSCs can migrate to peripheral sites where they are able to develop into resident dendritic cells (Massberg et al. 2007) and transplanted stem cells can form colonies composed of mixed cell populations in the spleen (Becker et al. 1963; Till and Mc 1961). Within the BM, the idea of a specific "stem cell niche" was developed from the observations that, after irradiation, histological sections revealed that hematopoietic repopulation first appeared along the endosteal bone surface (Schofield 1978).

In addition, subcutaneous transplantation of BM results in development of bone and vascularized BM (Maloney and Patt 1975; Patt and Maloney 1972) with the frequency of primitive progenitors highest close to the endosteal surface (Lord and Hendry 1972; Lord et al. 1975). The BM endosteum also appears to harbor quiescent HSCs (Zhang et al. 2003) and regulates expansion and differentiation through a complex interplay of cellular communication between osteoblasts and HSCs (Rizo et al. 2006; Wilson and Trumpp 2006). Early multipotent progenitors can also be found near the sinusoidal endothelium in the BM (Kiel et al. 2005) but in contrast to the cells in the endosteal area, the HSCs in the vascular niche appear to be more actively dividing (Kopp et al. 2005; Wilson and Trumpp 2006) suggesting that this might be the more active site for early differentiation. The HSC niche is discussed in detail by Wilson and Trumpp in Chap. 3.

While the site of blood cell development in the adult is fairly well defined, the development of progenitor cells in the embryo involves a complex interplay between several different anatomical sites. During mouse embryogenesis, the three germ layers ectoderm, endoderm, and mesoderm are formed in the blastocyst around 6.5 days post coitum/conception (dpc) through a process referred to as gastrulation. Hematopoietic cells originate from the mesoderm where progenitor cells arise that likely possesses a combined endothelial/hematopoietic potential (hemangioblasts, HE) (Ferkowicz and Yoder 2005; Ueno and Weissman 2006). Even though the issue of the existence of a HE in mouse is somewhat controversial, the idea is supported by in vitro differentiation studies using mouse ES cells (Choi et al. 1998; Fehling et al. 2003). In addition, several other model systems for studies of early blood cell development support the existence of such a common precursor. For instance, gene expression analysis in zebrafish embryos has revealed that one and the same progenitor express markers for both hematopoietic and endothelial cells (Gering et al. 1998). The first wave of blood cell development in mice is found in the extraembryonic yolk sac. In this tissue, blood islands primarily generating fetal erythrocytes are detected around day 7.5 dpc (Cumano et al. 2001; Cumano and Godin 2001). The first signs of intraembryonic blood cell development are found around day 8.5 dpc in the para-aorta splanchnopleura (P-Sp/PAS), which later develops into the aorta, gonads, and mesonephros (AGM) region (Medvinsky and Dzierzak 1996). This is half a day after the onset of circulation in the embryo (McGrath et al. 2003) and could therefore be a result of cell seeding from the blood islands in the yolk sac (Moore and Metcalf 1970). However, even though tracing experiments suggest that yolk sac (YS) derived cells contribute to adult hematopoiesis (Samokhvalov et al. 2007; Ueno and Weissman 2007) and that YS progenitors are involved in seeding of blood cells in the fetal liver (FL) around day 10 dpc (Godin et al. 1999; Palis et al. 2001), spatial and temporal analysis of HSCs has suggested that the AGM-derived HSCs are generated independently of the YS progenitors and that definitive hematopoiesis is largely derived from these cells (Cumano et al. 2001; Cumano and Godin 2001; Dieterlen-Lievre 1975; Medvinsky and Dzierzak 1996). Another site for de novo generation of hematopoietic cells is the placenta where myeloid progenitor cells can be found around day 9 dpc (Alvarez-Silva et al. 2003) and hematpoietic stem cells around day 11 dpc (Gekas et al. 2005;

Ottersbach and Dzierzak 2005). Even though the FL does not appear to be a site for de novo generation of HSCs, the number of cells in the FL is dramatically increased (Ema and Nakauchi 2000; Morrison et al. 1995a) before seeding of the BM around day 16 dpc (Morrison et al. 1995b). The appearance of primitive erythrocytes at day 7.5 dpc, while adult repopulating HSCs have not been identified until about three days later, argues that primitive erythropoiesis arises from cells that do not have full developmental potential. These observations indicate that several types of multipotent hematopoietic progenitors exist in the early embryo, whereas the earliest cell in the YS may have a limited developmental potential with a profound preference for development into erythroid cells.

Constructing the Hematopoietic Tree

Even though the gold standard for characterization of HSCs is direct functional investigation by means of transplantation experiments, the characterization of early multipotent cells has been largely facilitated by the development of cell sorting protocols that allow prospective purification and identification of HSCs. The development of highly complex sorting protocols has allowed for detailed investigation of the lineage potential of sorted cells and provided information about the structure of hematopoietic development. Several more or less independent protocols has been developed; one of the currently dominating schemes involves the purification of Lineage marker negative (Lin-), Sca1, and c-Kit high (LSK) cells (Ikuta and Weissman 1992; Li and Johnson 1995). This population represents approximately 0.1% of the cells in the mouse BM, contains about a 1,000-fold higher HSC activity than a whole BM preparation and appears to include the absolute majority of multipotent progenitors. However, the LSK population is heterogeneous and only about 1/30 of the cells represents functional stem cells (Bryder et al. 2006), which stresses the importance of using additional markers to achieve a high level of purity of true HSCs (Fig. 2.1). Expression of surface markers, such as CD150 (Slamf1) (Kiel et al. 2005), CD105 (Endoglin) (Chen et al. 2002), and in some mouse strains Thy1.1 (Spangrude et al. 1988) on HSCs, allows for the enrichment of functional HSCs, while CD34 (Osawa et al. 1996), Flt3 (Flk2) (Adolfsson et al. 2001; Christensen and Weissman 2001), and CD48 (Kiel et al. 2005) expression are markers for LSK cells which lack long-term repopulating HSC activity (LT-HSC) (Please see Chapter 1 and 3 for discussion about long-term (LT-HSC) and short-term "stem cell" (ST-HSC) populations). Expression of Flt3 not only marks loss of long-term reconstitution capacity but also serves as an indicator of a lineage restriction event associated with a dramatic reduction of megakaryocyte and erythrocyte (MkE) potential (Adolfsson et al. 2005) (see below). In addition to surface marker expression, efflux of dyes like Hoechst 33342 (Wolf et al. 1993) can be utilized to isolate the so-called side population cells, a subset which is also highly enriched for functional HSC activity. While isolation protocols for mouse HSCs today have reached a very advanced level, where a high portion of prospectively isolated candidate HSCs can be functionally established as representing true HSCs, the methods for isolation of human cells are slightly less well

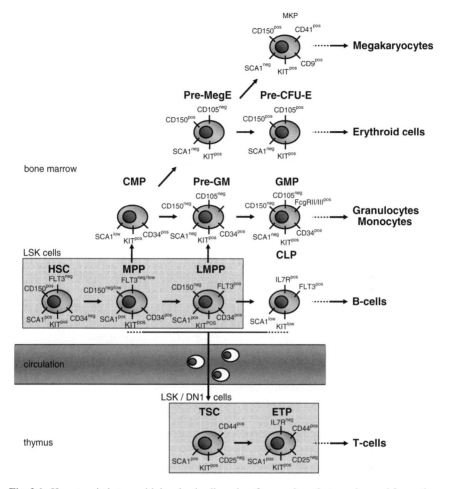

Fig. 2.1 Hematopoietic tree with involved cells and surface markers that may be used for sorting of the defined progenitors. *HSC* hematopoietic stem cell, *MPP* multipotent progenitor, *LMPP* lymphoid myeloid primed progenitor, *CLP* common lymphoid progenitor, *CMP* common myeloid progenitor, *GMP* granulocyte monocyte progenitor, *MkP* megakaryocyte progenitor, *CFU-E* colony forming unit – erythrocytes, *TSC* thymus seeding cell, *ETP* early thymic progenitors

developed. However, isolation of Lin-CD34+CD38– rhodamine low cord blood cells enables enrichment to about 1 HSC among 30 cells as judged by xenograft transplantation assays using nonobese diabetic/severe combined immunodeficiency (NOD/SCID) mice (Dick et al. 1997; McKenzie et al. 2007).

Several of the surface markers used to identify HSC can also be used to identify lineage-restricted progenitor cells (Fig. 2.1), such as the $Lin^{-}Sca^{Int}Kit^{Int}Il7r\alpha^{+}$ common lymphoid progenitor (CLP), a cell type which can transiently give rise to B, T, NK, and dendritic cells after transplantation (Kondo et al. 1997), while at the same time having limited myeloid potential. A second type of lymphoid progenitor, early thymic progenitor (ETP), which is characterized by high expression of c-Kit and Sca1 has been identified amongst a CD4⁻CD8⁻CD44⁺CD25⁻ thymocyte population (Allman et al. 2003). ETPs possess both B and T lineage potential as well as a residual myeloid potential and lack expression of IL7Rα (Allman et al. 2003), which suggests that ETPs might represent a complementary differentiation path into the T cell lineage, rather than all T cell development being the result of the differentiation from BM residing CLPs. Common myeloid progenitors (CMPs) are defined by CD34, c-kit, and low levels of FcRgII/III expression while simultaneously lacking expression of Sca1 and lineage markers (Akashi et al. 2000). High expression of FcRgII/III marks progenitors restricted to granulucyte/monocyte generation (GMPs) while lack of FcRgII/III and CD34 expression defines a population of megakaryocyte/erythroid (MkE)-restricted progenitor cells (MEP). Early studies of CMPs reported that these cells, at the clonal level, can give rise to granulucyte (G), monocyte (M), megakaryocytes (Mk) as well as erythroid (E) cells, but the low frequency of mixed clones in in vitro differentiation assays (Akashi et al. 2000) indicates that few cells in the CMP population have the full lineage potential of all myeloid lineages. Subfractionation of the CMP compartment based on differential expression of CD105 and CD150 revealed that this population can be divided into four subsets, each one with defined and restricted lineage potentials (Pronk et al. 2007). Thus, a rather substantial arsenal of markers can be used for positive and negative selection to allow for enrichment of hematopoietic progenitor cells at defined stages of development, before the onset of expression of more mature lineage-associated surface markers. The resolution of different cell populations is continuously increasing by the combinatorial use of surface markers as well as more modern approaches that include transgenic expression of reporter genes under the control of stage- and lineage-restricted control elements (see below).

FACS has proved extremely useful for the prospective identification and isolation of specific subpopulations; however, the information is rather limited unless the functional potential of the purified cells can be analyzed. In the mouse model system with several inbred strains, progenitor cell potential can be investigated by transplantation into congenic mice, similar to the approaches used to define functional HSC activity. One such system is based on the expression of different forms of the leukocyte antigen CD45 where even single cells from CD45.1 mice can be easily followed after transplantation into irradiated CD45.2 mice or vice versa (Osawa et al. 1996). Donor contribution to specific lineages can be followed by FACS analysis of the combined expression of the specific CD45 form and lineage markers, to provide a picture of the lineage potential in vivo. Whereas this assay has been very successfully applied to HSC biology, it has a limitation for other cell types in that the functional readout will at least in part be dependent of the ability of the transplanted progenitors cells to enter the bone marrow (homing) and analysis of single cells are only possible for cells whose progeny have an enormous proliferative potential (such as a LT-HSC). Thus, in order to investigate lineage potentials of single cells, several in vitro differentiation assays has been developed, many of which allow for the investigation of several lineage potentials simultaneously. MkE and GM cells can rather easily develop from progenitors in vitro after addition of hematopoietic stimulating cytokines (Adolfsson et al. 2005). The nature of the generated cells can then be investigated by morphology, cell surface antigen profiles as well as molecular techniques. Even though B-lymphocytes also can be generated by the addition of hematopoietic growth factors, including Flt3 ligand and IL7 (Veiby et al. 1996), the efficiency of cell differentiation is dramatically improved when candidate progenitor cells with B cell competence are grown on a stromal cell feeder layer. Here, the stromal cell line OP-9 has received extra attention both due to its ability to effectively support B-cell development and also due to the fact that an OP-9-derived cell line engineered to express the Notch ligand Delta 1 is highly capable of supporting the development of T-cells (Schmitt and Zuniga-Pflucker 2002). Functional investigations of human progenitor cells are largely performed in vitro, even though xenograft transplantation models in mice have been developed (Dick et al. 1997).

Pathways to Lineage Restriction

The prospective isolation of a CLP (Kondo et al. 1997) and a CMP (Akashi et al. 2000) suggested that lymphoid restriction would reflect the initial loss of multipotency in an upstream multipotent hematopoietic progenitor. This model presumes that lymphoid cell fate is established in close proximity to the HSC compartment, data in part supported by the isolation of a lymphoid-restricted LSK population (ELP) in mice that express GFP under the control of Rag-1 regulatory elements (Igarashi et al. 2002). However, subsequent purification of lymphoid-primed multipotent progenitors (LMPP), with reduced MkE but preserved GM potential (Adolfsson et al. 2005), has provided support for a revised view of how and when lymphoid cell fate is established. Investigations of early progenitor cells from mice transgenic for a GFP gene under the control of regulatory sequences of the Pu.1 transcription factor gene revealed that, in contrast to what is observed for GATA1 reporter mice, GFP$^+$CD34$^+$LSK cells maintain lymphoid and myeloid potential but display a loss of MkE potential (Arinobu et al. 2007; Nutt et al. 2005). Thus, the balance between GATA1 and Pu.1 expression in early progenitors appear to correlate with restrictions in cell fate already in LSK cells. In addition, the dose of Pu.1 may be of importance in the myeloid versus lymphoid cell fate in the LMPP, possibly by modulation of the function of downstream transcription factors such as EBF (Medina et al. 2004). Reduced MkE potential could also be observed in GFP$^+$ LSK cells from mice expressing GFP under the control of the regulatory elements from the IKAROS gene (Yoshida et al. 2006). This transcription factor is crucial for normal lymphoid development (Wang et al. 1996) and for the development of Flt3^{high} LSK cells (Yoshida et al. 2006) but the use of the reporter mouse on an IKAROS null background revealed that a GFP-positive population with reduced MkE potential resembling LMPPs develops, although without the high expression of Flt3 Rag1, or IL7Rα transcripts found in the wild-type reporter positive cells (Yoshida et al. 2006). These results indicate that IKAROS is critical for the lymphoid priming thought to be reflected in expression of lymphoid-restricted genes observed in a subfraction of the LMPPs (Adolfsson et al. 2005; Mansson et al. 2007).

Such an early lymphoid priming appears to involve a restricted set of genes including TdT, Rag-1, Rag-2, and sterile immunoglobulin heavy-chain transcripts, while the expression of T- or B-cell-restricted genes is absent (Adolfsson et al. 2005; Mansson et al. 2007). Interestingly, several of these genes have been proposed to be targets for the transcription factor E47 (Choi et al. 1996) encoded by the E2A gene, which is crucial for normal early lymphoid development (Bain et al. 1994; Zhuang et al. 1994). A function for E2A proteins in the establishment of the lymphoid program in the LMPP is also supported from the observation that LMPPs from E2A-deficient mice display a dramatic reduction in the expression of these genes (Dias et al. 2008). Thus, it appears as if priming toward lymphoid cell fate is regulated by the concerted activities of several transcription factors, whose roles are just beginning to be unraveled. The hallmark of the LMPP subcompartment of LSK cells is the high expression of the tyrosine kinase receptor Flt3 (Adolfsson et al. 2005). This receptor itself has been attributed a role in lymphoid restriction, since mice lacking this receptor display a pronounced defect in the generation of CLPs (Sitnicka et al. 2002). However, expression of Flt3 does not appear to mark or be sufficient for full lymphoid lineage restriction, since these cells have a maintained Granulocyte/Monocyte (GM) potential (Adolfsson et al. 2005). A more lymphoid-restricted surface expression is observed for IL7R alpha. The IL7R signal transduction pathway appears to be of critical importance for the development of the earliest B cell progenitors, possibly by induction of key transcription factors such as EBF (Dias et al. 2005; Kikuchi et al. 2005). The Flt3 ligand has been shown to act in collaboration with IL7, since mice lacking both these signaling pathways display a more dramatic and synergistic phenotype with regard to lymphoid development, than either of the single mutants (Sitnicka et al. 2003, 2007). The combined expression of these receptors can be observed in a small subpopulation of LMPPs (Adolfsson et al. 2005), but even if this can be taken as a sign of predisposition toward the lymphoid cell fate, functional single cell analysis of LMPPs suggested that a combined B/GM or T/GM potential may be preserved even after the combined B/T potential is lost (Mansson et al. 2007). An early separation between the pathways for B and T lymphocyte development is also supported by the observation that CLPs are rarely found in the circulation (Schwarz and Bhandoola 2004), a necessary step in order to seed the thymus. Instead, it has been suggested that the thymus-seeding cell represents a circulating LSK population (Schwarz and Bhandoola 2004) (Fig. 2.1) with a reduced B-lineage potential, which upon thymic entry is driven into T cell differentiation by Notch signaling (Sambandam et al. 2005). The close relationship between GM cells and lymphocytes is further supported by the observation that deletion or overexpression of single transcription factor in CLPs, early B-lymphocyte progenitors, or even B-cell lines can result in a loss of lineage identity and expression of GM features (Laiosa et al. 2006; Nutt et al. 1999; Pongubala et al. 2008; Xie et al. 2004).

Even though GM potential appears to be maintained in the early branch of the lymphoid hematopoietic tree, the identification of the CMP population, which lacks lymphoid potential (Akashi et al. 2000), suggests that GM lineages are generated through two different developmental pathways. The CMP population is capable of generating MkE-restricted MEPs that can give rise to CD9[+] megakaryocyte lineage-committed progenitors (MkPs) (Nakorn et al. 2003) and erythroid progenitor cells. CMPs are also able to generate the GM-restricted GMPs that are able to develop into even more lineage-restricted granulocytes and macrophages (Akashi et al. 2000). Bipotent progenitors that generate macrophages and dendritic cells could be identified in mice carrying a GFP reporter gene inserted into the Cx3cr1 locus (Fogg et al. 2006), while eosinophil-committed progenitors could be isolated based on the expression of the Il-5 receptor alpha chain (Iwasaki et al. 2005a). The common basophil/mast cell progenitor (Arinobu et al. 2005) can be isolated from the spleen based on the expression of the β7-integrin important for the homing of mast cells to the intestine (Gurish et al. 2001), while the restricted basophil progenitor (BaP) can be identified in the BM based on high expression of FceR1a (Arinobu et al. 2005). Mast cell-committed progenitors are mainly found in the intestine where they can be identified based on the expression of FcgRII/III. Additional insight into the lineage relationships of multipotent myeloid progenitors was gained when the CMP compartment was subfractionated based on differential expression of Endoglin (CD105) and Slamf1 (CD150) (Pronk et al. 2007). Endoglin high cells were found to be highly primed toward erythroid lineage development, while cell surface expression of Slamf1 was associated with cells displaying erythroid, megakaryocyte (MkE), or both of these lineage potentials (Pronk et al. 2007). However, CD150[high] or CD105[high] progenitor populations did not contain a substantial fraction of cells with granulucyte/monocyte (GM) potential, an activity which, by contrast, was associated within the Slamf1[-]Endoglin[-] fraction of CMPs (Pronk et al. 2007). These data strongly argue that the previously assigned CMP fraction, for the largest part, represents a combination of already lineage-restricted cells, an idea supported by the finding that conditional mutation of the Pu.1 gene results in loss of CLPs, GMPs, and CMPs while the development of functional MEPs and granulocytes was slightly increased (Dakic et al. 2005; Iwasaki et al. 2005b). Even though these findings could be explained, by a direct loss of specific surface markers, it adds further support for the idea that the CMP population is a complex mixture of cell types, leaving the issue of how these earliest steps in MkE/GM and lymphoid restriction occur open for further investigations.

The segregation of MkE and lymphoid potential with retained GM potential in both these branches is interesting from an evolutionary perspective. Even though the full spectra of mature blood cell types is restricted to vertebra phyla, several examples of blood cell-like cells (amebocytes, interstitial cell, or neoblasts) can be found even in diploblastic animals lacking a developed mesoderm (reviewed in Hartenstein 2006). One example of this is in sponges, considered to represent one of the most basic organizations of metazoa, where a gelatinous matrix fills the space between the ectoderm and endoderm, a structure denoted mesoglea (reviewed in Harrison and De Vos 1991). This matrix contains motile cells, amebocytes, presumed to take up, digest, and transport nutrients within the body of the animal. The mesoglea of the slightly evolutionarily younger Cnidarians contains interstitial cells that may act as phagocytes, but such cells also possess stem cell characteristics since they can replace injured cells (Bode 1996; Miller et al. 2000). Thus, by functional criteria, these blood cell ancestors appear to be most similar to GM cells in vertebras. GM-like cells, hyaline hemocytes and granular hemocytes, can also be found in the vascular system of annelids (segmented worms) where these cells are suggested to participate in immune responses as well as in processes to prevent bleeding (reviewed in Hartenstein 2006). Clotting functions are even more refined in terrestrial arthropods, like drosophila, where injury results in release of clotting factors from storage granules and subsequent aggregation of hemocytes (Kanost et al. 2004; Theopold et al. 2002). Thus, the function of blood cells in the maintenance of the integrity of the circulatory system appears to be an early step in the evolution of the hematopoietic tree. Even if the blood/hemolymph in several invertebra taxa contains oxygene-carrying proteins such as hemoglobin or hemocyanin, dedicated oxygen-carrying cells represent the exception among invertebrates (reviewed in Hartenstein 2006). Therefore, from an evolutionary perspective, the generation of cells with a function resembling erythrocytes would appear to be more restricted than that of thrombocytes. However, the appearance of oxygen-carrying cells in some annelids, sipuncolides, lophophorates, and echinoderms (reviewed in Hartenstein 2006) could indicate that these cells have existed in early common ancestors, and that the refinement and specialization of these cells has been restricted to vertebra phyla. Vertebraes also represent the only species that has developed lymphoid cells capable of mounting specific immune responses, suggesting that the lymphoid lineage represent the most recent development in vertebras. Thus, based on evolutionary considerations, it appears reasonable that GM cells represent the founding class of hematopoietic cells and that all other blood cell lineages arise from GM-competent cells. It would also be reasonable that Mk and E-lineage cells develop from GM-competent cells, a view compatible with the model of a combined MkEGM progenitor (CMP). However, this would demand that lymphoid cells must develop as an independent branch directly from the multipotent progenitor, rather than from a separate branch of GM-competent cells. Even though the latter at this point cannot be excluded, it seems more reasonable that lymphoid cells develop from another branch of GM-competent cells, in a manner predicted by the presence of an LMPP that have a dramatically reduced MkE potential (Adolfsson et al. 2005).

Events Guiding the Control and Establishment of Hematopoietic Progenitors

To maintain appropriate homeostasis and to cope with varying physiological demands, the HSC has several options including migration to other tissues, apoptosis, entry into a quiescent G0 stage, symmetric cell division in order to give rise to a new HSC (self renewal), and asymmetric division, resulting in one LT-HSC and one progenitor cell that initiates differentiation into mature blood cell lineages (Fig. 2.2a). Events governing these early HSC fate decisions (Fig. 2.2) are largely unknown, although they are thought to be closely connected to signals emanating from their proximal environment (Attar and Scadden 2004; Wilson and Trumpp 2006). Even though it is formally possible that the HSC itself may become externally activated to develop along a given lineage, this would be a dangerous scenario as it could rather rapidly result in a depletion of the HSC pool upon excessive differentiation signals. Thus, it is reasonable to assume that initiation of differentiation in most cases is associated with cell division (Morrison and Kimble 2006). This process can either result in the generation of two identical cells, symmetric cell division, on which the environment could next induce inductive cues instructing cell fate (Fig. 2.2b). In this scenario, both cells can either lose HSC potential and initiate differentiation or they can both persist as HSCs. Alternatively, the cells could divide by asymmetric cell division, resulting in one HSC and one progenitor cell destined to differentiation (Morrison and Kimble 2006). To be able to cope with changes in demand and environment, it would be unlikely that HSC would divide only by asymmetric cell division since this would not allow for the expansion of the HSC pool observed after transplantation. However, the environmental conditions in a steady state scenario could well result in a process resembling asymmetric cell division since this would also simplify the regulation of the size of the HSC pool (Dick 2003; Morrison and Kimble 2006). Asymmetric cell divisions can be achieved in different ways and in the case of the HSC it is likely to involve either divisional asymmetry, where the unequal distribution of fate determining-factors in daughter cells results in differential phenotypes, or environmental asymmetry,

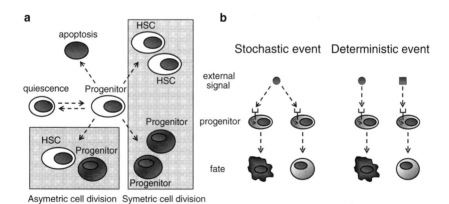

Fig. 2.2 (a) Possible stem cell fates used to balance HSC homeostasis and output of mature blood cells. Symmetric cell division leads to either expansion or depletion of HSCs, while asymmetric divisions lead to maintenance of HSC numbers. (b) Determination of HSC fate by stochastic and deterministic events. If fate is determined stochastically, the HSC randomly commits to a fate because of intrinsic events regardless of external signal. Whereas, if fate is regulated deterministically, HSC will adopt different fates as dictated by the extrinsic signal

where the daughter cells are exposed to different environmental cues in conjunction with the division, turning a symmetric division into an apparently asymmetric event resulting in one HSC and one multipotent progenitor destined to proliferation and subsequently differentiate into specific lineages (Morrison and Kimble 2006).

The cause of lineage restriction appears to be the result of a combination of cell extrinsic and intrinsic events and two major conceptual theories about cytokine function have been established (Fig. 2.2b). The first theory states that the actions of cytokines is permissive such that the central role for extrinsic signals is to give the cell the possibility to develop along a given lineage. This could be mediated through activation of survival factors like BCL-2 or through induction of proliferation. An alternative theory about how extrinsic signals impact the development of early hematopoietic progenitors is through instructive activity, meaning that cytokines directly guide the lineage decision events. These two theories are not mutually exclusive, and we favor the hypothesis that cytokines possess both permissive and instructive activity. Evidence for the former has been obtained from transgenic mice lacking functional genes-encoding cytokines associated with certain lineages such as EPO (Wu et al. 1995), G-CSF (Lieschke et al. 1994a), or GM-CSF (Stanley et al. 1994), mice which display rather mild steady-state phenotypes. This could, to some degree, be contributed to by functional redundancy and overlapping receptor expression; however, the combined inactivation of GM-CSF and G-CSF (Seymour et al. 1997) or GM-CSF and M-CSF (Lieschke et al. 1994b) does not result in an enhanced disturbance of hematopoiesis as compared to the single knock-out mice. Expression of CSF1-R (McArthur et al. 1994) or a constitutively active EpoR (Pharr et al. 1994) in early progenitor cells did not result in any significant changes when investigating the lineage choices of the multipotent progenitors. Further support for a predominant permissive action of cytokines was obtained through experiments utilizing chimeric receptors, for example studies of G-CSFR-EpoR fusion receptors, which result in a receptor that is still competent to promote development of granulocytes upon stimulation with G-CSF (Semerad et al. 1999). Further, a targeted knock-in of the G-CSFR intracellular domain into the Tpo-R locus is sufficient to rescue the thrombocytopenic phenotype of TpoR-deficient mice (Stoffel et al. 1999). Even if these data argue that the action of cytokines in early hematopoiesis is permissive, experiments where IL2Rβ or GM-CSFR were expressed in lymphoid-restricted CLPs resulted in the generation of GM cells after incubation with the corresponding cytokine (Kondo et al. 2000). These findings demonstrate that cytokine signaling, at least under these conditions, can directly instruct lineage choice. In addition to cytokines, other environmental cues such as Delta/Notch signaling may act in an instructive manner since Notch signaling is able to suppress myeloid conversion of pro-T cells by Pu.1 (Franco et al. 2006) and B-lymphoid development of CLPs (Schmitt and Zuniga-Pflucker 2002). Therefore, even though cytokines clearly influence the outcome of cell differentiation processes in the hematopoietic system, the precise understanding of these activities is obscured by the huge number of combinatorial activities achieved by their concerted action as well as the action of other regulatory factors in the BM microenvironment.

Even though external signals are clearly involved in the development of early progenitor cells, their action is likely to be modulated by the transcription factor networks active in a given cell at a defined stage of development both due to regulation of cytokine receptor expression and through differential responses to given signals. One functional antagonism found in the earliest progenitor involves the transcription factors GATA1 and Pu.1. In the absence of GATA1, embryos die from anemia due to an early differentiation block in erythropoiesis (Fujiwara et al. 1996) revealing a crucial role for this transcription factor in erythroid development. Furthermore, mutant mice carrying a functional modification of a DNAse1 hypersensitive region 5′ of the Gata1 gene displayed Mk lineage-specific phenotype with severe disruption of platelet production (Shivdasani et al. 1997). An instructive role of GATA1 has been suggested from the findings that GATA1 expression is able to revert the GM fate of avian MEPE26 cells (Kulessa et al. 1995) and to drive development of MkE cells from CLPs, GMPs as well as c-kit$^+$CD34$^+$ progenitor cells (Iwasaki et al. 2003, 2006). These data indicate that GATA1 expression dictates MkE fate in progenitor cells. However, the use of a transgenic mouse where GFP was put under the control of Gata1 regulatory sequences allowed for the identification of GFP$^+$CD34$^+$LSK cells with a dramatic reduction in their lymphoid potential but with a preserved or increased ability to generate MkE and to some extent also capable of generating GM cells (Arinobu et al. 2007). Thus, even if GATA1 appears to be a crucial factor in the specification of the MkE lineage, expression of GATA1 in vivo does not appear to be a definitive marker for MkE commitment. One potential explanation for these findings has evolved from experiments revealing that GATA1 can interact with and antagonize the function of Pu.1 (Nerlov and Graf 1998; Zhang et al. 1999), a positive regulator of GM fate (Nerlov and Graf 1998) and vice versa. Pu.1 is able to repress GATA1 function via direct protein/protein interactions, resulting in a repressor complex including both factors as well as the retinoblastoma protein (Rb), bound to the regulatory region of erythroid-associated genes (Rekhtman et al. 1999). Both factors are controlled by autoregulatory loops (Chen et al. 1995) and Pu.1 also appears to act in a feed forward loop by inducing the transcription factor CEBPα, a potent cofactor for GM development and by itself also an inducer of Pu.1 expression. The expression of both GATA1 and PU.1 in the LSK compartment indicate that this may be one of the earliest regulatory networks in blood cell differentiation but other regulatory loops are involved in downstream lineage choices. Taken together, these recent studies demonstrate that transcription factors are crucial regulators of hematopoiesis.

Epigenetic Modulation and Lineage Priming in Early Hematopoiesis

One of the favored explanations for how lineage potential is gradually lost during the differentiation of the HSC is that progressive epigenetic changes result in permanent silencing of genes that could participate in the development into alternative cell fates.

Epigenetic alterations are reversible changes in DNA structure in the cell mediated for instance by DNA methylation or histone modifications. This is to be compared to somatic changes that involves permanent loss of genetic material, for instance by chromosomal deletions in tumor cells. While loss of genetic material in normal somatic cells essentially is restricted to lymphoid cells that recombine their surface antigen receptors, epigenetic changes compose a crucial part of all developmental programs by generating specialized effector cells from immature progenitors. The most dramatic epigenetic changes are associated with early embryogenesis where the level of CpG methylation (mC) transiently drops from the 70–80% observed in normal somatic cells (Ehrlich et al. 1982), to less than one third of that level (Kafri et al. 1992). The normal mC levels are restored by de novo methylation by the time of implantation (Kafri et al. 1992). A striking feature of methylation patterns is the presence of 5'CpG islands potentially involved in gene regulation in about 60% of human genes (Antequera and Bird 1993). DNA methylation is believed to be of key importance for epigenetic memory after cell division, since the hemimethylated DNA generated after DNA synthesis is a preferred substrate for the DNA-methylating enzyme DNMT1 (Pradhan et al. 1999). De novo methylation in somatic cells in adults is not as common as in the developing embryo, but there is ample evidence of progressive methylation of CpG islands during aging and in abnormal cells such as cancers (Lund and van Lohuizen 2004). Even though gene methylation appears to be a powerful mechanism for stable silencing of a gene, most experimental evidence points to this as secondary event, resulting from gene silencing rather then being the primary mechanisms of gene repression. Thus, to penetrate the issue of regulation of developmentally and disease-associated genes, there is a need to consider the other factors involved in the formation of higher order of chromatin structures. Such factors include modifications of histones and histone-associated proteins (Giadrossi et al. 2007; Spivakov and Fisher 2007). The composition of the histones is in some instances linked to the methylation status of the DNA but, in contrast to the slow process of de novo methylation, histone proteins can be rapidly posttranslationally modified by transcription factors and their coactivator/repressor proteins, thereby resulting in dramatic and rapid altera-tions in the DNA structure. Histone modifications include, amongst others, acetylation, phosphorylation, methylation, sumoylation and ubiquitination, modifications that correlate with distinct states of transcriptional competence (Giadrossi et al. 2007; Spivakov and Fisher 2007). In addition, there can be differences in the composition of the histone core protein complex in active and silent genes (Ahmad and Henikoff 2002). This heterogeneity has been utilized to investigate epigenetic changes in developing cells and provided some evidence for the idea that chromatin displays a more "active" configuration in early as compared to later lineage-restricted progenitor cells (Giadrossi et al. 2007; Spivakov and Fisher 2007). Links between multipo-tency and epigenetic status have been suggested from studies of embryonic stem (ES) cells, which display a unique pattern of histone modifications and also display a generally higher turnover of nucleosomes than differentiated cells (Meshorer et al. 2006). Increased nucleosome turnover is believed to be a result of high chromatin accessibility, which is in line with the finding that undifferentiated cells have a higher content of acetylated histone 3 (H3), histone 4 (H4), and di- and tri-methylated H3K4, all modifications associated with active genes (Giadrossi et al. 2007; Spivakov and Fisher 2007). However, even though several developmentally regulated genes are associated with these forms of activation markers in ES cells, their expression is silenced possibly due to the combined presence of repressive histone modifications such as H3K27 trimethylation placing the gene in a poised, but not active, state (Bernstein et al. 2006). Another indication that developmentally important genes are kept in a special configuration in ES cells derives from studies of replication timing of defined genes. Replication timing is believed to be highly dependent on chromatin structure, and when the timing of replication of the neuronal restricted genes Pax6, Sox2, and Math1 was investigated in ES cells, it became clear that they were replicated early in S-phase but in HSCs, that have lost neuronal potential, these genes are replicated later (Azuara et al. 2006). Direct evidence for a need for proper control of epigenetic modifications in early hematopoietic progenitors comes from studies of the polycomb protein Bmi1. This factor is part of the Polycomb repressor complex 1 (PRC1) that contains several subunits believed to act in concert to maintain genes in a silenced state during proliferation (Rajasekhar and Begemann 2007). Targeted deletion of the Bmi1 gene results in a loss of long-term reconstitution capacity of BM as well as fetal liver HSCs (Park et al. 2003) indicating that this factor is critical for the self-renewal capacity of HSCs. This idea is further supported by studies where Bmi1 was retrovirally overexpressed in LSK cells, which results in an increase in in vivo reconstitution activity and increased numbers of symmetric cell divisions in vitro (Iwama et al. 2004). These results support the idea that polycomb-containing complexes are involved in repressing genes involved in HSC differentiation during proliferation. Further support for a generalized role of polycomb in such processes is derived from studies where ectopic expression of the PRC2 complex-associated histone methyl transferase enhancer of zeste 2 (Ezh2) resulted in increased ability of LSK cells to repopulate hosts during serial transplantations (Kamminga et al. 2006). The PRC2 complex is believed to be involved in de novo silencing of genes due to its ability to introduce H3K27 trimethylation, thereby creating a substrate for associated DNA methylation activities by DNMT1 (Vire et al. 2006). Direct evidence for epigenetic changes in association with blood cell differentiation has been obtained since changes in the histone and methylation patterns of lineage-associated genes in HSCs as well as lineage-restricted progenitors have been characterized (Attema et al. 2007). Thus, several lines of investigations support the idea that epigenetic changes are associated with and may be crucial for normal hematopoiesis; however, the understanding of these processes are currently not as developed as those involved in the differentiation of ES cells.

In totipotent ES cells, the transcription factors Nanog and Oct4, both involved in the preservation of multipotency, bind directly to the regulatory regions of key genes involved in the differentiation of specific lineages, resulting in silencing of these target genes (Giadrossi et al. 2007; Spivakov and Fisher 2007). Thus, it appears that in ES cells multipotency is associated with an active silencing of genes involved in the development of lineage-restricted progenitors. A similar mechanism could also explain the multipotency associated with HSCs; however, when investigating the phenotypically defined CD34⁻LSK HSC compartment, such cells express both cytokine receptors and transcription factors known to be key regulators of downstream lineage specification (Hu et al. 1997; Mansson et al. 2007; Miyamoto et al. 2002). This data has been taken as support for the hypothesis that lineage priming occurs already at the level of HSCs. However, singe cell PCR analysis revealed that one and the same CD34⁻LSK cell can express genes associated with Mk, E, and GM but not lymphoid programs (Mansson et al. 2007; Miyamoto et al. 2002) even though the HSC clearly possess lymphoid potential. The notion that transcription of lineage-restricted genes in HSCs is not associated with reduced potency is supported from studies using mice transgenic for a CRE recombinase gene controlled by the lysosome M regulatory elements and a CRE-dependent ROSA26GFP gene (Ye et al. 2003). In these mice, only cells that derive from cells that have expressed the lysozyme M-driven CRE will express GFP. Importantly, green cells could be found among lymphoid cells indicating that the expression of this myeloid gene in HSCs or multipotent progenitors was not associated with myeloid lineage restriction (Ye et al. 2003). It remains to be determined whether these transcripts are merely a consequence of maintenance of genes in a primed state or whether they produce sufficient amounts of protein to set up regulatory networks that may have a direct effect on the fate of the HSC. Nonetheless, the presence of these transcripts suggests that HSC maintain multipotency by different means as compared to ES cells.

Concluding Remarks

Even though there still remains a discussion regarding whether the earliest lineage restriction event downstream of HSCs involves a CMP or CLP fate (Forsberg et al. 2006), accumulating evidence supports the idea of a MkE/lymphoid separation with preserved GM potential in both pathways (Adolfsson et al. 2005; Arinobu et al. 2007) that represent the first line of lineage restriction events in adult mice. The situation may be different in fetal development since progenitor cells differ with regard to gene and marker expression patterns and while transcription factors such as Scl/Tal1, Aml1/Runx1, and Sox17 are crucial for the development of fetal HSCs, these genes all appear to be dispensable for the function and maintenance of adult HSCs (Dzierzak and Speck 2008; Kim et al. 2007). The nature of the adult HSCs also appears to undergo a phenotypic change associated with aging. Even though the actual numbers of cells with a stem cell surface phenotype increase dramatically in aged mice, the differentiation potential is altered as compared to adult progenitor cells, with a high degree of development of myeloid progeny at the expense of lymphopoiesis, and B cell production in particular (Rossi et al. 2007, 2008). Even though this effect may be in part dependent on changes in the BM microenvironment, transplantation of HSCs from aged mice into young hosts revealed the same alterations in development as observed when studying aged mice in steady-state suggesting that these effects have a degree of cell autonomy (Rossi et al. 2007, 2008). The understanding of changes in progenitor compartments with age may be of large importance also for the understanding of human leukemias since many of these are age associated (Rossi et al. 2008). Thus, the work focused to resolve the order of events and the molecular mechanisms regulating hematopoietic development in health and disease is likely to proceed until a complete understanding is achieved.

References

Adolfsson, J., Borge, O. J., Bryder, D., Theilgaard-Monch, K., Astrand-Grundstrom, I., Sitnicka, E., Sasaki, Y., and Jacobsen, S. E. (2001). Upregulation of Flt3 expression within the bone marrow Lin(-)Sca1(+)c-kit(+) stem cell compartment is accompanied by loss of self-renewal capacity. Immunity 15, 659–669.

Adolfsson, J., Mansson, R., Buza-Vidas, N., Hultquist, A., Liuba, K., Jensen, C. T., Bryder, D., Yang, L., Borge, O. J., Thoren, L. A., et al. (2005). Identification of Flt3+ lympho-myeloid stem cells lacking erythro-megakaryocytic potential: a revised road map for adult blood lineage commitment. Cell 121, 295–306.

Ahmad, K., and Henikoff, S. (2002). The histone variant H3.3 marks active chromatin by replication-independent nucleosome assembly. Mol Cell 9, 1191–1200.

Akashi, K., Traver, D., Miyamoto, T., and Weissman, I. L. (2000). A clonogenic common myeloid progenitor that gives rise to all myeloid lineages. Nature 404, 193–197.

Allman, D., Sambandam, A., Kim, S., Miller, J. P., Pagan, A., Well, D., Meraz, A., and Bhandoola, A. (2003). Thymopoiesis independent of common lymphoid progenitors. Nat Immunol 4, 168–174.

Alvarez-Silva, M., Belo-Diabangouaya, P., Salaun, J., and Dieterlen-Lievre, F. (2003). Mouse placenta is a major hematopoietic organ. Development 130, 5437–5444.

Antequera, F., and Bird, A. (1993). Number of CpG islands and genes in human and mouse. Proc Natl Acad Sci USA 90, 11995–11999.

Arinobu, Y., Mizuno, S., Chong, Y., Shigematsu, H., Iino, T., Iwasaki, H., Graf, T., Mayfield, R., Chan, S., Kastner, P., and Akashi, K. (2007). Reciprocal activation of Gata-1 and PU.1 marks initial specification of hematopoietic stem cells into myeloerythroid and myelolymphoid lineages. Cell Stem Cell 1, 416–427.

Arinobu, Y., Iwasaki, H., Gurish, M. F., Mizuno, S., Shigematsu, H., Ozawa, H., Tenen, D. G., Austen, K. F., and Akashi, K. (2005). Developmental checkpoints of the basophil/mast cell lineages in adult murine hematopoiesis. Proc Natl Acad Sci USA 102, 18105–18110.

Attar, E. C., and Scadden, D. T. (2004). Regulation of hematopoietic stem cell growth. Leukemia 18, 1760–1768.

Attema, J. L., Papathanasiou, P., Forsberg, E. C., Xu, J., Smale, S. T., and Weissman, I. L. (2007). Epigenetic characterization of hematopoietic stem cell differentiation using miniChIP and bisulfite sequencing analysis. Proc Natl Acad Sci USA 104, 12371–12376.

Azuara, V., Perry, P., Sauer, S., Spivakov, M., Jorgensen, H. F., John, R. M., Gouti, M., Casanova, M., Warnes, G., Merkenschlager, M., and Fisher, A. G. (2006). Chromatin signatures of pluripotent cell lines. Nat Cell Biol 8, 532–538.

Bain, G., Maandag, E. C. R., Izon, D. J., Amsen, D., Kruisbeek, A. M., Weintraub, B. C., Kroop, I., Schlissel, M. S., Feeney, A. J., van Roon, M., et al. (1994). E2A proteins are required for proper B cell development and initiation of immunoglobulin gene rearrangements. Cell 79, 885–892.

Becker, A. J., Mc, C. E., and Till, J. E. (1963). Cytological demonstration of the clonal nature of spleen colonies derived from transplanted mouse marrow cells. Nature 197, 452–454.

Bernstein, B. E., Mikkelsen, T. S., Xie, X., Kamal, M., Huebert, D. J., Cuff, J., Fry, B., Meissner, A., Wernig, M., Plath, K., et al. (2006). A bivalent chromatin structure marks key developmental genes in embryonic stem cells. Cell *125*, 315–326.

Bode, H. R. (1996). The interstitial cell lineage of hydra: a stem cell system that arose early in evolution. J Cell Sci *109* (Pt 6) , 1155–1164.

Bradford, G. B., Williams, B., Rossi, R., and Bertoncello, I. (1997). Quiescence, cycling, and turnover in the primitive hematopoietic stem cell compartment. Exp Hematol *25*, 445–453.

Bryder, D., Rossi, D. J., and Weissman, I. L. (2006). Hematopoietic stem cells: the paradigmatic tissue-specific stem cell. Am J Pathol *169*, 338–346.

Chen, C. Z., Li, M., de Graaf, D., Monti, S., Gottgens, B., Sanchez, M. J., Lander, E. S., Golub, T. R., Green, A. R., and Lodish, H. F. (2002). Identification of endoglin as a functional marker that defines long-term repopulating hematopoietic stem cells. Proc Natl Acad Sci USA *99*, 15468–15473.

Chen, H., Ray-Gallet, D., Zhang, P., Hetherington, C. J., Gonzalez, D. A., Zhang, D. E., Moreau-Gachelin, F., and Tenen, D. G. (1995). PU.1 (Spi-1) autoregulates its expression in myeloid cells. Oncogene *11*, 1549–1560.

Cheng, T. (2004). Cell cycle inhibitors in normal and tumor stem cells. Oncogene *23*, 7256–7266.

Cheng, T., Rodrigues, N., Shen, H., Yang, Y., Dombkowski, D., Sykes, M., and Scadden, D. T. (2000). Hematopoietic stem cell quiescence maintained by p21cip1/waf1. Science *287*, 1804–1808.

Cheshier, S. H., Morrison, S. J., Liao, X., and Weissman, I. L. (1999). In vivo proliferation and cell cycle kinetics of long-term self-renewing hematopoietic stem cells. Proc Natl Acad Sci USA *96*, 3120–3125.

Choi, J. K., Shen, C.-P., Radomska, H. S., Eckhardt, L. A., and Kadesch, T. (1996). E47 activates the Ig-heavy chain and TdT loci in non-B cells. EMBO J *15*, 5014–5021.

Choi, K., Kennedy, M., Kazarov, A., Papadimitriou, J. C., and Keller, G. (1998). A common precursor for hematopoietic and endothelial cells. Development *125*, 725–732.

Christensen, J. L., and Weissman, I. L. (2001). Flk-2 is a marker in hematopoietic stem cell differentiation: a simple method to isolate long-term stem cells. Proc Natl Acad Sci USA *98*, 14541–14546.

Cumano, A., Ferraz, J. C., Klaine, M., Di Santo, J. P., and Godin, I. (2001). Intraembryonic, but not yolk sac hematopoietic precursors, isolated before circulation, provide long-term multilineage reconstitution. Immunity *15*, 477–485.

Cumano, A., and Godin, I. (2001). Pluripotent hematopoietic stem cell development during embryogenesis. Curr Opin Immunol *13*, 166–171.

Dakic, A., Metcalf, D., Di Rago, L., Mifsud, S., Wu, L., and Nutt, S. L. (2005). PU.1 regulates the commitment of adult hematopoietic progenitors and restricts granulopoiesis. J Exp Med *201*, 1487–1502.

Dias, S., Silva, H., Jr., Cumano, A., and Vieira, P. (2005). Interleukin-7 is necessary to maintain the B cell potential in common lymphoid progenitors. J Exp Med *201*, 971–979.

Dias S, Månsson R, Gurbuxani S, Sigvardsson M, Kee BL (2008). E2A proteins promote development of lymphoid-primed multipotent progenitors. Immunity. Aug;29(2):217–27.

Dick, J. E. (2003). Stem cells: self-renewal writ in blood. Nature *423*, 231–233.

Dick, J. E., Bhatia, M., Gan, O., Kapp, U., and Wang, J. C. (1997). Assay of human stem cells by repopulation of NOD/SCID mice. Stem Cells *15* Suppl 1, 199–203; discussion 204–207.

Dick, J. E., Magli, M. C., Huszar, D., Phillips, R. A., and Bernstein, A. (1985). Introduction of a selectable gene into primitive stem cells capable of long-term reconstitution of the hemopoietic system of W/Wv mice. Cell *42*, 71–79.

Dieterlen-Lievre, F. (1975). On the origin of haemopoietic stem cells in the avian embryo: an experimental approach. J Embryol Exp Morphol *33*, 607–619.

Dzierzak, E., and Speck, N. A. (2008). Of lineage and legacy: the development of mammalian hematopoietic stem cells. Nat Immunol *9*, 129–136.

Ehrlich, M., Gama-Sosa, M. A., Huang, L. H., Midgett, R. M., Kuo, K. C., McCune, R. A., and Gehrke, C. (1982). Amount and distribution of 5-methylcytosine in human DNA from different types of tissues of cells. Nucleic Acids Res *10*, 2709–2721.

Ema, H., and Nakauchi, H. (2000). Expansion of hematopoietic stem cells in the developing liver of a mouse embryo. Blood *95*, 2284–2288.

Fehling, H. J., Lacaud, G., Kubo, A., Kennedy, M., Robertson, S., Keller, G., and Kouskoff, V. (2003). Tracking mesoderm induction and its specification to the hemangioblast during embryonic stem cell differentiation. Development *130*, 4217–4227.

Ferkowicz, M. J., and Yoder, M. C. (2005). Blood island formation: longstanding observations and modern interpretations. Exp Hematol *33*, 1041–1047.

Fogg, D. K., Sibon, C., Miled, C., Jung, S., Aucouturier, P., Littman, D. R., Cumano, A., and Geissmann, F. (2006). A clonogenic bone marrow progenitor specific for macrophages and dendritic cells. Science *311*, 83–87.

Forsberg, E. C., Serwold, T., Kogan, S., Weissman, I. L., and Passegue, E. (2006). New evidence supporting megakaryocyte-erythrocyte potential of flk2/flt3+ multipotent hematopoietic progenitors. Cell *126*, 415–426.

Franco, C. B., Scripture-Adams, D. D., Proekt, I., Taghon, T., Weiss, A. H., Yui, M. A., Adams, S. L., Diamond, R. A., and Rothenberg, E. V. (2006). Notch/Delta signaling constrains reengineering of pro-T cells by PU.1. Proc Natl Acad Sci USA *103*, 11993–11998.

Fujiwara, Y., Browne, C. P., Cunniff, K., Goff, S. C., and Orkin, S. H. (1996). Arrested development of embryonic red cell precursors in mouse embryos lacking transcription factor GATA-1. Proc Natl Acad Sci USA *93*, 12355–12358.

Gekas, C., Dieterlen-Lievre, F., Orkin, S. H., and Mikkola, H. K. (2005). The placenta is a niche for hematopoietic stem cells. Dev Cell *8*, 365–375.

Gering, M., Rodaway, A. R., Gottgens, B., Patient, R. K., and Green, A. R. (1998). The SCL gene specifies haemangioblast development from early mesoderm. EMBO J *17*, 4029–4045.

Giadrossi, S., Dvorkina, M., and Fisher, A. G. (2007). Chromatin organization and differentiation in embryonic stem cell models. Curr Opin Genet Dev *17*, 132–138.

Godin, I., Garcia-Porrero, J. A., Dieterlen-Lievre, F., and Cumano, A. (1999). Stem cell emergence and hemopoietic activity are incompatible in mouse intraembryonic sites. J Exp Med *190*, 43–52.

Gurish, M. F., Tao, H., Abonia, J. P., Arya, A., Friend, D. S., Parker, C. M., and Austen, K. F. (2001). Intestinal mast cell progenitors require CD49dbeta7 (alpha4beta7 integrin) for tissue-specific homing. J Exp Med *194*, 1243–1252.

Harrison, F., and De Vos, L. (1991). Porifera. In: Microscopic Anatomy of Invertebrates. Wiley, New York, pp. 29–90.

Hartenstein, V. (2006). Blood cells and blood cell development in the animal kingdom. Annu Rev Cell Dev Biol *22*, 677–712.

Hu, M., Krause, D., Greaves, M., Sharkis, S., Dexter, M., Heyworth, C., and Enver, T. (1997). Multilineage gene expression precedes commitment in the hemopoietic system. Genes Dev *11*, 774–785.

Igarashi, H., Gregory, S. C., Yokota, T., Sakaguchi, N., and Kincade, P. W. (2002). Transcription from the RAG1 locus marks the earliest lymphocyte progenitors in bone marrow. Immunity *17*, 117–130.

Ikuta, K., and Weissman, I. L. (1992). Evidence that hematopoietic stem cells express mouse c-kit but do not depend on steel factor for their generation. Proc Natl Acad Sci USA *89*, 1502–1506.

Iwama, A., Oguro, H., Negishi, M., Kato, Y., Morita, Y., Tsukui, H., Ema, H., Kamijo, T., Katoh-Fukui, Y., Koseki, H., et al. (2004). Enhanced self-renewal of hematopoietic stem cells mediated by the polycomb gene product Bmi-1. Immunity *21*, 843–851.

Iwasaki, H., Mizuno, S., Arinobu, Y., Ozawa, H., Mori, Y., Shigematsu, H., Takatsu, K., Tenen, D. G., and Akashi, K. (2006). The order of expression of transcription factors directs hierarchical specification of hematopoietic lineages. Genes Dev *20*, 3010–3021.

Iwasaki, H., Mizuno, S., Mayfield, R., Shigematsu, H., Arinobu, Y., Seed, B., Gurish, M. F., Takatsu, K., and Akashi, K. (2005a). Identification of eosinophil lineage-committed progenitors in the murine bone marrow. J Exp Med *201*, 1891–1897.

Iwasaki, H., Mizuno, S., Wells, R. A., Cantor, A. B., Watanabe, S., and Akashi, K. (2003). GATA-1 converts lymphoid and myelomonocytic progenitors into the megakaryocyte/erythrocyte lineages. Immunity *19*, 451–462.

Iwasaki, H., Somoza, C., Shigematsu, H., Duprez, E. A., Iwasaki-Arai, J., Mizuno, S., Arinobu, Y., Geary, K., Zhang, P., Dayaram, T., et al. (2005b). Distinctive and indispensable roles of PU.1 in maintenance of hematopoietic stem cells and their differentiation. Blood *106*, 1590–1600.

Kafri, T., Ariel, M., Brandeis, M., Shemer, R., Urven, L., McCarrey, J., Cedar, H., and Razin, A. (1992). Developmental pattern of gene-specific DNA methylation in the mouse embryo and germ line. Genes Dev *6*, 705–714.

Kamminga, L. M., Bystrykh, L. V., de Boer, A., Houwer, S., Douma, J., Weersing, E., Dontje, B., and de Haan, G. (2006). The Polycomb group gene Ezh2 prevents hematopoietic stem cell exhaustion. Blood *107*, 2170–2179.

Kanost, M. R., Jiang, H., and Yu, X. Q. (2004). Innate immune responses of a lepidopteran insect, Manduca sexta. Immunol Rev *198*, 97–105.

Keller, G., Paige, C., Gilboa, E., and Wagner, E. F. (1985). Expression of a foreign gene in myeloid and lymphoid cells derived from multipotent haematopoietic precursors. Nature *318*, 149–154.

Kiel, M. J., Yilmaz, O. H., Iwashita, T., Yilmaz, O. H., Terhorst, C., and Morrison, S. J. (2005). SLAM family receptors distinguish hematopoietic stem and progenitor cells and reveal endothelial niches for stem cells. Cell *121*, 1109–1121.

Kikuchi, K., Lai, A. Y., Hsu, C. L., and Kondo, M. (2005). IL-7 receptor signaling is necessary for stage transition in adult B cell development through up-regulation of EBF. J Exp Med *201*, 1197–1203.

Kim, I., Saunders, T. L., and Morrison, S. J. (2007). Sox17 dependence distinguishes the transcriptional regulation of fetal from adult hematopoietic stem cells. Cell *130*, 470–483.

Kondo, M., Scherer, D. C., Miyamoto, T., King, A. G., Akashi, K., Sugamura, K., and Weissman, I. L. (2000). Cell-fate conversion of lymphoid-committed progenitors by instructive actions of cytokines. Nature *407*, 383–386.

Kondo, M., Weissman, I. L., and Akashi, K. (1997). Identification of clonogenic common lymphoid progenitors in mouse bone marrow. Cell *91*, 661–672.

Kopp, H. G., Avecilla, S. T., Hooper, A. T., and Rafii, S. (2005). The bone marrow vascular niche: home of HSC differentiation and mobilization. Physiology (Bethesda) *20*, 349–356.

Kulessa, H., Frampton, J., and Graf, T. (1995). GATA-1 reprograms avian myelomonocytic cell lines into eosinophils, thromboblasts, and erythroblasts. Genes Dev *9*, 1250–1262.

Laiosa, C. V., Stadtfeld, M., Xie, H., de Andres-Aguayo, L., and Graf, T. (2006). Reprogramming of committed T cell progenitors to macrophages and dendritic cells by C/EBP alpha and PU.1 transcription factors. Immunity *25*, 731–744.

Lemischka, I. R., Raulet, D. H., and Mulligan, R. C. (1986). Developmental potential and dynamic behavior of hematopoietic stem cells. Cell *45*, 917–927.

Li, C. L., and Johnson, G. R. (1995). Murine hematopoietic stem and progenitor cells: I. Enrichment and biologic characterization. Blood *85*, 1472–1479.

Lieschke, G. J., Grail, D., Hodgson, G., Metcalf, D., Stanley, E., Cheers, C., Fowler, K. J., Basu, S., Zhan, Y. F., and Dunn, A. R. (1994a). Mice lacking granulocyte colony-stimulating factor have chronic neutropenia, granulocyte and macrophage progenitor cell deficiency, and impaired neutrophil mobilization. Blood *84*, 1737–1746.

Lieschke, G. J., Stanley, E., Grail, D., Hodgson, G., Sinickas, V., Gall, J. A., Sinclair, R. A., and Dunn, A. R. (1994b). Mice lacking both macrophage- and granulocyte-macrophage colony-stimulating factor have macrophages and coexistent osteopetrosis and severe lung disease. Blood *84*, 27–35.

Lord, B. I., and Hendry, J. H. (1972). The distribution of haemopoietic colony-forming units in the mouse femur, and its modification by X-rays. Br J Radiol *45*, 110–115.

Lord, B. I., Testa, N. G., and Hendry, J. H. (1975). The relative spatial distributions of CFUs and CFUc in the normal mouse femur. Blood *46*, 65–72.

Lorenz, E., Uphoff, D., Reid, T. R., and Shelton, E. (1951). Modification of irradiation injury in mice and guinea pigs by bone marrow injections. J Natl Cancer Inst *12*, 197–201.

Lund, A. H., and van Lohuizen, M. (2004). Epigenetics and cancer. Genes Dev *18*, 2315–2335.

Maloney, M. A., and Patt, H. M. (1975). On the origin of hematopoietic stem cells after local marrow extirpation. Proc Soc Exp Biol Med 149, 94–97.

Mansson, R., Hultquist, A., Luc, S., Yang, L., Anderson, K., Kharazi, S., Al-Hashmi, S., Liuba, K., Thoren, L., Adolfsson, J., et al. (2007). Molecular evidence for hierarchical transcriptional lineage priming in fetal and adult stem cells and multipotent progenitors. Immunity 26, 407–419.

Massberg, S., Schaerli, P., Knezevic-Maramica, I., Kollnberger, M., Tubo, N., Moseman, E. A., Huff, I. V., Junt, T., Wagers, A. J., Mazo, I. B., and von Andrian, U. H. (2007). Immunosurveillance by hematopoietic progenitor cells trafficking through blood, lymph, and peripheral tissues. Cell 131, 994–1008.

McArthur, G. A., Rohrschneider, L. R., and Johnson, G. R. (1994). Induced expression of c-fms in normal hematopoietic cells shows evidence for both conservation and lineage restriction of signal transduction in response to macrophage colony-stimulating factor. Blood 83, 972–981.

McGrath, K. E., Koniski, A. D., Malik, J., and Palis, J. (2003). Circulation is established in a stepwise pattern in the mammalian embryo. Blood 101, 1669–1676.

McKenzie, J. L., Takenaka, K., Gan, O. I., Doedens, M., and Dick, J. E. (2007). Low rhodamine 123 retention identifies long-term human hematopoietic stem cells within the Lin-CD34+CD38– population. Blood 109, 543–545.

Medina, K. L., Pongubala, J. M., Reddy, K. L., Lancki, D. W., Dekoter, R., Kieslinger, M., Grosschedl, R., and Singh, H. (2004). Assembling a gene regulatory network for specification of the B cell fate. Dev Cell 7, 607–617.

Medvinsky, A., and Dzierzak, E. (1996). Definitive hematopoiesis is autonomously initiated by the AGM region. Cell 86, 897–906.

Meshorer, E., Yellajoshula, D., George, E., Scambler, P. J., Brown, D. T., and Misteli, T. (2006). Hyperdynamic plasticity of chromatin proteins in pluripotent embryonic stem cells. Dev Cell 10, 105–116.

Miller, M. A., Technau, U., Smith, K. M., and Steele, R. E. (2000). Oocyte development in Hydra involves selection from competent precursor cells. Dev Biol 224, 326–338.

Miyamoto, T., Iwasaki, H., Reizis, B., Ye, M., Graf, T., Weissman, I. L., and Akashi, K. (2002). Myeloid or lymphoid promiscuity as a critical step in hematopoietic lineage commitment. Dev Cell 3, 137–147.

Moore, M. A., and Metcalf, D. (1970). Ontogeny of the haemopoietic system: yolk sac origin of in vivo and in vitro colony forming cells in the developing mouse embryo. Br J Haematol 18, 279–296.

Morrison, S. J., Hemmati, H. D., Wandycz, A. M., and Weissman, I. L. (1995a). The purification and characterization of fetal liver hematopoietic stem cells. Proc Natl Acad Sci USA 92, 10302–10306.

Morrison, S. J., and Kimble, J. (2006). Asymmetric and symmetric stem-cell divisions in development and cancer. Nature 441, 1068–1074.

Morrison, S. J., Uchida, N., and Weissman, I. L. (1995b). The biology of hematopoietic stem cells. Annu Rev Cell Dev Biol 11, 35–71.

Nakorn, T. N., Miyamoto, T., and Weissman, I. L. (2003). Characterization of mouse clonogenic megakaryocyte progenitors. Proc Natl Acad Sci USA 100, 205–210.

Nerlov, C., and Graf, T. (1998). PU.1 induces myeloid lineage commitment in multipotent hematopoietic progenitors. Genes Dev 12, 2403–2412.

Nutt, S. L., Heavey, B., Rolink, A. G., and Busslinger, M. (1999). Commitment to the B-lymphoid lineage depends on the transcription factor Pax5 [see comments]. Nature 401, 556–562.

Nutt, S. L., Metcalf, D., D'Amico, A., Polli, M., and Wu, L. (2005). Dynamic regulation of PU.1 expression in multipotent hematopoietic progenitors. J Exp Med 201, 221–231.

Ogawa, M. (1993). Differentiation and proliferation of hematopoietic stem cells. Blood 81, 2844–2853.

Osawa, M., Hanada, K., Hamada, H., and Nakauchi, H. (1996). Long-term lymphohematopoietic reconstitution by a single CD34-low/negative hematopoietic stem cell. Science 273, 242–245.

Ottersbach, K., and Dzierzak, E. (2005). The murine placenta contains hematopoietic stem cells within the vascular labyrinth region. Dev Cell 8, 377–387.

Palis, J., Chan, R. J., Koniski, A., Patel, R., Starr, M., and Yoder, M. C. (2001). Spatial and temporal emergence of high proliferative potential hematopoietic precursors during murine embryogenesis. Proc Natl Acad Sci USA 98, 4528–4533.

Park, I. K., Qian, D., Kiel, M., Becker, M. W., Pihalja, M., Weissman, I. L., Morrison, S. J., and Clarke, M. F. (2003). Bmi-1 is required for maintenance of adult self-renewing haematopoietic stem cells. Nature *423*, 302–305.

Patt, H. M., and Maloney, M. A. (1972). Bone formation and resorption as a requirement for marrow development. Proc Soc Exp Biol Med *140*, 205–207.

Pharr, P. N., Ogawa, M., Hofbauer, A., and Longmore, G. D. (1994). Expression of an activated erythropoietin or a colony-stimulating factor 1 receptor by pluripotent progenitors enhances colony formation but does not induce differentiation. Proc Natl Acad Sci USA *91*, 7482–7486.

Pongubala, J. M., Northrup, D. L., Lancki, D. W., Medina, K. L., Treiber, T., Bertolino, E., Thomas, M., Grosschedl, R., Allman, D., and Singh, H. (2008). Transcription factor EBF restricts alternative lineage options and promotes B cell fate commitment independently of Pax5. Nat Immunol *9*, 203–215.

Pradhan, S., Bacolla, A., Wells, R. D., and Roberts, R. J. (1999). Recombinant human DNA (cytosine-5) methyltransferase. I. Expression, purification, and comparison of de novo and maintenance methylation. J Biol Chem *274*, 33002–33010.

Pronk, C., Rossi, D., Månsson, R., Attema, J., Norddahl, G., Chan, C., Sigvardsson, M., Weissman, I., and Bryder, D. (2007). Elucidation of the phenotype, functional, and molecular topography of a myeloerythroid progenitor cell hierarchy. Cell Stem Cell *1*, 428–442.

Rajasekhar, V. K., and Begemann, M. (2007). Concise review: roles of polycomb group proteins in development and disease: a stem cell perspective. Stem Cells *25*, 2498–2510.

Randall, T. D., and Weissman, I. L. (1997). Phenotypic and functional changes induced at the clonal level in hematopoietic stem cells after 5-fluorouracil treatment. Blood *89*, 3596–3606.

Rekhtman, N., Radparvar, F., Evans, T., and Skoultchi, A. I. (1999). Direct interaction of hematopoietic transcription factors PU.1 and GATA-1: functional antagonism in erythroid cells. Genes Dev *13*, 1398–1411.

Rizo, A., Vellenga, E., de Haan, G., and Schuringa, J. J. (2006). Signaling pathways in self-renewing hematopoietic and leukemic stem cells: do all stem cells need a niche? Hum Mol Genet *15* Spec No 2, R210–R219.

Rossi, D. J., Bryder, D., and Weissman, I. L. (2007). Hematopoietic stem cell aging: mechanism and consequence. Exp Gerontol *42*, 385–390.

Rossi, D. J., Jamieson, C. H., and Weissman, I. L. (2008). Stems cells and the pathways to aging and cancer. Cell *132*, 681–696.

Sambandam, A., Maillard, I., Zediak, V. P., Xu, L., Gerstein, R. M., Aster, J. C., Pear, W. S., and Bhandoola, A. (2005). Notch signaling controls the generation and differentiation of early T lineage progenitors. Nat Immunol *6*, 663–670.

Samokhvalov, I. M., Samokhvalova, N. I., and Nishikawa, S. (2007). Cell tracing shows the contribution of the yolk sac to adult haematopoiesis. Nature *446*, 1056–1061.

Schmitt, T. M., and Zuniga-Pflucker, J. C. (2002). Induction of T cell development from hematopoietic progenitor cells by delta-like-1 in vitro. Immunity *17*, 749–756.

Schofield, R. (1978). The relationship between the spleen colony-forming cell and the haemopoietic stem cell. Blood Cells *4*, 7–25.

Schwarz, B. A., and Bhandoola, A. (2004). Circulating hematopoietic progenitors with T lineage potential. Nat Immunol 5, 953–960.

Semerad, C. L., Poursine-Laurent, J., Liu, F., and Link, D. C. (1999). A role for G-CSF receptor signaling in the regulation of hematopoietic cell function but not lineage commitment or differentiation. Immunity *11*, 153–161.

Seymour, J. F., Lieschke, G. J., Grail, D., Quilici, C., Hodgson, G., and Dunn, A. R. (1997). Mice lacking both granulocyte colony-stimulating factor (CSF) and granulocyte-macrophage CSF have impaired reproductive capacity, perturbed neonatal granulopoiesis, lung disease, amyloidosis, and reduced long-term survival. Blood *90*, 3037–3049.

Shivdasani, R. A., Fujiwara, Y., McDevitt, M. A., and Orkin, S. H. (1997). A lineage-selective knockout establishes the critical role of transcription factor GATA-1 in megakaryocyte growth and platelet development. EMBO J *16*, 3965–3973.

Sitnicka, E., Brakebusch, C., Martensson, I. L., Svensson, M., Agace, W. W., Sigvardsson, M., Buza-Vidas, N., Bryder, D., Cilio, C. M., Ahlenius, H., et al. (2003). Complementary signaling through flt3 and interleukin-7 receptor alpha is indispensable for fetal and adult B cell genesis. J Exp Med *198*, 1495–1506.

Sitnicka, E., Bryder, D., Theilgaard-Monch, K., Buza-Vidas, N., Adolfsson, J., and Jacobsen, S. E. (2002). Key role of flt3 ligand in regulation of the common lymphoid progenitor but not in maintenance of the hematopoietic stem cell pool. Immunity *17*, 463–472.

Sitnicka, E., Buza-Vidas, N., Ahlenius, H., Cilio, C. M., Gekas, C., Nygren, J. M., Mansson, R., Cheng, M., Jensen, C. T., Svensson, M., et al. (2007). Critical role of FLT3 ligand in IL-7 receptor-independent T lymphopoiesis and regulation of lymphoid-primed multipotent progenitors. Blood *110*, 2955–2964.

Smith, L. G., Weissman, I. L., and Heimfeld, S. (1991). Clonal analysis of hematopoietic stem-cell differentiation in vivo. Proc Natl Acad Sci USA *88*, 2788–2792.

Spangrude, G. J., Heimfeld, S., and Weissman, I. L. (1988). Purification and characterization of mouse hematopoietic stem cells. Science *241*, 58–62.

Spivakov, M., and Fisher, A. G. (2007). Epigenetic signatures of stem-cell identity. Nat Rev Genet *8*, 263–271.

Stanley, E., Lieschke, G. J., Grail, D., Metcalf, D., Hodgson, G., Gall, J. A., Maher, D. W., Cebon, J., Sinickas, V., and Dunn, A. R. (1994). Granulocyte/macrophage colony-stimulating factor-deficient mice show no major perturbation of hematopoiesis but develop a characteristic pulmonary pathology. Proc Natl Acad Sci USA *91*, 5592–5596.

Stoffel, R., Ziegler, S., Ghilardi, N., Ledermann, B., de Sauvage, F. J., and Skoda, R. C. (1999). Permissive role of thrombopoietin and granulocyte colony-stimulating factor receptors in hematopoietic cell fate decisions in vivo. Proc Natl Acad Sci USA *96*, 698–702.

Tan, B. T., Park, C. Y., Ailles, L. E., and Weissman, I. L. (2006). The cancer stem cell hypothesis: a work in progress. Lab Invest *86*, 1203–1207.

Theopold, U., Li, D., Fabbri, M., Scherfer, C., and Schmidt, O. (2002). The coagulation of insect hemolymph. Cell Mol Life Sci *59*, 363–372.

Till, J. E., and Mc, C. E. (1961). A direct measurement of the radiation sensitivity of normal mouse bone marrow cells. Radiat Res *14*, 213–222.

Ueno, H., and Weissman, I. L. (2006). Clonal analysis of mouse development reveals a polyclonal origin for yolk sac blood islands. Dev Cell *11*, 519–533.

Ueno, H., and Weissman, I. L. (2007). Stem cells: blood lines from embryo to adult. Nature 446, 996–997.

Wang, J. H., Nichogiannopoulou, A., Wu, L., Sun, L., Sharpe, A. H., Bigby, M., and Georgopoulos, K. (1996). Selective defects in the development of the fetal and adult lymphoid system in mice with an Ikaros null mutation. Immunity *5*, 537–549.

Veiby, O. P., Lyman, S. D., and Jacobsen, S. E. (1996). Combined signaling through interleukin-7 receptors and flt3 but not c-kit potently and selectively promotes B-cell commitment and differentiation from uncommitted murine bone marrow progenitor cells. Blood *88*, 1256–1265.

Wilson, A., and Trumpp, A. (2006). Bone-marrow haematopoietic-stem-cell niches. Nat Rev Immunol 6, 93–106.

Vire, E., Brenner, C., Deplus, R., Blanchon, L., Fraga, M., Didelot, C., Morey, L., Van Eynde, A., Bernard, D., Vanderwinden, J. M., et al. (2006). The Polycomb group protein EZH2 directly controls DNA methylation. Nature *439*, 871–874.

Wolf, N. S., Kone, A., Priestley, G. V., and Bartelmez, S. H. (1993). In vivo and in vitro characterization of long-term repopulating primitive hematopoietic cells isolated by sequential Hoechst 33342-rhodamine 123 FACS selection. Exp Hematol *21*, 614–622.

Wu, H., Liu, X., Jaenisch, R., and Lodish, H. F. (1995). Generation of committed erythroid BFU-E and CFU-E progenitors does not require erythropoietin or the erythropoietin receptor. Cell *83*, 59–67.

Xie, H., Ye, M., Feng, R., and Graf, T. (2004). Stepwise reprogramming of B cells into macrophages. Cell *117*, 663–676.

Ye, M., Iwasaki, H., Laiosa, C. V., Stadtfeld, M., Xie, H., Heck, S., Clausen, B., Akashi, K., and Graf, T. (2003). Hematopoietic stem cells expressing the myeloid lysozyme gene retain long-term, multilineage repopulation potential. Immunity *19*, 689–699.

Yoshida, T., Ng, S. Y., Zuniga-Pflucker, J. C., and Georgopoulos, K. (2006). Early hematopoietic lineage restrictions directed by Ikaros. Nat Immunol *7*, 382–391.

Zhang, J., Niu, C., Ye, L., Huang, H., He, X., Tong, W. G., Ross, J., Haug, J., Johnson, T., Feng, J. Q., et al. (2003). Identification of the haematopoietic stem cell niche and control of the niche size. Nature *425*, 836–841.

Zhang, P., Behre, G., Pan, J., Iwama, A., Wara-Aswapati, N., Radomska, H. S., Auron, P. E., Tenen, D. G., and Sun, Z. (1999). Negative cross-talk between hematopoietic regulators: GATA proteins repress PU.1. Proc Natl Acad Sci USA *96*, 8705–8710.

Zhuang, Y., Soriano, P., and Weintraub, H. (1994). The helix-loop-helix gene E2A is required for B cell formation. Cell *79*, 875–884.

Chapter 3
Hematopoietic Stem Cell Niches

Anne Wilson and Andreas Trumpp

Introduction

The hematopoietic stem cell (HSC) is probably the best characterized somatic stem cell and is still the only one regularly used in clinical practice. Nevertheless, expansion of HSCs in vitro has been surprisingly unsuccessful, limiting their full therapeutic potential. During homeostasis, the vast majority of HSCs are found in the bone marrow (BM) localized to specific microenvironments called stem cell "niches." Over the last few years our knowledge of cellular niche components and the signaling molecules that coordinate the crosstalk between HSCs and niche cells has dramatically increased. Here we review the two main niche types found in the BM: the endosteal and the vascular niches, and provide an overview of the different signaling and cell adhesion molecules that form the HSC–niche synapse. Signals from BM niches not only control HSC dormancy, but also regulate the balance between self-renewal and differentiation. In the future, successful expansion of HSCs for therapeutic use will require three-dimensional reconstruction of a stem cell-niche unit.

The Hematopoietic Stem Cell

Regenerating tissues such as the gut, the skin, and the hematopoietic system are maintained under homeostatic conditions by a small number of immature cells called stem cells. Stem cells are functionally defined as cells that at the clonal level can both self-renew (maintain their numbers at a constant level), and give rise to all mature cells in the tissue in which they reside. Identification and characterization of this minor immature population of cells has long-occupied researchers in many

A. Trumpp(✉)
Deutsches Krebsforschungszentrum (DKFZ), DKFZ-ZMBH Alliance, Im Neuenheimer Feld 280, D-69120 Heidelberg, Germany
e-mail: a.trumpp@dkfz.de

A. Wickrema and B. Kee (eds.), *Molecular Basis of Hematopoiesis*,
DOI: 10.1007/978-0-387-85816-6_1, © Springer Science+Business Media, LLC 2009

biological disciplines (Weissman 2000). Although stem cells present in the gut mucosa and skin epidermis have recently been identified in situ, by far the most well-characterized stem cell in the mouse is the HSC (Shizuru et al. 2005; Barker et al. 2007; Blanpain et al. 2007).

The presence of HSCs in the BM was first demonstrated in landmark experiments performed by two Canadian researchers in the 1960s. Till and McCulloch (1961) showed that an unidentified population of cells found within the BM were able to fully reconstitute all lineages of the hematopoietic system of a mouse after lethal irradiation. However, elucidation of the precise phenotype of cells with this self-renewal capacity did not occur until more than 40 years later. The increasing availability of stem-cell-specific monoclonal antibodies combined with multiparameter fluorescent activated cell sorting (FACS) have revolutionized technology for isolating and characterizing minor subpopulations of the BM. It is now possible to purify nearly homogeneous, functional adult mouse HSCs using a specific combination of monoclonal antibodies together with FACS sorting. Indeed, long-term reconstitution of the hematopoietic system of a lethally irradiated recipient by transplantation of a single purified BM HSC has been achieved by several groups, thus providing functional proof that bona fide adult stem cells exist (Osawa et al. 1996; Kiel et al. 2005; Camargo et al. 2006; Dykstra et al. 2007).

In the adult mouse, all functional HSC activity is found in the Lin^-Sca1^+ $c-Kit^+$ (LSK) population that represents only about 0.05% of total BM cells. However, LSK cells remain heterogeneous, as only less than one out of ten actually has repopulation capacity. This minor BM subset comprises at least five subsets based on the differential expression of cell surface antigens such as CD34, CD135 (Flk2), CD150, and CD48 (the latter two being members of the large signal lymphocyte activation marker (SLAM) family) (Wilson et al. 2007). Independent single cell reconstitution studies have shown that all functional stem-cell activity is found within the two highly overlapping LSK populations: $CD34^-$ LSK and $CD150^+$ $CD48^-$ LSK (Osawa et al. 1996; Kiel et al. 2005).

Functional stem-cell activity has independently also been associated with physical properties of stem cells such as relative quiescence and side population (SP) activity. Loss of long-term self-renewal capacity and differentiation into more committed progenitors is concurrent with exit from the resting (predominantly G_0) state and upregulation of metabolic activity (Passegue et al. 2005; Arai and Suda 2007; Wilson et al. 2007; Orford and Scadden 2008). A more detailed discussion of relative quiescence and niche activity is provided below. In addition, several groups have used SP activity (the ability of stem cells to actively efflux fluorescent dyes such as Hoechst 33342 or Rhodamine-123 via multidrug resistant pumps), for isolation of functional HSC activity (Camargo et al. 2006; Dykstra et al. 2007).

In summary, the most current model of developmental progression starts with a dormant/quiescent CD34 negative HSC that most likely does not contribute to normal homeostasis of the hematopoietic system, but may function as a reserve pool of stem cells, only to be activated upon injury (Wilson et al., in press). In contrast, during normal homeostasis CD34 positive HSCs are those that are responsible for the day-to-day maintenance of hematopoiesis, providing both a constant source of progenitors that give rise to all mature blood lineages, as well as to self-renewing HSCs to maintain the stem-cell pool (Wilson et al. 2007).

Stem Cells and Asymmetry; or Two Ways to Divide

One of the unique capacities of a stem cell is its ability to generate two daughter cells with different fates: a new stem cell and a progenitor cell committed to terminal differentiation. This generation of asymmetric offspring is used during steady-state homeostasis in all regenerative adult tissues. It permits maintenance of a constant number of stem cells while simultaneously producing differentiated progeny to replace cells continuously lost during normal cellular turnover (i.e., erythrocytes and neutrophils, skin keratinocytes, mucosal enterocytes, and others). To achieve such asymmetry during cell division two distinct mechanisms have been proposed. The major difference between these mechanisms is whether asymmetry is achieved before, (intrinsic or divisional asymmetry), or after (extrinsic or environmental asymmetry) cell division (Fig.3.1).

Asymmetry Based on Cell Division

In divisional asymmetry (also called intrinsic asymmetry), specific cell fate determinants in the cytoplasm (mRNA and/or proteins) redistribute unequally into only one daughter cell prior to the onset of division. In this situation, stem cells set up an axis of polarity during interphase and use it to asymmetrically localize cell fate determinants during mitosis (Knoblich 2008). Thus two nonidentical daughter cells are produced, one retaining the stem-cell fate while the other initiates differentiation (Fig.3.1a). The molecular basis for this complex process has mostly been elucidated in Drosophila neuroblasts and sensory organ precursor cells (reviewed in Knoblich (2008)). However, a few further examples of divisional asymmetry have been documented in higher vertebrates such as brain progenitors (Konno et al. 2008), mouse T lymphocytes (Chang et al. 2007) and skin epithelial cells (Lechler and Fuchs 2005). For example, in the mammalian fetal epidermis, basal cells not only divide symmetrically to allow two-dimensional epidermal expansion, but also divide asymmetrically to promote stratification and differentiation of the skin. In the latter case, a protein complex that includes Par3, LGN, and murine inscutable forms an apical crescent that dictates the directionality of the ensuing cell division (Lechler and Fuchs 2005). However, although the mechanism for divisional asymmetry is conserved between invertebrates and mammals, so far such a mechanism has not yet been demonstrated in vertebrate stem cells in vivo. Nevertheless, in vitro studies have suggested that HSCs can undergo some type of asymmetric division as

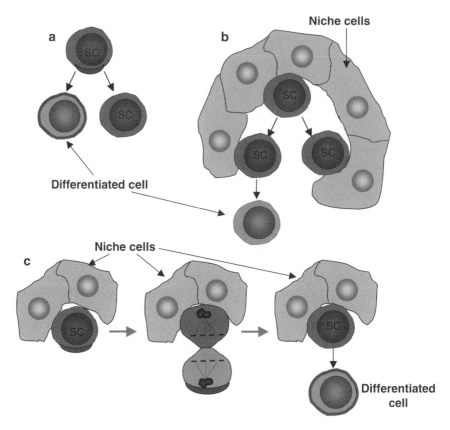

Fig. 3.1 Three ways to achieve asymmetric cell division. (**a**) Divisional or intrinsic asymmetry. Specific cell fate determinants redistribute unequally into the differentiating daughter cell prior to the onset of cell division. (**b**) Environmental or extrinsic asymmetry. After undergoing symmetric cell division, one of the two identical daughter cells remains in the niche and is maintained as a stem cell while the other eventually leaves the niche and therefore initiates a differentiation program. (**c**) A combination of niche signals and intrinsic cues. Niche signals are important to asymmetrically polarize cell fate determinants prior to division. The cell division plane is parallel to the niche so that the daughter cell receiving cell fate determinants is distant from the niche and therefore differentiation is initiated. *SC* stem cell (*see Color Plates*)

analysis of the ability of the two daughter cells of a single HSC to long-term reconstitute lethally irradiated recipients showed that ~20% of HSCs produced nonidentical daughter cells (Suda et al. 1984; Takano et al. 2004; Ho 2005). However, these studies neither provide a mechanism for the observed asymmetry, nor show if it occurs precell or postcell division. Moreover, whether these in vitro studies reflect the situation of BM HSCs remains unclear, and future studies will still need to address whether divisional or intrinsic asymmetry is indeed used by HSCs in vivo.

Asymmetry Based on the Environment: The Stem Cell Niche Concept

Environmental (or extrinsic) asymmetry can be achieved by exposing the two daughter stem cells to different extrinsic signals (soluble or cellular) provided by the distinct local microenvironment (Fig.3.1b). In this scenario, after first undergoing a symmetric self-renewing division, one of the two identical daughter cells that remains in the niche microenvironment would conserve a stem-cell fate. However, the other identical daughter cell would make contact (passively or actively), with a different microenvironment that would provide signals initiating differentiation, thus its stem-cell phenotype would no longer be preserved and differentiation would be initiated (Spradling et al. 2001; Ohlstein et al. 2004; Murphy et al. 2005; Wilson and Trumpp 2006). As with divisional asymmetry, the final product would also be nonidentical daughter cells, but this would have been achieved postcell division and not precell division (Fig.3.1b).

Asymmetry Achieved by a Combination of Niche Interactions and Intrinsic Cues

Extrinsic (niche-based) and intrinsic mechanisms of achieving asymmetry after cell division are not necessarily mutually exclusive. They may in fact cooperate in order to assure asymmetry, thus fine tuning the process. In this scenario, the stem cell is attached to the niche via cell adhesion molecules, and niche cell-derived signals at the contact site or HSC–niche synapse (Fig.3.2b) would then direct the asymmetric segregation of cell fate determinants that orient the mitotic spindle perpendicular to the niche surface. Consequently, whereas one daughter cell would retain its contact with the niche cells, the second daughter cell would be released and thus become committed to differentiate (Fig.3.1c). In summary, with any of the three proposed mechanisms governing asymmetric cell division (intrinsic, extrinsic, or a combination of the two), two nonidentical daughter cells, each with its own distinct fate (self-renewal or differentiation) are generated.

Stem Cell Niches

A stem cell niche can be defined as a spatial structure in which stem cells are housed and maintained by keeping them quiescent or allowing self-renewal in the absence of differentiation. The term "niche" comes from the French word "nicher" or "nichier" – to nest, or from the latin "nidus" – nest. In classical architecture, a niche has been described as an exedra or an apse, reduced in size and forming part of a wall.

a

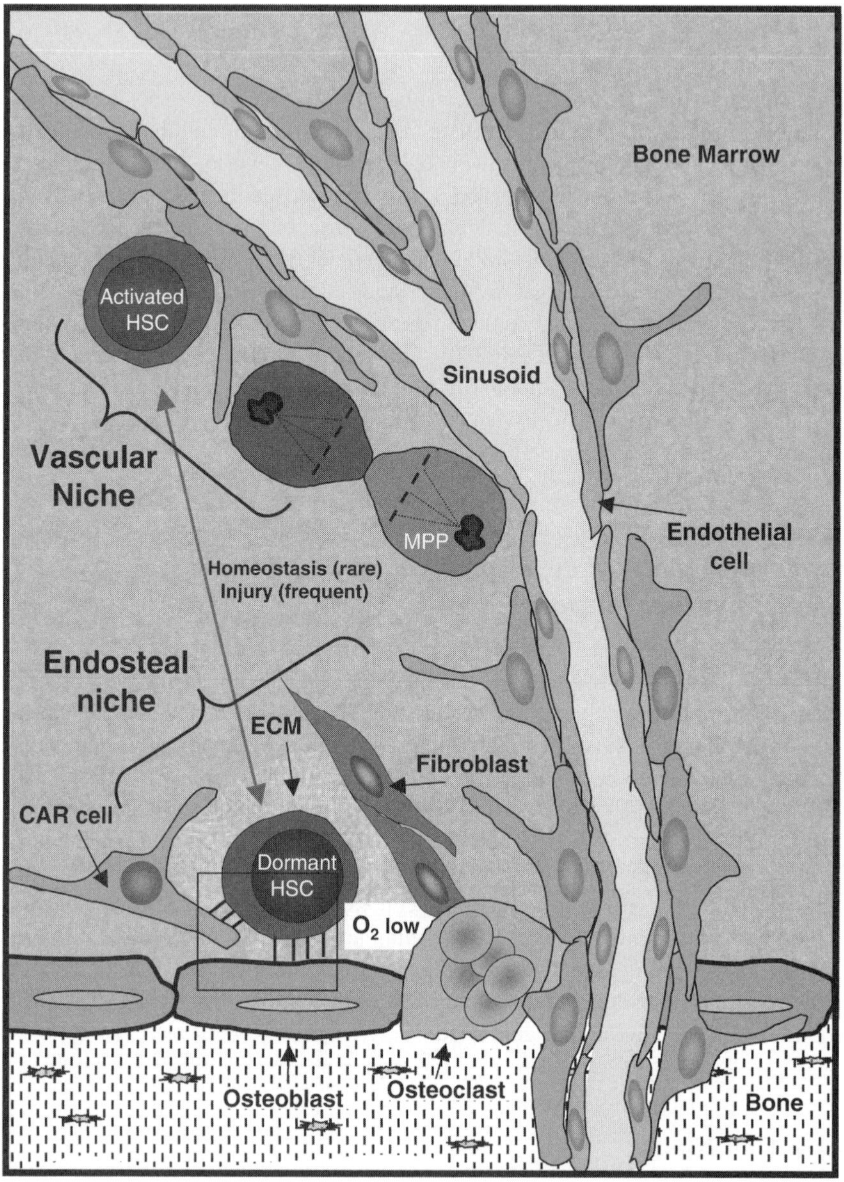

Fig. 3.2 (**a**) Two types of bone marrow niche. At the endosteum, specialized niche stromal cells such as osteoblasts, CAR cells, fibroblasts, and osteoclasts create an hypoxic local environment. Dormant HSCs are retained in this storage niche by cell:cell adhesion. The *box* indicates the enlarged area shown in (**b**). In the center of the BM, close to fenestrated sinusoids, CAR cells and endothelium create a second type of vascular niche near to blood vessels. Activated HSCs undergo extrinsic asymmetric cell division, move out of the niche and initiate differentiation programs in preparation for migration to the periphery via the circulation. *CAR* CXCL12-abundant reticular cells,

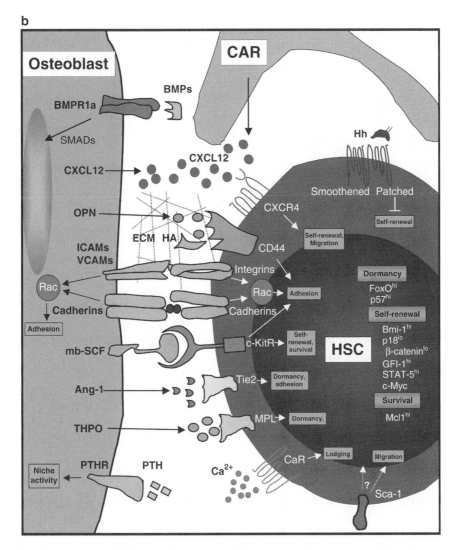

Fig. 3.2 (continued) *ECM* extracellular matrix, *HSCs* hematopoietic stem cells, *MPP* multipotent progenitor. (**b**) The HSC: niche synapse. In the space between HSCs and niche stromal cells (osteoblasts and CAR cells), a large number of receptor: receptor and ligand: receptor interactions take place. Signals emanating from these interactions influence integral stem cell processes such as adhesion, dormancy, lodging, migration, self-renewal, and survival. Some of the transcription factors mediating these processes are listed in the nucleus of the HSC (*right*). *Ang-1* Angeopoitin-1, *BMP* bone morphogenic protein, *BMPR1a* BMP receptor 1a, *CaR* Ca2+ receptor, *CAR* CXCL12-abundant reticular cell, *CXCR4* CXC-receptor-4, *CXCL12* CXC-Ligand-12, *ECM* extracellular matrix, *HA* Hyalurin, *Hh* Hedgehog, *HSC* hematopoietic stem cell, *mb-SCF* membrane bound stem cell factor, *MPL* THPO receptor, *OPN* osteopontin, *PTH* parathyroid hormone, *PTHR* PTH receptor, *THPO* thrombopoietin, *VCAM* vascular cell adhesion molecule (*see Color Plates*)

It usually has a half-dome shape and is often filled with a statue. It may also be a hollow, crack or crevice. In a biological context the term "niche" was first used by the naturalist Joseph Grinnell in 1917, and the first functional definition of a niche was provided 10 years later by Charles Sutherland Elton, an ecologist in Britain. However, it was not until 1958 that the niche concept was popularized by the zoologist G. Evelyn Hutchinson (Niche 2008). Needless to say, in all of the above situations, the term "niche" was used within the context of ecological systems where it has a much more complex function than just a location or type of home. In this case, niche refers to the relational position of a population within its natural habitat, and regulates how an organism or population responds to the resources provided by the surrounding environment. Although in those days, this description was coined with only ecological systems in mind, in many respects these parameters can easily be applied to stem cell niches. Moreover, the competitive exclusion principle, which states that only one population may occupy any given niche at any one time, might just as well have been written about stem cell niches.

More specifically, in biological terms a stem cell niche has been described as a three-dimensional spatial structure containing one or more stem cells. The niche provides a specialized microenvironment in which stem cells are maintained through self-renewal, while differentiation is inhibited. The balance between self-renewal and differentiation is thought to be controlled by interactions between the stem cells and either soluble factors produced by niche-forming stromal cells or the stromal cells themselves (Spradling et al. 2001; Wilson and Trumpp 2006). Several different types of niche may exist depending on the composition of stromal cells making up the niche environment, or the physical location of the niche within the tissue.

The Bone Marrow Niche

After the initial discovery that multipotent cells capable of self-renewal were present in the BM, several groups observed a close relationship between hematopoiesis (marrow formation) and osteogenesis (development of bone). These studies showed that upon subcutaneous transfer of total, unmanipulated BM, first bone and then vascularized BM developed (Patt and Maloney 1972; Maloney and Patt 1975). Prompted by these discoveries, Schofield first used the term niche to describe a putative HSC-specific BM microenvironment in 1978 (Schofield 1978). Although the main focus of studies at the time was in elucidating the nature of the hematopoietic cell responsible for the recently discovered stem-cell activity, curiosity about the underlying mechanisms involved in maintenance of this activity had arisen. Thus Schofield proposed that HSCs and bone would be in close contact, and that the resulting signals/interactions were responsible for the apparently unlimited proliferative capacity (or as we now know, life-long capacity to proliferate – self-renew), and at the same time to inhibit differentiation and maturation of HSCs (Schofield 1978).

When the delicate balance between HSCs and niche components is perturbed, homeostasis of the hematopoietic system is often affected. Studies in which primary defects in bone development or remodeling have been observed have provided support for the concept that both osteoblasts (bone forming cells of mesenchymal origin) and osteoclasts (bone remodeling, macrophage like cells of hematopoietic origin) may be involved in formation and function of the BM stem cell niche (Wilson and Trumpp 2006; Kollet et al. 2007). In particular, defective BM hematopoiesis due to defects in osteoblast differentiation and the subsequent failure to form bone were reported in mice lacking Cbfa1/Runx-2 (one of the earliest osteoblast specific transcription factors) (Deguchi et al. 1999; Ducy et al. 2000). However, whether this hematopoietic deficiency is a secondary effect caused by the lack of formation of a suitable BM cavity or whether the defect in *Cbfa1/Runx-2* directly affects hematopoiesis remains unclear. To date, several other mouse mutants with defects in bone development and/or remodeling have been also been described (Karsenty and Wagner 2002), however any potential effects on hematopoiesis have not yet been documented.

Physical Evidence for the BM Niche

In contrast to other regenerating tissues such as the skin and the gut mucosa, providing physical evidence for HSC stem cell niches has been difficult due to the technical difficulties in obtaining sections of solid bones while at the same time preserving marrow structure. Nevertheless, HSCs were first localized to trabecular bone nearly 40 years ago (Lord et al. 1975), and more recently, several groups have postulated a close association of niches with the endosteal lining of marrow cavities in the trabecular region of long bones (Gong 1978; Nilsson et al. 2001; Askenasy and Farkas 2003). The relatively simple homing assay relies on the previously characterized ability of HSCs to home to and engraft into BM niches after i.v. transfer into the blood of irradiated recipients. To visualize the HSCs and the sites to which they home, FACS-enriched HSC/progenitor cells are labeled with fluorescence dyes (i.e., CFSE) and then transferred into recipient mice. Fifteen hours posttransfer, a significant enrichment of CFSE-labeled stem/progenitor cells is seen at, or near the endosteum of trabecular bone on histological sections (Nilsson et al. 2001; Wilson et al. 2004). The advantage of this assay is that it is relatively rapid compared to classical, functional stem-cell assays. In addition, putative niche stromal cells can be identified using markers expressed on niche stromal cells.

Nevertheless, it is important to note that the localization of HSCs using this technique may be unspecific as homing of stem cells after i.v. transfer may not reflect endogenous niche location, given that irradiation damage may alter the BM vasculature thereby facilitating or altering the homing process. Further studies have refined this localization to show that HSCs are most likely physically interacting with various cell types including osteoblasts, osteoclasts, stromal fibroblasts, and endothelial cells located in close proximity to the bone surface (endosteal niche)

(Zhang et al. 2003; Arai et al. 2004; Wilson et al. 2004). Evidence for the participation of these cell types in BM niche function is discussed below.

Cellular Components of the BM Niche

Osteoblastic Cells

HSC-supporting activity from cells involved in bone formation was first provided by in vitro studies where cytokines promoting the expansion of hematopoietic cells were secreted by both murine and human osteoblast cell lines. Moreover, many stromal cell lines that can maintain HSCs in vitro also have bone formation activity and many long-term BM cultures have been found to contain osteoblasts (Taichman and Emerson 1998; Oostendorp et al. 2002; Taichman 2005). In addition, homed CFSE[+] HSC/progenitors have also been shown to be in close contact to osteopontin[+], N-cadherin[+] and bone morphogenic protein (BMP) receptor-1A[+] osteoblasts located at the endosteum (Wilson et al. 2004). These immunolocalization data provide further evidence that specialized osteoblasts are part of the endosteal niche.

Indeed, a direct role for osteoblasts in HSC regulation and/or maintenance in vivo has also recently been observed in studies in which osteoblast numbers were experimentally manipulated. First (Calvi et al. 2003), using the type 1 collagen α1 (Colα1) promoter, expression of a constitutively activated form of parathyroid hormone (PTH) or the PTH-related protein (PTHrP) receptor (PPR), was targeted to osteoblasts. Constitutive expression of this important regulator of calcium homeostasis and thus bone formation and resorption in osteoblasts, resulted in increased numbers of both osteoblasts and HSCs in the BM. In an independent study (Zhang et al. 2003), mice in which BM stromal cells were specifically deficient in BMPR1A, provided confirmation of the close association of HSC numbers and osteoblast numbers. Lack of BMPR1A, which is normally expressed on osteoblasts lining the endosteum, also results in a simultaneous increase in the number of both osteoblasts and repopulating HSCs. Moreover, specialized spindle-shaped N-cadherin expressing osteoblasts (SNOs) located in the endosteum were shown to be in contact with Lin[-] label-retaining cells (LRC) via N-cadherin. When the number of osteoblasts increased, a corresponding increase in HSC numbers was observed, leading to the hypothesis that osteoblasts (and thus SNO cells) were essential components of the HSC BM niche (Calvi et al. 2003; Zhang et al. 2003).

Support for this concept was provided by experiments in which osteoblasts were conditionally ablated by targeting thymidine kinase expression (which induces cell death in response to ganciclovir) to the osteoblast lineage (Visnjic et al. 2001, 2004). The progressive bone loss observed in these mice correlated with a decrease in both total BM and the number of LSK-HSCs. It is important to note that this loss of osteoblasts was accompanied by an increase in progenitor cells (and HSCs) in extramedullary sites such as the liver, spleen and peripheral blood, suggesting that the BM niches were lost. As osteoblast depletion in this situation is dependent on continued thymidine kinase activity, removal of ganciclovir caused the re-establishment of BM hematopoiesis (Visnjic et al. 2004). However, when osteoblasts are genetically ablated using the osteocalcin promoter (which is active at a later stage of osteoblastogenesis compared to Col1a1), to drive thymidine kinase expression, no effects on hematopoiesis were observed (Corral et al. 1998), suggesting that niches may comprise immature osteoblasts. Finally, deletion of the retinoblastoma protein (RB) in the BM causes a dramatic increase in osteoclasts followed by depletion of osteoblasts and eventually loss of trabecular bone. As a result, RB-deficient HSCs are mobilized to the spleen, initiating extramedullary hematopoiesis. Most interestingly, these mice develop myeloid proliferative disease (MPD), a pre-leukemic stage. In addition, normal HSCs are lost most probably due to the eliminated osteoblastic BM niche. These data confirm that the endosteal niche is required to maintain HSCs long term and also suggest that defects in niche cells can cause MPD (Walkley et al. 2007b).

CAR Cells

In addition to specialized osteoblasts (SNO cells), CXC-Ligand (CXCL)-12-abundant reticular (CAR) cells that are often closely associated with the microvasculature have also been localized in close proximity to HSCs in the BM (Sugiyama et al. 2006). However, as CAR cells also serve as a niche for BM B cells (Tokoyoda et al. 2004), it is unlikely that these cells alone are sufficient to define HSC niches. Interestingly, while CAR cells are mostly found near the sinusoids (low-pressure vessels with a fenestrated endothelium), a significant number are also located near to, and at the endosteum together with osteoblasts, osteoclasts, and stromal fibroblasts (Sugiyama et al. 2006; Kollet et al. 2007), raising the possibility that CAR cells may provide a cellular, and possible molecular link between vascular and endosteal niches. These data suggest that different types of niches may be present in the BM: one at the endosteum (inner cellular lining of the bone cavities interfacing with the BM) of trabecular bone, and the other, in more vascularized regions in the center of the BM.

Osteoclasts

Osteoblasts and osteoclasts are instrumental in controlling the amount of bone tissue: osteoblasts form bone and osteoclasts are macrophage-like cells that resorb bone. Both bone-remodeling cell types maintain bone equilibrium and act in the endosteum in the vicinity of HSCs. Although osteoclasts have not yet been shown to be physically part of the endosteal niche unit, there is increasing evidence that signals from osteoclasts directly affect HSC behavior by altering niche activity (Kollet et al. 2007). For example, mouse receptor activator of nuclear factor-κB ligand (RANKL)-stimulated osteoclasts are actively involved in stress-induced mobilization of HSCs into the circulation, in a CXCR4- and matrix metalloproteinase-9

(MMP-9)-dependent manner. In parallel, expression of stromal derived factor (SDF)-1, stem cell factor (SCF), and osteopontin are reduced in response to osteoclast activity, suggesting that signals derived from osteoclasts can mediate HSC exit and mobilization by altering niche activity (Kollet et al. 2007).

The Endosteal Niche Maintains HSCs by Keeping Them Quiescent

It is becoming evident that one of the main functions of the niche is to keep the most immature HSCs in a quiescent or even dormant status. Most functional HSCs are thought to divide infrequently and may remain quiescent for weeks or even months (Bradford et al. 1997; Cheshier et al. 1999; Passegue et al. 2005; Arai and Suda 2007). They are also considered metabolically inactive (Suda et al. 2005). In agreement, the adult stem-cell pool is largely resistant to classical chemotherapeutic agents that target cycling cells (Lerner and Harrison 1990). In addition, quiescent HSCs engraft more efficiently after transplantation into patients or irradiated mice than those that are cycling (Spangrude and Johnson 1990; Uchida et al. 1997; Passegue et al. 2005).

Label-retaining assays have been used to locate quiescent stem cells in situ in many tissues (Cotsarelis et al. 1990; Potten and Loeffler 1990; Zhang et al. 2003; Arai et al. 2004; Waghmare et al. 2008). In this assay, cellular DNA is labeled during expansion by nucleotide analogues (such as BrdU), or by an inducible transgene expressing histone H2B-GFP, and can then be retained for months by quiescent cells, including stem cells. In the BM, LRCs that fail to express differentiation markers are highly enriched for HSCs. On sections, HSC-LRCs are found in trabecular bone (Zhang et al. 2003; Arai et al. 2004; Yoshihara et al. 2007). Thus, long-term quiescent HSCs that cannot significantly contribute to normal homeostasis are present in the BM. Consequently it has been proposed that these are stored in a "quiescent storage" niche (Ohlstein et al. 2004; Wilson and Trumpp 2006). The ability of the HSC niche to maintain their quiescent state is important, as the self-renewal capacity of HSCs appears to be enormous, but not unlimited. Evidence for this comes from a number of mutant mice in which HSC proliferation is increased (see below and reviewed in Orford and Scadden (2008)).

The HSC–Niche Synapse

Similar to what has been found in invertebrate niches, cell adhesion molecules play an important role in the establishment of a stable HSC–niche unit (Spradling et al. 2001; Wilson and Trumpp 2006). It is thought that a significant number of cell adhesion molecules form a network that, together with secreted extracellular matrix (ECM), bring HSCs into close proximity to the different niche cell components (particularly osteoblasts and CAR cells). Tight adhesion and juxtaposition of HSCs to niche cells leads to the formation of an intercellular space in which efficient ligand–receptor mediated crosstalk can occur. As most secreted signaling molecules are bound to the cell surface or to the ECM, and consequently do not diffuse very far, this setup allows efficient signaling in both directions. In analogy to the neuronal and immunological synapses (Friedl and Storim 2004), the space between HSCs and BM niche elements has been proposed to be named as the "HSC–niche synapse" (Fig.3.2b). Below we review some of the signaling pathways and cell adhesion components that have been shown to be functionally involved in controlling HSC–niche interactions.

Stem Cell Factor-Kit Signaling

One of the first molecules identified as playing a role in HSCs were the genes mutated in the mouse mutants Steel (encoding SCF) and Kit encoding its receptor c-Kit. The Steel (Sl) locus encodes both membrane bound-SCF (mb-SCF) and secreted-SCF (s-SCF) (Flanagan et al. 1991). SCF binds and activates c-Kit, which is expressed at high levels by HSCs as well as other stem cells. Mutations at either of these loci affect migration and differentiation of primordial germ cells, neural crest-derived melanoblasts, and hematopoietic cells (Lyman and Jacobsen 1998). Analysis of the different SCF and c-Kit mutants revealed that although not essential for the generation and initial expansion of HSCs in the embryo and in the fetal liver, they are crucial for long-term maintenance and self-renewal of adult HSCs in the BM. Importantly, mb-SCF is expressed on osteoblasts and has a higher and more sustained capacity to activate c-Kit on HSCs than s-SCF (Miyazawa et al. 1995; Lyman and Jacobsen 1998). In addition, mb-SCF is a potent stimulator of adhesion of HSCs/progenitors to stromal cells as it can activate very-late-antigen (VLA)-4 and VLA-5 (Kinashi and Springer 1994), suggesting that mb-SCF can affect adhesive properties of the niche by modifying the functional states of specific integrins (Kovach et al. 1995). Importantly, BM of young Sl/Sl^d mice, which only produce s-SCF, but not mb-SCF, have normal HSC-activity when transplanted into lethally irradiated recipients, whereas BM from old Sl/Sl^d mice has greatly reduced HSC-activity suggesting a progressive loss of stem-cell activity over time, potentially due to reduced niche activity (Barker 1994, 1997). This indicates that osteoblastic mb-SCF activates c-Kit expressed on HSCs and that this signaling is critical to maintain BM stem cells. In agreement with this notion, studies on Kit$^{W41/W41}$ mice, which have a partial loss of c-Kit function, show that c-Kit signaling in HSCs is required for HSC maintenance since they are progressively lost in chimeric settings. Moreover, decreased c-Kit signaling induces apoptosis and increased HSC proliferation suggesting that this signaling pathway is crucial for survival and quiescence of HSCs (Thoren et al. 2008).

Finally, treatment of mice with ACK-2, an antibody which blocks c-Kit function leads to the transient removal of almost all HSCs, but allows engraftment of transplanted HSCs without irradiation or conditioning other than immunosuppression

(Czechowicz et al. 2007). These data not only suggest that SCF-c-Kit signaling is important for HSC function and niche localization, but also suggest that the availability of niches is a limiting factor to transplantation.

Thrombopoietin–MPL Signaling

One of the key signaling pathways recently identified to be required to keep HSCs in a quiescent status is the thrombopoietin (THPO)–MPL pathway (Solar et al. 1998). The cytokine THPO has a double role, as it not only regulates production and differentiation of megakaryocytes and platelets, but also regulates HSC quiescence. Since thrombopoiesis is critical for the immediate survival of the organism, to guarantee sufficient numbers of platelets during homeostasis and after injury, production of THPO appears to induce a feedback signal from platelets to HSCs. Most interestingly, HSCs lacking the THPO receptor MPL, which signals through the Jak-STAT, Ras-Raf-MAPK, and PI3K pathways (Chanprasert et al. 2006), progressively decline over time and are 150-fold decreased in mutants more than 1-year-old (Qian et al. 2007). This loss of HSCs over time correlates with increased HSC cycling and reduced expression of the CDK inhibitors $p57^{Kip2}$ and $p19^{INK4D}$. These data make it likely that HSCs are exhausted over time due to overproliferation and loss of quiescence. Most relevant to the niche, THPO is expressed by the osteoblasts lining the endosteal surface suggesting that the THPO-MPL signaling axis both anatomically and functionally links quiescent HSCs to their niche (Yoshihara et al. 2007). Similar to MPL mutants, inhibition of THPO in osteoblasts reduces the number of quiescent HSCs, while exogenous THPO inhibits HSC proliferation. In addition, increased expression of β1-integrin most likely further increases the attachment of HSCs to niche osteoblasts. Curiously THPO is included in most cocktails which are used to grow HSCs in vitro. However, none of these cytokine cocktails have yet allowed significant expansion of HSCs in vitro. This suggests that while THPO-MPL signaling is required for HSC maintenance, other factors and probably adhesion to the correct niche cell type are essential for the complex balance between HSC quiescence and self-renewal.

Angiopoietin-1 (Ang1)-Tie2 Signaling

Similar to THPO-MPL signaling, the Ang1-Tie2 pathway also appears to be critical for maintenance of the quiescent status of HSCs (Arai et al. 2004). Genetic evidence for the requirement of Tie2 in HSC–niche interactions has been obtained from chimeric mice comprised of wild-type and Tie1/Tie2 double-mutant morulae (Puri and Bernstein 2003). Although Tie1 and Tie2 are not required for the development and differentiation of fetal HSCs, double mutant HSCs fail to be maintained in the adult microenvironment. In adult BM, Tie2 (which is expressed specifically by quiescent long-term-HSCs) is activated by Ang1, which is secreted by osteoblasts. Most interestingly, Tie2 signaling leads to the upregulation of N-cadherin (see below) on HSCs, providing the first example of a secreted niche factor promoting HSC–osteoblast adhesion. Similar to THPO, the Ang1–Tie2 signaling pathway prevents HSC cell division and maintains HSC quiescence, both in vitro and in vivo (Arai et al. 2004; Arai and Suda 2007). Since blocking of THPO decreases Tie2 expression it appears that MPL signaling is upstream of Tie2 (Yoshihara et al. 2007).

N-Cadherin

One of the most studied but still controversial cell adhesion molecules proposed to be critical for HSC–osteoblast interaction is N-cadherin (Wilson and Trumpp 2006; Kiel et al. 2007; Haug et al. 2008). N-cadherin is expressed on both osteoblasts and on Flk2⁻LSK-HSCs and it has been postulated that homotypic N-cadherin interactions play a major role in anchoring HSCs to the endosteal niche (Zhang et al. 2003; Wilson et al. 2004; Haug et al. 2008). In support of this hypothesis, ectopic expression of N-cadherin in the OP9 stromal cell line significantly increases its capacity to maintain murine HSCs in vitro (Arai et al. 2004). FACS analysis in combination with functional reconstitution assays revealed that HSCs that drive day-to-day hematopoiesis, as well as mobilized HSCs express only low levels of N-cadherin. In contrast, quiescent HSCs appear to express higher levels of N-cadherin in agreement with the notion that they would be present in the osteoblastic, quiescent niche (Haug et al. 2008). Unfortunately, genetic evidence for an essential role for N-cadherin in HSC–osteoblast adhesion and/or signaling is still lacking, since N-cadherin mutant embryos fail to develop past mid-gestation (Radice et al. 1997). Nevertheless, indirect support for the importance of N-cadherin has also been obtained from studies showing that c-Myc represses, and Ang1–Tie2 signaling increases N-cadherin expression on HSCs. The antagonistic effects of c-Myc and Tie-2 on N-cadherin expression correlates with the model that Myc activity promotes HSCs to exit the niche (Wilson et al. 2004; Murphy et al. 2005), while Tie2 preserves them in a quiescent state and therefore retains them in the niche (Arai et al. 2004; Arai and Suda 2007). In summary, homotypic N-cadherin interactions are a critical part of the adhesion network used to anchor quiescent HSCs in the endosteal niche (Haug et al. 2008).

CXCL12–CXCR4 Signaling

A crucial chemokine involved in maintenance, retention, and mobilization of HSCs during homeostasis and after injury is CXCL12 (or SDF-1). CXCL12 is highly expressed not only by CAR cells, but also by osteoblasts, and vascular endothelial cells (Ponomaryov et al. 2000; Ara et al. 2003; Sugiyama et al. 2006) (Fig.3.2b). Similar to SCF, CXCL12 expression and secretion is induced in response to hematopoietic cell loss due to irradiation, chemotherapy or hypoxia, and purified HSCs migrate specifically toward CXCL12 but not to any other single chemokine (Wright et al. 2002). The biological effects of CXCL12 are mediated by its capacity to induce motility, chemotaxis, and adhesion, as well as to induce secretion of MMPs and angiogenic factors (such as vascular endothelial growth factor (VEGF) in cells expressing its receptor, CXC-chemokine receptor 4 (CXCR4)). Mice lacking either CXCL12 or CXCR4 show similar embryonic lethal defects, including impaired myeloid and B cell hematopoiesis (Nagasawa et al. 1996; Zou et al. 1998; Lapidot and Petit 2002). Importantly, CXCL12 is not essential for HSC generation or expansion in the fetal liver, but is crucial for colonization of the BM during late fetal development. Interestingly, engraftment of CXCR4 deficient fetal liver HSCs in normal adult recipients is unaffected, suggesting that the BM niche needs to provide CXCL12 to ensure HSC engraftment (Ara et al. 2003). Moreover, conditional knockouts revealed that this signaling pathway is also required for maintaining the quiescence of adult HSCs, and in its absence long-term HSCs are progressively lost (Sugiyama et al. 2006). The most important source of CXCL12 is CAR cells which are found next to niche osteoblasts and are associated with the microvasculature at the vascular niche (see below), suggesting that the CXCR4 pathway is activated in HSCs present in both niches.

Rac

Cell adhesion molecules cooperate with the cytoskeleton to regulate migration and adhesion, which is essential for HSC homing, engraftment and mobilization. Critical components mediating this interaction in HSCs are the Rho GTPases Rac1 and Rac2. For example, Lin^-c-Kit^+ cells lacking Rac1 not only fail to engraft (Gu et al. 2003), but also exhibit reduced homing efficiency to the BM and the endosteum. Moreover, deletion of both Rac1 and Rac2 causes massive defects in stem/progenitor cell proliferation, survival, adhesion to VLA-4 and/or VLA-5 receptors, and migration towards CXCL12 in vitro. Deletion of both genes in engrafted HSCs in vivo leads to massive mobilization of HSC/progenitors to the peripheral blood (Gu et al. 2003; Cancelas et al. 2005). Interestingly, Rac is activated by adhesion via $\beta1$-integrin and via stimulation of CXCL12–CXCR4 and by SCF-c-Kit signaling (Williams et al. 2008). Together, these data suggest that Rac1 and Rac2 have essential roles in homing, lodging, and retention of HSCs in the BM niche (Gu et al. 2003; Cancelas et al. 2005; Williams et al. 2008).

Osteopontin

The acidic glycoprotein Osteopontin (Opn) is secreted into the bone matrix. Opn synthesis is stimulated by calcitriol (1,25-dihydroxy-vitamin D3) and binds to $\alpha V\beta3$ integrin and to CD44 and extracellular matrix components (Wai and Kuo 2008). Mice lacking Opn have a twofold increase in HSCs and since the same effect was observed by transplanting wild-type HSCs into lethally irradiated Opn mutants, Opn production by osteoblasts has a negative effect on HSC number (Nilsson et al. 2005; Stier et al. 2005). Since apoptosis is induced in cultured Lin⁻Sca1⁺ BM in response to soluble Opn, the increase of HSCs in Opn deficient mice was postulated to be a result of decreased apoptosis (Stier et al. 2005).

Hedgehog Signaling

Treatment of cultured HSC/progenitor cells with sonic hedgehog (Shh) induced expansion that was dependent on BMP signaling, suggesting that hedgehog (Hh) signaling promotes HSC proliferation (Bhardwaj et al. 2001). In agreement with these data, it was shown that HSCs in mice lacking one copy of the Hedgehog receptor Patched (ptc) and therefore displaying increased Hh signaling show increased proliferation (Trowbridge et al. 2006). Ptc⁺/⁻ HSCs have an excellent short-term capacity to expand, as observed in an injury response, but as a consequence mutant HSCs exhaust and are eventually lost. These data suggest that while Hh signaling is used to promote self-renewal divisions of HSCs, prolonged signaling needs to be strictly controlled in order to prevent HSC exhaustion. Furthermore, as treatment with the inhibitor cyclopamine (a general inhibitor of Hh signaling) is well tolerated it seems unlikely that Hh signaling is absolutely critical for HSC maintenance and function, although it remains to be demonstrated that HSC signaling is completely blocked by this drug in vivo (Trowbridge et al. 2006).

Calcium Signaling

An increase in calcium concentration is found next to bone surfaces and freshly remodeled bone, suggesting that HSCs at the endosteum are in a calcium-rich environment, which may in turn influence the niche microenvironment. Indeed, HSCs normally express the calcium-sensing receptor (CaR), and HSCs lacking this receptor fail to home and engraft in the BM suggesting that high calcium concentrations contribute to the activity of the endosteal stem cell niche (Adams et al. 2006).

Wnt and Notch Signaling

Although canonical Wnt-β-catenin and Notch-RBPjK signaling have been frequently implicated in HSC function (Stier et al. 2002; Reya et al. 2003), recent data analyzing various genetic models make it unlikely that either of these two pathways play a major role in HSC maintenance or activity (Jeannet et al. 2008; Koch et al. 2008; Maillard et al. 2008).

The Vascular Bone Marrow Niche

Phenotypic identification of HSCs is normally performed using a combination of monoclonal antibodies labeled with multiple fluorescent dyes – a combination too complex for localization on bone sections using fluorescence microscopy. However, the recent discovery that HSCs can also be highly purified based on the expression of two SLAM family receptors CD150 and CD48 (CD150$^+$ CD48$^-$) has made immunolocalization more feasible (Kiel et al. 2005). Interestingly, about two-thirds of CD150$^+$ CD48$^-$ HSCs were found to be attached to sinusoidal endothelium, while only a minority were present at the endosteum (Kiel et al. 2005). These data provided the first evidence that the BM may harbor more than one type of stem cell niche.

As HSCs and endothelial cells arise from a common embryonic precursor (the hemangioblast) (Huber et al. 2004), interactions between these two cell types are not unexpected. Indeed, embryonic primary endothelial cells or cell lines have been shown to promote maintenance and/or expansion of adult LSK-HSCs *in culture* (Ohneda et al. 1998; Li et al. 2003), while adult, nonhematopoietic vascular endothelial cells only very inefficiently maintain HSCs in vitro (Li et al. 2004). Thus, BM sinusoidal endothelial cells (BMECs) are functionally and phenotypically distinct from microvasculature endothelial cells of other organs. Moreover cytokines important for HSC mobilization, homing, and engraftment, such as CXCL12 and adhesion molecules such as E-selectin and vascular cell-adhesion molecule-1 (VCAM-1), are constitutively expressed by BMECs (Rafii et al. 1997; Avecilla et al. 2004) thus providing further support for an important role of the vascular niche. In agreement with this concept, during HSC mobilization after myeloid ablation, hematopoiesis is re-established after quiescent HSCs detach from the endosteal niche and migrate toward the vascular zone in the center of the bones (Heissig et al. 2002; Avecilla et al. 2004).

In the BM, dormant/quiescent stem cells, as defined by label retaining assays (LRC-HSCs), have been shown to be almost exclusively confined to the endosteal regions of bone (Zhang et al. 2003; Oser et al. unpublished observations). These data show that most dormant HSCs are found in the endosteal niche, consistent with the idea that it may serve as a storage niche, or a self-renewing niche comprising both quiescent and self-renewing HSCs. In contrast, the CD150$^+$ HSC population which is principally localized to vascular niches, would comprise mostly activated self-renewing HSCs (Kiel et al. 2005; Wilson and Trumpp 2006). Their close proximity to sinusoids thus enables CD150$^+$ HSCs to respond rapidly and robustly to situations of BM stress. Thus the two BM stem cell niches (endosteal versus vascular) may differ in the HSC populations they support (quiescent vs. self-renewing). However, because deletion of osteoblasts results in extramedullary hematopoiesis (Visnjic et al. 2004), the vascular niche alone may not be sufficient to maintain long-term hematopoiesis, and may require influx of HSCs from the primary endosteal niche.

Reactive Oxygen Species

Studies in a number of mouse mutants showed that loss of either ataxia telangiectasia mutated (ATM) or forkhead-related FOXO transcription factors in HSCs result in elevated levels of reactive oxygen species (ROS) and disrupted self-renewal activity. In response to oxidative stress, HSCs specifically activate the p38 MAPK-p16^{Ink4a} pathway initially promoting HSC cycling, but eventually leading to stem cell exhaustion (Ito et al. 2006; Tothova et al. 2007). Most interestingly, ROS suppresses N-cadherin expression on HSCs possibly by activation of metalloproteases (Hosokawa et al. 2007). These results suggest that ROS-induced cycling and exhaustion may be caused by the enforced exit of HSCs from the niche. Importantly, these data also indicate that quiescent HSCs located in the endosteal niche are in a rather hypoxic environment. In contrast, vascular niches next to the oxygen-rich microvasculature are more likely to harbor actively cycling HSCs, which would run the day-to-day hematopoiesis (Wilson et al. 2007). In summary, these data also suggest that endosteal and vascular niches differ in the type of HSC (quiescent with long-term self-renewal activity vs. proliferating, short-term self-renewing) and that this may be at least in part be controlled by the level of ROS (Fig.3.2a).

Concluding Remarks

Stem-cell-specific microenvironments had already been proposed in the 1960s and 1970s; however, significant progress in our understanding of the cellular and molecular composition of the BM niches only occurred after 2000. The endosteal niche comprising osteoblasts and CAR cells has now been characterized and most likely preserves HSCs by keeping them quiescent (Fig.3.2a). In this location, HSCs can even be dormant long term, and are kept as a reserve pool which can be activated in response to hematopoietic injury (Wilson et al., in press). Destruction or absence of the endosteal niche causes HSCs to exhaust most likely due to overproliferation of stem cells (Visnjic et al. 2004; Walkley et al. 2007a; Orford and Scadden 2008). The vascular niche is less well characterized than the endosteal niche, and it is currently thought that this niche would harbor activated HSCs that run day-to-day hematopoiesis. Despite the progress made in recent years, which also includes identification of a number of biochemical pathways regulating HSC–niche function (Fig.3.2b), several important questions remain.

Although osteoblasts have been shown to be rate limiting for the BM-HSC niche, their exact phenotype and differentiation status remains unclear. Are these niche cells derived from the same lineage as the mesenchymal stem-cell-derived osteoblasts that continue to differentiate into osteocytes, or have they branched off to generate a distinct "niche–osteoblast"? If the latter is true, do they differentiate in response to signals derived from an attaching HSC indicating that HSCs and osteoblastic cells would form a self-organizing unit? How dynamic or stable are such HSC–niche interactions?

Moreover, the first data suggesting that altered niche function may be causative for disease have emerged. For example, microenvironment specific elimination of the retinoblastoma (Rb) or retinoic acid receptor γ (RARγ) genes causes the development of myeloproliferative disease (Walkley et al. 2007a, b). In addition, it has been proposed that leukemic stem cells (LSC) are also located in specific niches (Flynn and Kaufman 2007; Trumpp and Wiestler 2008), and indeed disruption of LSC–niche attachment by an antibody targeting CD44 has recently been successful in some leukemias (Jin et al. 2006).

Finally, in order to eventually be able to significantly expand HSCs in vitro, not only specific cytokine cocktails, but also a three-dimensional reconstruction of the "niche–stem cell synapse" with appropriate cell types and ECM will be required. The molecular and cellular biology of stem cell niches will soon receive even more attention in the stem cell field as niches not only control stem cell function, but also appear to be involved in disease and thus new knowledge could soon be translated into novel therapies.

Acknowledgments   We thank the Trumpp laboratory members for helpful discussions. This work was supported in part by grants to A.T. from the Swiss National Science Foundation, the Swiss Cancer League, the EU-FP6 Program "INTACT" and the EU FP7 EuroSyStem Integrated Project.

References

Adams, G.B., Chabner, K.T., Alley, I.R., Olson, D.P., Szczepiorkowski, Z.M., Poznansky, M.C., Kos, C.H., Pollak, M.R., Brown, E.M., and Scadden, D.T. 2006. Stem cell engraftment at the endosteal niche is specified by the calcium-sensing receptor. Nature 439(7076): 599–603.

Ara, T., Tokoyoda, K., Sugiyama, T., Egawa, T., Kawabata, K., and Nagasawa, T. 2003. Long-term hematopoietic stem cells require stromal cell-derived factor-1 for colonizing bone marrow during ontogeny. Immunity 19(2): 257–267.

Arai, F. and Suda, T. 2007. Maintenance of quiescent hematopoietic stem cells in the osteoblastic niche. Ann N Y Acad Sci 1106: 41–53.

Arai, F., Hirao, A., Ohmura, M., Sato, H., Matsuoka, S., Takubo, K., Ito, K., Koh, G.Y., and Suda, T. 2004. Tie2/angiopoietin-1 signaling regulates hematopoietic stem cell quiescence in the bone marrow niche. Cell 118(2): 149–161.

Askenasy, N. and Farkas, D.L. 2003. In vivo imaging studies of the effect of recipient conditioning, donor cell phenotype and antigen disparity on homing of haematopoietic cells to the bone marrow. Br J Haematol 120(3): 505–515.

Avecilla, S.T., Hattori, K., Heissig, B., Tejada, R., Liao, F., Shido, K., Jin, D.K., Dias, S., Zhang, F., Hartman, T.E., et al. 2004. Chemokine-mediated interaction of hematopoietic progenitors with the bone marrow vascular niche is required for thrombopoiesis. Nat Med 10(1): 64–71.

Barker, J.E. 1994. Sl/Sld hematopoietic progenitors are deficient in situ. Exp Hematol 22(2): 174–177.

Barker, J.E. 1997. Early transplantation to a normal microenvironment prevents the development of Steel hematopoietic stem cell defects. Exp Hematol 25(6): 542–547.

Barker, N., van Es, J.H., Kuipers, J., Kujala, P., van den Born, M., Cozijnsen, M., Haegebarth, A., Korving, J., Begthel, H., Peters, P.J., et al. 2007. Identification of stem cells in small intestine and colon by marker gene Lgr5. Nature 449(7165): 1003–1007.

Bhardwaj, G., Murdoch, B., Wu, D., Baker, D.P., Williams, K.P., Chadwick, K., Ling, L.E., Karanu, F.N., and Bhatia, M. 2001. Sonic hedgehog induces the proliferation of primitive human hematopoietic cells via BMP regulation. Nat Immunol 2(2): 172–180.

Blanpain, C., Horsley, V., and Fuchs, E. 2007. Epithelial stem cells: Turning over new leaves. Cell 128(3): 445–458.

Bradford, G.B., Williams, B., Rossi, R., and Bertoncello, I. 1997. Quiescence, cycling, and turnover in the primitive hematopoietic stem cell compartment. Exp Hematol 25(5): 445–453.

Calvi, L.M., Adams, G.B., Weibrecht, K.W., Weber, J.M., Olson, D.P., Knight, M.C., Martin, R.P., Schipani, E., Divieti, P., Bringhurst, F.R., et al. 2003. Osteoblastic cells regulate the haematopoietic stem cell niche. Nature 425(6960): 841–846.

Camargo, F.D., Chambers, S.M., Drew, E., McNagny, K.M., and Goodell, M.A. 2006. Hematopoietic stem cells do not engraft with absolute efficiencies. Blood 107(2): 501–507.

Cancelas, J.A., Lee, A.W., Prabhakar, R., Stringer, K.F., Zheng, Y., and Williams, D.A. 2005. Rac GTPases differentially integrate signals regulating hematopoietic stem cell localization. Nat Med 11(8): 886–891.

Chang, J.T., Palanivel, V.R., Kinjyo, I., Schambach, F., Intlekofer, A.M., Banerjee, A., Longworth, S.A., Vinup, K.E., Mrass, P., Oliaro, J., et al. 2007. Asymmetric T lymphocyte division in the initiation of adaptive immune responses. Science 315(5819): 1687–1691.

Chanprasert, S., Geddis, A.E., Barroga, C., Fox, N.E., and Kaushansky, K. 2006. Thrombopoietin (TPO) induces c-myc expression through a PI3K- and MAPK-dependent pathway that is not mediated by Akt, PKCzeta or mTOR in TPO-dependent cell lines and primary megakaryocytes. Cell Signal 18(8): 1212–1218.

Cheshier, S.H., Morrison, S.J., Liao, X., and Weissman, I.L. 1999. In vivo proliferation and cell cycle kinetics of long-term self-renewing hematopoietic stem cells. Proc Natl Acad Sci USA 96(6): 3120–3125.

Corral, D.A., Amling, M., Priemel, M., Loyer, E., Fuchs, S., Ducy, P., Baron, R., and Karsenty, G. 1998. Dissociation between bone resorption and bone formation in osteopenic transgenic mice. Proc Natl Acad Sci USA 95(23): 13835–13840.

Cotsarelis, G., Sun, T.T., and Lavker, R.M. 1990. Label-retaining cells reside in the bulge area of pilosebaceous unit: Implications for follicular stem cells, hair cycle, and skin carcinogenesis. Cell 61(7): 1329–1337.

Czechowicz, A., Kraft, D., Weissman, I.L., and Bhattacharya, D. 2007. Efficient transplantation via antibody-based clearance of hematopoietic stem cell niches. Science 318(5854): 1296–1299.

Deguchi, K., Yagi, H., Inada, M., Yoshizaki, K., Kishimoto, T., and Komori, T. 1999. Excessive extramedullary hematopoiesis in Cbfa1-deficient mice with a congenital lack of bone marrow. Biochem Biophys Res Commun 255(2): 352–359.

Ducy, P., Schinke, T., and Karsenty, G. 2000. The osteoblast: A sophisticated fibroblast under central surveillance. Science 289(5484): 1501–1504.

Dykstra, B., Kent, D., Bowie, M., McCaffrey, L., Hamilton, M., Lyons, K., Lee, S.-J., Brinkman, R., and Eaves, C.J. 2007. Long-term propagation of distinct hematopoietic differentiation programs in vivo. Cell Stem Cell 1: 218–229.

Flanagan, J.G., Chan, D.C., and Leder, P. 1991. Transmembrane form of the kit ligand growth factor is determined by alternative splicing and is missing in the Sld mutant. Cell 64(5): 1025–1035.

Flynn, C.M. and Kaufman, D.S. 2007. Donor cell leukemia: Insight into cancer stem cells and the stem cell niche. Blood 109(7): 2688–2692.

Friedl, P. and Storim, J. 2004. Diversity in immune–cell interactions: States and functions of the immunological synapse. Trends Cell Biol 14(10): 557–567.

Gong, J.K. 1978. Endosteal marrow: A rich source of hematopoietic stem cells. Science 199(4336): 1443–1445.

Gu, Y., Filippi, M.D., Cancelas, J.A., Siefring, J.E., Williams, E.P., Jasti, A.C., Harris, C.E., Lee, A.W., Prabhakar, R., Atkinson, S.J., et al. 2003. Hematopoietic cell regulation by Rac1 and Rac2 guanosine triphosphatases. Science 302(5644): 445–449.

Haug, J.S., He, X.C., Grindley, J.C., Wunderlich, J.P., Gaudenz, K., Ross, J.T., Paulson, A., Wagner, K.P., Xie, Y., Zhu, R., et al. 2008. N-cadherin expression level distinguishes reserved versus primed states of hematopoietic stem cells. Cell Stem Cell 2(4): 367–379.

Heissig, B., Hattori, K., Dias, S., Friedrich, M., Ferris, B., Hackett, N.R., Crystal, R.G., Besmer, P., Lyden, D., Moore, M.A., et al. 2002. Recruitment of stem and progenitor cells from the bone marrow niche requires MMP-9 mediated release of kit-ligand. Cell 109(5): 625–637.

Ho, A.D. 2005. Kinetics and symmetry of divisions of hematopoietic stem cells. Exp Hematol 33(1): 1–8.

Hosokawa, K., Arai, F., Yoshihara, H., Nakamura, Y., Gomei, Y., Iwasaki, H., Miyamoto, K., Shima, H., Ito, K., and Suda, T. 2007. Function of oxidative stress in the regulation of hematopoietic stem cell–niche interaction. Biochem Biophys Res Commun 363(3): 578–583.

Huber, T.L., Kouskoff, V., Fehling, H.J., Palis, J., and Keller, G. 2004. Haemangioblast commitment is initiated in the primitive streak of the mouse embryo. Nature 432(7017): 625–630.

Ito, K., Hirao, A., Arai, F., Takubo, K., Matsuoka, S., Miyamoto, K., Ohmura, M., Naka, K., Hosokawa, K., Ikeda, Y., et al. 2006. Reactive oxygen species act through p38 MAPK to limit the lifespan of hematopoietic stem cells. Nat Med 12(4): 446–451.

Jeannet, G., Scheller, M., Scarpellino, L., Duboux, S., Gardiol, N., Back, J., Kuttler, F., Malanchi, I., Birchmeier, W., Leutz, A., et al. 2008. Long-term, multilineage hematopoiesis occurs in the combined absence of beta-catenin and gamma-catenin. Blood 111(1): 142–149.

Jin, L., Hope, K.J., Zhai, Q., Smadja-Joffe, F., and Dick, J.E. 2006. Targeting of CD44 eradicates human acute myeloid leukemic stem cells. Nat Med 12(10): 1167–1174.

Karsenty, G. and Wagner, E.F. 2002. Reaching a genetic and molecular understanding of skeletal development. Dev Cell 2(4): 389–406.

Kiel, M.J., Yilmaz, O.H., Iwashita, T., Terhorst, C., and Morrison, S.J. 2005. SLAM family receptors distinguish hematopoietic stem and progenitor cells and reveal endothelial niches for stem cells. Cell 121(7): 1109–1121.

Kiel, M.J., Radice, G.L., and Morrison, S.J. 2007. Lack of evidence that hematopoietic stem cells depend on N-cadherin-mediated adhesion to osteoblasts for their maintenance. Cell Stem Cell 1(2): 204–217.

Kinashi, T. and Springer, T.A. 1994. Steel factor and c-kit regulate cell–matrix adhesion. Blood 83(4): 1033–1038.

Knoblich, J.A. 2008. Mechanisms of asymmetric stem cell division. Cell 132(4): 583–597.

Koch, U., Wilson, A., Cobas, M., Kemler, R., Macdonald, H.R., and Radtke, F. 2008. Simultaneous loss of beta- and gamma-catenin does not perturb hematopoiesis or lymphopoiesis. Blood 111(1): 160–164.

Kollet, O., Dar, A., and Lapidot, T. 2007. The multiple roles of osteoclasts in host defense: Bone remodeling and hematopoietic stem cell mobilization. Annu Rev Immunol 25: 51–69.

Konno, D., Shioi, G., Shitamukai, A., Mori, A., Kiyonari, H., Miyata, T., and Matsuzaki, F. 2008. Neuroepithelial progenitors undergo LGN-dependent planar divisions to maintain self-renewability during mammalian neurogenesis. Nat Cell Biol 10(1): 93–101.

Kovach, N.L., Lin, N., Yednock, T., Harlan, J.M., and Broudy, V.C. 1995. Stem cell factor modulates avidity of alpha 4 beta 1 and alpha 5 beta 1 integrins expressed on hematopoietic cell lines. Blood 85(1): 159–167.

Lapidot, T. and Petit, I. 2002. Current understanding of stem cell mobilization: The roles of chemokines, proteolytic enzymes, adhesion molecules, cytokines, and stromal cells. Exp Hematol 30(9): 973–981.

Lechler, T. and Fuchs, E. 2005. Asymmetric cell divisions promote stratification and differentiation of mammalian skin. Nature 437: 275–280.

Lerner, C. and Harrison, D.E. 1990. 5-Fluorouracil spares hemopoietic stem cells responsible for long-term repopulation. Exp Hematol 18(2): 114–118.

Li, W., Johnson, S.A., Shelley, W.C., Ferkowicz, M., Morrison, P., Li, Y., and Yoder, M.C. 2003. Primary endothelial cells isolated from the yolk sac and para-aortic splanchnopleura support the expansion of adult marrow stem cells in vitro. Blood 102(13): 4345–4353.

Li, W., Johnson, S.A., Shelley, W.C., and Yoder, M.C. 2004. Hematopoietic stem cell repopulating ability can be maintained in vitro by some primary endothelial cells. Exp Hematol 32(12): 1226–1237.

Lord, B.I., Testa, N.G., and Hendry, J.H. 1975. The relative spatial distributions of CFUs and CFUc in the normal mouse femur. Blood 46(1): 65–72.

Lyman, S.D. and Jacobsen, S.E. 1998. c-kit ligand and Flt3 ligand: Stem/progenitor cell factors with overlapping yet distinct activities. Blood 91(4): 1101–1134.

Maillard, I., Koch, U., Dumortier, A., Shestova, O., Xu, L., Sai, H., Pross, S.E., Aster, J.C., Bhandoola, A., Radtke, F., et al. 2008. Canonical notch signaling is dispensable for the maintenance of adult hematopoietic stem cells. Cell Stem Cell 2(4): 356–366.

Maloney, M.A. and Patt, H.M. 1975. On the origin of hematopoietic stem cells after local marrow extirpation. Proc Soc Exp Biol Med 149(1): 94–97.

Miyazawa, K., Williams, D.A., Gotoh, A., Nishimaki, J., Broxmeyer, H.E., and Toyama, K. 1995. Membrane-bound steel factor induces more persistent tyrosine kinase activation and longer life span of c-kit gene-encoded protein than its soluble form. Blood 85(3): 641–649.

Murphy, M.J., Wilson, A., and Trumpp, A. 2005. More than just proliferation: Myc function in stem cells. Trends Cell Biol 15(3): 128–137.

Nagasawa, T., Hirota, S., Tachibana, K., Takakura, N., Nishikawa, S., Kitamura, Y., Yoshida, N., Kikutani, H., and Kishimoto, T. 1996. Defects of B-cell lymphopoiesis and bone-marrow myelopoiesis in mice lacking the CXC chemokine PBSF/SDF-1. Nature 382(6592): 635–638.

Niche, W. Accession May 2008. Wikipedia http://en.wikipedia.org/wiki/niche.

Nilsson, S.K., Johnston, H.M., and Coverdale, J.A. 2001. Spatial localization of transplanted hemopoietic stem cells: Inferences for the localization of stem cell niches. Blood 97(8): 2293–2299.

Nilsson, S.K., Johnston, H.M., Whitty, G.A., Williams, B., Webb, R.J., Denhardt, D.T., Bertoncello, I., Bendall, L.J., Simmons, P.J., and Haylock, D.N. 2005. Osteopontin, a key component of the hematopoietic stem cell niche and regulator of primitive hematopoietic progenitor cells. Blood 106(4): 1232–1239.

Ohlstein, B., Kai, T., Decotto, E., and Spradling, A. 2004. The stem cell niche: Theme and variations. Curr Opin Cell Biol 16(6): 693–699.

Ohneda, O., Fennie, C., Zheng, Z., Donahue, C., La, H., Villacorta, R., Cairns, B., and Lasky, L.A. 1998. Hematopoietic stem cell maintenance and differentiation are supported by embryonic aorta-gonad-mesonephros region-derived endothelium. Blood 92(3): 908–919.

Oostendorp, R.A., Harvey, K.N., Kusadasi, N., de Bruijn, M.F., Saris, C., Ploemacher, R.E., Medvinsky, A.L., and Dzierzak, E.A. 2002. Stromal cell lines from mouse aorta-gonads-mesonephros subregions are potent supporters of hematopoietic stem cell activity. Blood 99(4): 1183–1189.

Orford, K.W. and Scadden, D.T. 2008. Deconstructing stem cell self-renewal: Genetic insights into cell-cycle regulation. Nat Rev Genet 9(2): 115–128.

Osawa, M., Hanada, K., Hamada, H., and Nakauchi, H. 1996. Long-term lymphohematopoietic reconstitution by a single CD34-low/negative hematopoietic stem cell. Science 273(5272): 242–245.

Passegue, E., Wagers, A.J., Giuriato, S., Anderson, W.C., and Weissman, I.L. 2005. Global analysis of proliferation and cell cycle gene expression in the regulation of hematopoietic stem and progenitor cell fates. J Exp Med 202(11): 1599–1611.

Patt, H.M. and Maloney, M.A. 1972. Bone formation and resorption as a requirement for marrow development. Proc Soc Exp Biol Med 140(1): 205–207.

Ponomaryov, T., Peled, A., Petit, I., Taichman, R.S., Habler, L., Sandbank, J., Arenzana-Seisdedos, F., Magerus, A., Caruz, A., Fujii, N., et al. 2000. Induction of the chemokine stromal-derived factor-1 following DNA damage improves human stem cell function. J Clin Invest 106(11): 1331–1339.

Potten, C.S. and Loeffler, M. 1990. Stem cells: Attributes, cycles, spirals, pitfalls and uncertainties. Lessons for and from the crypt. Development 110(4): 1001–1020.

Puri, M.C. and Bernstein, A. 2003. Requirement for the TIE family of receptor tyrosine kinases in adult but not fetal hematopoiesis. Proc Natl Acad Sci USA 100(22): 12753–12758.

Qian, H., Buza-Vidas, N., Hyland, C.D., Jensen, C.T., Antonchuk, J., Mansson, R., Thoren, L.A., Ekblom, M., Alexander, W.S., and Jacobsen, S.E. 2007. Critical role of thrombopoietin in maintaining adult quiescent hematopoietic stem cells. Cell Stem Cell 1(6): 671–684.

Radice, G.L., Rayburn, H., Matsunami, H., Knudsen, K.A., Takeichi, M., and Hynes, R.O. 1997. Developmental defects in mouse embryos lacking N-cadherin. Dev Biol 181(1): 64–78.

Rafii, S., Mohle, R., Shapiro, F., Frey, B.M., and Moore, M.A. 1997. Regulation of hematopoiesis by microvascular endothelium. Leuk Lymphoma 27(5–6): 375–386.

Reya, T., Duncan, A.W., Ailles, L., Domen, J., Scherer, D.C., Willert, K., Hintz, L., Nusse, R., and Weissman, I.L. 2003. A role for Wnt signalling in self-renewal of haematopoietic stem cells. Nature 423(6938): 409–414.

Schofield, R. 1978. The relationship between the spleen colony-forming cell and the haemopoietic stem cell. Blood Cells 4(1–2): 7–25.

Shizuru, J.A., Negrin, R.S., and Weissman, I.L. 2005. Hematopoietic stem and progenitor cells: Clinical and preclinical regeneration of the hematolymphoid system. Annu Rev Med 56: 509–538.

Solar, G.P., Kerr, W.G., Zeigler, F.C., Hess, D., Donahue, C., de Sauvage, F.J., and Eaton, D.L. 1998. Role of c-mpl in early hematopoiesis. Blood 92(1): 4–10.

Spangrude, G.J. and Johnson, G.R. 1990. Resting and activated subsets of mouse multipotent hematopoietic stem cells. Proc Natl Acad Sci USA 87(19): 7433–7437.

Spradling, A., Drummond-Barbosa, D., and Kai, T. 2001. Stem cells find their niche. Nature 414(6859): 98–104.

Stier, S., Cheng, T., Dombkowski, D., Carlesso, N., and Scadden, D.T. 2002. Notch1 activation increases hematopoietic stem cell self-renewal in vivo and favors lymphoid over myeloid lineage outcome. Blood 99(7): 2369–2378.

Stier, S., Ko, Y., Forkert, R., Lutz, C., Neuhaus, T., Grunewald, E., Cheng, T., Dombkowski, D., Calvi, L.M., Rittling, S.R., et al. 2005. Osteopontin is a hematopoietic stem cell niche component that negatively regulates stem cell pool size. J Exp Med 201(11): 1781–1791.

Suda, J., Suda, T., and Ogawa, M. 1984. Analysis of differentiation of mouse hemopoietic stem cells in culture by sequential replating of paired progenitors. Blood 64(2): 393–399.

Suda, T., Arai, F., and Hirao, A. 2005. Hematopoietic stem cells and their niche. Trends Immunol 26(8): 426–433.

Sugiyama, T., Kohara, H., Noda, M., and Nagasawa, T. 2006. Maintenance of the hematopoietic stem cell pool by CXCL12-CXCR4 chemokine signaling in bone marrow stromal cell niches. Immunity 25(6): 977–988.

Taichman, R.S. 2005. Blood and bone: Two tissues whose fates are intertwined to create the hematopoietic stem-cell niche. Blood 105(7): 2631–2639.

Taichman, R.S. and Emerson, S.G. 1998. The role of osteoblasts in the hematopoietic microenvironment. Stem Cells 16(1): 7–15.

Takano, H., Ema, H., Sudo, K., and Nakauchi, H. 2004. Asymmetric division and lineage commitment at the level of hematopoietic stem cells: Inference from differentiation in daughter cell and granddaughter cell pairs. J Exp Med 199(3): 295–302.

Thoren, L.A., Liuba, K., Bryder, D., Nygren, J.M., Jensen, C.T., Qian, H., Antonchuk, J., and Jacobsen, S.E. 2008. Kit regulates maintenance of quiescent hematopoietic stem cells. J Immunol 180(4): 2045–2053.

Till, J.E. and McCulloch, E.A. 1961. A direct measurement of the radiation sensitivity of normal mouse bone marrow cells. Radiat Res 14: 213–222.

Tokoyoda, K., Egawa, T., Sugiyama, T., Choi, B.I., and Nagasawa, T. 2004. Cellular niches controlling B lymphocyte behavior within bone marrow during development. Immunity 20(6): 707–718.

Tothova, Z., Kollipara, R., Huntly, B.J., Lee, B.H., Castrillon, D.H., Cullen, D.E., McDowell, E.P., Lazo-Kallanian, S., Williams, I.R., Sears, C., et al. 2007. FoxOs are critical mediators of hematopoietic stem cell resistance to physiologic oxidative stress. Cell 128(2): 325–339.

Trowbridge, J.J., Scott, M.P., and Bhatia, M. 2006. Hedgehog modulates cell cycle regulators in stem cells to control hematopoietic regeneration. Proc Natl Acad Sci USA 103(38): 14134–14139.

Trumpp, A. and Wiestler, O.D. 2008. Mechanisms of disease: Cancer stem cells-targeting the evil twin. Nat Clin Pract Oncol 5(6): 337–347.

Uchida, N., He, D., Friera, A.M., Reitsma, M., Sasaki, D., Chen, B., and Tsukamoto, A. 1997. The unexpected G0/G1 cell cycle status of mobilized hematopoietic stem cells from peripheral blood. Blood 89(2): 465–472.

Visnjic, D., Kalajzic, I., Gronowicz, G., Aguila, H.L., Clark, S.H., Lichtler, A.C., and Rowe, D.W. 2001. Conditional ablation of the osteoblast lineage in Col2.3deltatk transgenic mice. J Bone Miner Res 16(12): 2222–2231.

Visnjic, D., Kalajzic, Z., Rowe, D.W., Katavic, V., Lorenzo, J., and Aguila, H.L. 2004. Hematopoiesis is severely altered in mice with an induced osteoblast deficiency. Blood 103(9): 3258–3264.

Waghmare, S.K., Bansal, R., Lee, J., Zhang, Y.V., McDermitt, D.J., and Tumbar, T. 2008. Quantitative proliferation dynamics and random chromosome segregation of hair follicle stem cells. EMBO J 27(9):1309–1320.

Wai, P.Y. and Kuo, P.C. 2008. Osteopontin: Regulation in tumor metastasis. Cancer Metastasis Rev 27(1): 103–118.

Walkley, C.R., Olsen, G.H., Dworkin, S., Fabb, S.A., Swann, J., McArthur, G.A., Westmoreland, S.V., Chambon, P., Scadden, D.T., and Purton, L.E. 2007a. A microenvironment-induced myeloproliferative syndrome caused by retinoic acid receptor gamma deficiency. Cell 129(6): 1097–1110.

Walkley, C.R., Shea, J.M., Sims, N.A., Purton, L.E., and Orkin, S.H. 2007b. Rb regulates interactions between hematopoietic stem cells and their bone marrow microenvironment. Cell 129(6): 1081–1095.

Weissman, I.L. 2000. Stem cells: Units of development, units of regeneration, and units in evolution. Cell 100(1): 157–168.

Williams, D.A., Zheng, Y., and Cancelas, J.A. 2008. Rho GTPases and regulation of hematopoietic stem cell localization. Methods Enzymol 439: 365–393.

Wilson, A. and Trumpp, A. 2006. Bone-marrow haematopoietic-stem-cell niches. Nat Rev Immunol 6: 93–106.

Wilson, A., Murphy, M.J., Oser, G.M., Oskarsson, T., Kaloulis, K., Bettess, M.D., Pasche, A.C., Knabenhans, C., MacDonald, H.R., and Trumpp, A. 2004. c-Myc controls the balance between hematopoietic stem cell self-renewal and differentiation. Genes Dev 18(22): 2747–2763.

Wilson, A., Oser, G.M., Jaworski, M., Blanco-Bose, W.E., Laurenti, E., Adolphe, C., Essers, M.A., Macdonald, H.R., and Trumpp, A. 2007. Dormant and self-renewing hematopoietic stem cells and their niches. Ann N Y Acad Sci 1106: 64–75.

Wilson, A., Laurenti, E., Oser, G.M., van der Wath, R.C., Blanco-Bose,, W., Jaworski, M., Offner, S., Dunant, C., Eshkind, L., Bockamp, E., Lio, P., MacDonald, H.R., and Trumpp, A. in press. Hematopoietic stem cells reversibly switch from dormancy to self-renewal during homeostasis and repair. Cell.

Wright, D.E., Bowman, E.P., Wagers, A.J., Butcher, E.C., and Weissman, I.L. 2002. Hematopoietic stem cells are uniquely selective in their migratory response to chemokines. J Exp Med 195(9): 1145–1154.

Yoshihara, H., Arai, F., Hosokawa, K., Hagiwara, T., Takubo, K., Nakamura, Y., Gomei, Y., Iwasaki, H., Matsuoka, S., Miyamoto, K., et al. 2007. Thrombopoietin/MPL signaling regulates hematopoietic stem cell quiescence and interaction with the osteoblastic niche. Cell Stem Cell 1(6): 685–697.

Zhang, J., Niu, C., Ye, L., Huang, H., He, X., Tong, W.G., Ross, J., Haug, J., Johnson, T., Feng, J.Q., et al. 2003. Identification of the haematopoietic stem cell niche and control of the niche size. Nature 425(6960): 836–841.

Zou, Y.R., Kottmann, A.H., Kuroda, M., Taniuchi, I., and Littman, D.R. 1998. Function of the chemokine receptor CXCR4 in haematopoiesis and in cerebellar development. Nature 393(6685): 595–599.

Chapter 4
Molecular Biology of Erythropoiesis

James Palis

Abstract Red cells, comprising the most abundant cell type in the body, are uniquely designed to withstand the vicissitudes of the microcirculation to deliver oxygen to the tissues. Red cells are produced in the bone marrow where they undergo progressive maturation from unilineage progenitors to morphologically defined precursors to enucleated erythrocytes. Two distinct erythroid lineages exist during mammalian ontogeny. The first "primitive" erythroid lineage originates during early embryogenesis in the yolk sac and generates a transient wave of maturing erythroid cells. The second "definitive" erythroid lineage exists in the fetus and throughout postnatal life. Erythropoietin is the primary cytokine regulating erythroid cell maturation by signaling through its receptor to activate multiple intracellular signaling cascades. Erythropoiesis is also regulated by transcriptional complexes containing GATA-1, SCL, EKLF, and multiple other factors. These complexes assist in the creation of transcriptionally active chromatin regions and upregulate erythroid-specific genes. MicroRNAs have recently been found in erythroid cells and raise the possibility that gene downregulation is also important for lineage maturation. A better understanding of the regulation of the globin genes expressed in the embryo, fetus, and adult will ultimately lead to improved therapies for people with hemoglobinopathies, including sickle cell disease and thalassemia.

Introduction: RBC Structure and Function

Red cells are the most abundant cell type in the human body, with more than 25 trillion in the bloodstream. These circulating erythrocytes contain no nucleus or organelles, but are filled with hemoglobin to carry oxygen from the lungs to all

J. Palis
Department of Pediatrics, Center for Pediatric Biomedical Research,
University of Rochester Medical Center,
601 Elmwood Ave.,
Rochester, NY14642, USA
e-mail: james_palis@urmc.rochester.edu

A. Wickrema and B. Kee (eds.), *Molecular Basis of Hematopoiesis*,
DOI: 10.1007/978-0-387-85816-6_1, © Springer Science+Business Media, LLC 2009

tissues throughout the body. Tissue oxygenation is critically dependent upon the hemoglobin content and the viscosity of blood, both of which increase as red cell numbers go up. The optimum red cell content of blood is approximately 40–45%, above which the increasing viscosity of blood begins to hinder flow and limit tissue oxygenation. Hemoglobin is composed of four protein (globin) chains, each containing a heme ring with an iron molecule that binds and releases oxygen molecules without changing valence. Consistent with the central role of red cells in tissue oxygenation, our red cell mass is carefully regulated by oxygen levels in the body. Tissue hypoxia leads to increased production of red cells through the hormone erythropoietin (EPO), which is produced primarily in the kidneys but acts in the bone marrow where red cells are synthesized. A decrease in circulating red cell mass (i.e., anemia) leads to a loss of wellbeing and, if severe or rapid, red cell loss can be life-threatening.

Mammalian red cells are biconcave discs with unique cell membranes. The outer component of the cell membrane consists of a lipid bilayer with embedded proteins including band 3 and glycophorin. The inner component consists of a cytoskeleton composed primarily of α and β spectrin molecules that are attached to the lipid bilayer through anchoring proteins. The cytoskeleton helps to maintain the shape of the red cell, regulates the lateral mobility of membrane proteins, and provides structural support for the lipid bilayer. Cellular deformability, due in part to the excess membrane that comes with a biconcave shape, provides the red cell with the capability of navigating capillary networks where gas exchange primarily occurs. Mutations in the genes encoding cytoskeletal components can cause a preferential loss of cell membrane leading to decreased deformability and increased red cell destruction. This hemolysis shortens the normal 120-day red cell lifespan.

Erythroid Lineage Differentiation

In the adult, all mature blood cells are ultimately derived from hematopoietic stem cells. Differentiation of hematopoietic stem cells is characterized by progressive restriction in potential and proliferative capacity leading to hematopoietic progenitors committed to single lineages. The production of red blood cells (erythropoiesis) in adult mammals occurs in the bone marrow and is characterized by three distinct stages (Fig. 4.1).

The first stage consists of lineage-committed progenitors defined by their ability to form colonies in semisolid media. The earliest recognizable erythroid-specific progenitor is the burst-forming unit erythroid (BFU-E), which from the mouse and the human give rise in vitro to large colonies of red cells in 7 and 14 days, respectively. BFU-E generate more mature erythroid-committed progenitors termed colony-forming units erythroid (CFU-E). CFU-E from the mouse and the human give rise in vitro to small colonies of red cells in 2 and 7 days of culture, respectively (Heath et al. 1976; Stephenson et al. 1971). CFU-E have an extremely limited proliferative potential compared to BFU-E. Several cytokines have been shown to enhance the

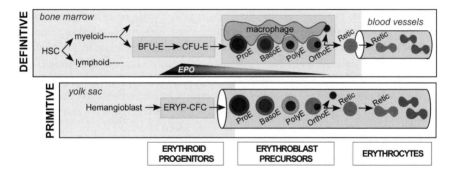

Fig. 4.1 Schema of definitive and primitive erythropoiesis. Erythroid lineage maturation is characterized by three progressive cellular compartments: Committed progenitors (BFU-E, CFU-E, EryP-CFC), erythroblast precursors (ProE, BasoE, PolyE, OrthoE), and enucleated erythrocytes. Definitive erythropoiesis in the adult arises from hematopoietic stem cells (HSC) in the bone marrow. Primitive erythropoiesis arises from hemangioblast precursors in the yolk sac. Primitive erythroid precursors enter the bloodstream where they mature as a semi-synchronous cohort. Erythropoietin (EPO) is postulated to primarily effect CFU-E and immature erythroid precursors that express highest levels of the erythropoietin receptor (EPO-R). *BFU-E* burst-forming unit erythroid, *CFU-E* colony forming unit erythroid, *EryP-CFC* primitive erythroid colony forming cell (*see Color Plates*)

formation of colonies from BFU-E, including IL-3, SCF, and high concentrations of EPO. In contrast, CFU-E survival depends almost exclusively on low concentrations of EPO.

The second stage of erythroid differentiation consists of morphologically identifiable, nucleated erythroid precursors that progress from proerythroblast to basophilic, polychromatophilic, and orthochromatic forms (Fig. 4.1). In mammals, three distinctive processes characterize this stage of differentiation. The first process is progressive accumulation of hemoglobin, which constitutes more than 90% of the protein content of the mature red cell. The second process is an expansion of erythroblast numbers through a limited number of cell divisions. Thus, erythroid precursors continue to divide as they differentiate. The third process is progressive nuclear pyknosis and ultimately loss of the nucleus.

Erythroid precursors mature in association with macrophage cells in "erythroblast islands" which serve as a stromal microenvironment within the bone marrow cavity (Bessis et al. 1978). In fact, every erythroid precursor in the marrow is in physical contact with a macrophage cell (Mohandas and Prenant 1978). Several adhesive interactions have been described that facilitate erythroblast–macrophage interactions. The first occurs between $\alpha4\beta1$ integrin expressed on erythroblasts and its counterreceptor vascular adhesion molecule 1 (VCAM-1) expressed on macrophage cells. Monoclonal antibodies directed toward $\beta1$-integrin and VCAM-1 each, disrupt erythroblast islands in vitro (Sadahira et al. 1995). A second molecule mediating adhesive interactions between erythroblasts and macrophage cells is erythroblast macrophage protein (Emp). Emp is a 36-kD transmembrane protein expressed both by erythroblasts and by macrophage cells (Hanspal and Hanspal 1994).

Culture of human erythroblasts without macrophage cells or in the presence of macrophage cells and Emp blocking antibodies resulted in a marked decrease in erythroblast proliferation and enucleation and a marked increase in erythroblast apoptosis (Hanspal and Hanspal 1994). These results support the notion that erythroblast–macrophage interactions are important for erythroblast maturation, survival, and enucleation. Emp was subsequently cloned and recently targeted in the mouse. At late gestation, fetuses lacking Emp develop pallor and have an increase in circulating nucleated erythroid cells (Soni et al. 2006). In vitro erythroblast island reconstitution assays using Emp-null erythroblasts or Emp-null macrophage cells mixed with their wild-type counterparts indicates that Emp function is both erythroid and macrophage cell autonomous. A third erythroblast–macrophage adhesive interaction occurs between the erythroid-specific isoform of intercellular adhesion molecule 4 (ICAM-4) and αv integrin on macrophage cells (Spring et al. 2001). The addition of synthetic αv integrin peptides blocks this interaction and disrupts erythroblast island integrity in vitro (Lee et al. 2006). Targeted disruption of ICAM-4 causes a 50% reduction in the number of erythroblast islands in the marrow; however, steady-state erythropoiesis is not adversely affected in the adult (Lee et al. 2006).

Cell–cell interactions within erythroblast islands have been proposed to facilitate both early and late stages of erythroid maturation. Regarding early stages of erythroid maturation, electron microscopic studies of erythroblast islands have suggested that macrophages may "nurse" erythroblasts by supplying them with iron (reviewed by Bessis et al. 1978). This process, initially termed "rhopheocytosis," is now recognized to be micropinocytosis, a process by which immature erythroblasts accumulate iron through a specific acid ferritin receptor (Meyron-Holz et al. 1994). It has also been proposed, but not proven, that central macrophage cells may serve as an important source of cytokines, in particular EPO, that support erythroid maturation (Rich 1986). Furthermore, the attachment to macrophage cells brings erythroblasts into close physical proximity and thus facilitates erythroblast–erythroblast interactions. The importance of these interactions has been supported by the proposed regulation of erythroid cell numbers by Fas–FasL signaling system. Expression of FasL on late-stage erythroblasts can transmit a death signal to adjacent immature erythroblasts that express Fas (de Maria et al. 1999). High levels of EPO can protect immature erythroblasts from this signaling pathway and lead to increased erythroid cell survival. Furthermore, the co-culture of erythroblasts with macrophage cells prevents erythroblast apoptosis and increases their proliferation (Hanspal et al. 1998; Rhodes et al. 2008).

Enucleation is the hallmark of mammalian definitive erythropoiesis and distinguishes the red cells of adult mammals from the red cells of nonmammalian species (Gulliver 1875). Enucleation begins when the nucleus becomes freely movable within late-stage erythroblasts, an event that coincides with the loss of vimentin intermediate filaments that form a cage around the nucleus. Soon thereafter, the nucleus becomes acentric and coalescing vacuoles with new membranes form in the furrow between nucleus and the incipient reticulocyte (Simpson and Kling 1967; Skutelsky and Danon 1970). The nucleus is then extruded with a thin rim of cytoplasm and a surrounding plasma membrane. This "extruded nucleus" is actually an intact cell that has recently been named "pyrenocyte" (McGrath et al. 2008). During enucleation, an actin ring forms between the nascent pyrenocyte and reticulocyte and inhibition of actin polymerization significantly blocks enucleation (Koury et al. 1989; Ji et al. 2008). A role for microtubules in the enucleation process remains controversial (Koury et al. 1989; Chasis et al. 1989).

The process of erythroblast enucleation also involves the segregation of cytoskeletal and cell surface proteins between the plasma membrane of the nucleus and the incipient reticulocyte. Interestingly, most of the major cytoskeletal proteins, including band 3, spectrin, ankyrin, and 4.1, segregate to the nascent reticulocyte, while Emp, β1-integrin and glycoconjugates recognized by concanavalin A partition to the nascent pyrenocyte (Gaiduschek and Singer 1979; Koury et al. 1989; Lee et al. 2004). The pyrenocyte rapidly loses phosphatidylserine asymmetry in its plasma membrane, sending an "eat me" signal to central macrophages (Yoshida 2005). Macrophage cells have been shown by time-lapse videography to actively phagocytose pyrenocytes (reviewed by Bessis et al. 1978; Yoshida et al. 2005). This macrophage function has also been documented in vitro in long-term bone marrow cultures (Allen and Dexter 1984) and in vivo in mice lacking DNase II, where fetal liver macrophages become engorged with multiple ingested, but not digested, nuclei (Kawane et al. 2001). This defect causes severe anemia and in utero death in late gestation. A similar phenotype of severe late gestation anemia and perinatal mortality has also been described in one of three phosphatidylserine receptor knockout mouse models (Kunisaki et al. 2004) that presumably leads to the inability of macrophage cells to ingest pyrenocytes. These results suggest that engulfment and breakdown of pyrenocytes play an important role in the regulation of late stages of erythropoiesis.

The third stage of erythroid differentiation consists of reticulocyte maturation and the circulation of mature red cells in the circulation (Fig. 4.1). Transcription ceases with the dismantling of the ribosomal machinery as reticulocytes transition to terminally differentiated erythrocytes. An active loss of internal organelles including endoplasmic reticulum and mitochondria occurs through internal degradation, termed autophagy (Fader and Colombo 2006). It has recently been shown that mitochondrial clearance in reticulocytes requires the BCL2-related protein NIX (Schweers et al. 2007). As the reticulocyte matures from a multilobulated cell to a biconcave disc, there is a significant restructuring of the membrane cytoskeleton resulting in significant loss of cell membrane associated with the loss of specific cell surface proteins through the release of exosomes (Chasis et al. 1989; Johnstone et al. 1991). Reticulocytes enter the bloodstream by migrating through the sinusoidal endothelium of the marrow. Within 24–48 h after entering the circulation, human reticulocytes lose their remaining ribosomes and thus transition to mature erythrocytes that circulate for approximately 120 days. As red cells age, there is a progressive loss of cell membrane, hemoglobin content, and CD47 expression (Gifford et al. 2006; Khandelwal et al. 2007). Eventually, senescent erythrocytes express phosphotidylserine on their cell surface and are recognized and engulfed by macrophage cells of the reticuloendothelial system (Boas et al. 1998).

Erythroid Cell Ontogeny

In contrast to the enucleated erythrocytes of mammals, the circulating red cells of fish, amphibians and birds are nucleated (Gulliver 1875). Nearly 100 years ago, examination of mammalian embryos revealed the presence of distinct nucleated and enucleated red cells (Maximow 1909). The continuous circulation of small, enucleated red cells during fetal and postnatal life, termed "definitive" erythropoiesis, was distinguished from "primitive" erythropoiesis, characterized by the transient circulation of large, nucleated red cells that originate in the yolk sac. Because mammalian primitive erythroblasts circulate as nucleated cells and are confined to the embryo, they have been thought to be a "primitive" form of erythropoiesis that shares many characteristics with the nucleated red cells of nonmammalian vertebrates (Tavassoli 1991). However, it was originally shown (Kingsley et al. 2004) and recently confirmed (Fraser et al. 2007) that primitive erythroid cells in the mouse enucleate to form erythrocytes, indicating that primitive erythropoiesis in mammals is clearly "mammalian" in nature.

Primitive Erythropoiesis

Primitive erythroid progenitors (EryP-CFC) capable of generating colonies of primitive erythroid cells were first identified when murine yolk sac cells were cultured in semisolid media (Wong et al. 1986). Murine EryP-CFC form compact red colonies in 5 days comprised of large erythroid cells that express both embryonic and adult globins (Palis et al. 1999). In contrast, BFU-E and CFU-E generate colonies comprised of small red cells expressing only adult globins. EryP-CFC arise during early gastrulation in the mouse embryo (E7.25), expand in numbers in the yolk sac, and disappear within 48 h (Palis et al. 1999). These results indicate that the primitive erythroid lineage is transient and distinct from the definitive erythroid lineage.

Pools of primitive erythroid cells are evident in the mouse yolk sac as "blood islands" beginning at E7.5 (Haar and Ackerman 1971). Over the next 24 h, they differentiate into primitive proerythroblasts ensheathed by a primary vascular plexus (Ferkowicz and Yoder 2005). Primitive erythroblasts begin to circulate at E8.25 soon after the onset of cardiac contractions (McGrath et al. 2003; Ji et al. 2003). They continue to divide in the bloodstream until E13, as evidenced by the presence of circulating mitotic figures (Bethlenfalvay and Block 1970), thymidine incorporation (de la Chapelle et al. 1969), and cell cycle studies (Sangiorgi et al. 1990). Primitive erythroblasts accumulate increasing amounts of hemoglobin and become progressively less basophilic (Steiner and Vogel 1973; Sasaki and Matsamura 1986). Morphologic analysis indicates that primitive erythroblasts mature in a semisynchronous cohort as they circulate in the bloodstream (Kingsley et al. 2004; Fraser et al. 2007). Primitive red cells reach their steady-state hemoglobin content of 80–100 pg per cell, approximately six times the amount of hemoglobin found in adult murine erythrocytes (Steiner and Vogel 1973). Using immunohisto-chemical and transgene approaches to positively identify primitive erythroid cells, it was recognized that primitive erythroid cells in the mouse embryo enucleate between E11.5 and E17.5 of gestation and continue to circulate as erythrocytes for several days afterwards (Kingsley et al. 2004; Fraser et al. 2007). It was recently noted that primitive "pyrenocytes" are found transiently in the fetal bloodstream, consistent with the notion that primitive erythroid precursors, like their definitive erythroid counterparts, enucleate by nuclear extrusion (McGrath et al. 2008).

While primitive erythropoiesis has been most thoroughly investigated in the mouse, relatively little is known about primitive erythropoiesis in humans owing to ethical concerns and physical inaccessibility. Primitive erythroid cells circulate in the human embryo throughout the first trimester (Knoll 1927). However, it is not yet known if human primitive erythroid cells arise from a transient population of EryP-CFC or if they ultimately enucleate to form primitive erythrocytes in the fetal bloodstream.

Definitive Erythropoiesis

Primitive erythropoiesis fulfills the erythroid functions critical for early postim-plantation embryonic survival and growth, however, the fetus requires increasing numbers of red cells throughout gestation. Prior to formation of the bone marrow cavity, the liver serves as the site of maturation of definitive erythroid cells in the fetus. Hematopoietic progenitors colonize the liver of the mouse fetus between E9.5 and E10.0, soon after it begins to form as an organ (Johnson and Moore 1975; Palis et al. 1999). The source of these definitive erythroid progenitors is the yolk sac, since BFU-E are first found in the yolk sac prior to the onset of circulation and are subsequently found in increasing numbers within the bloodstream prior to liver colonization (Palis et al. 1999). Furthermore, Ncx-1-null mouse embryos that lack a heartbeat and fail to establish a functional circulation, have normal numbers of definitive erythroid progenitors emerging in the yolk sac (Lux et al. 2008). However, few BFU-E are found within the embryo proper of these mutant mice supporting the hypothesis that definitive hematopoietic progenitors are initially generated in the yolk sac and are redistributed to the embryo proper following the onset of embryonic circulation.

Definitive erythropoiesis in the fetal liver is initially characterized by exponentially expanding numbers of BFU-E and CFU-E that peak in numbers at E14.5–15.5 (Rich and Kubanek 1979; Kurata et al. 1998). As gestation proceeds, there is gradual transition of hematopoietic activity to the bone marrow cavity and the liver ceases to be hematopoietic organ in both the mouse and the human soon after birth. While fundamentally similar, there are some differences between fetal erythroid progenitors and their adult bone marrow counterparts. For example, CFU-E in the murine fetus are more sensitive than bone marrow-derived CFU-E to EPO (Rich and Kubanek 1976). Furthermore, fetal BFU-E have a greater proliferative capacity and mature more rapidly when compared to adult BFU-E. Finally, fetal liver-derived BFU-E, unlike adult marrow-derived BFU-E, are capable of proliferating in response to EPO alone (Migliaccio and Migliaccio 1988; Emerson et al. 1989).

Cytokines in Erythroid Differentiation

It has long been recognized that an inverse relationship exists between the oxygen delivery in the body and the production of red blood cells (Tinsley et al. 1949). The mechanism responsible for this relationship became clearer after the cloning of the glycoprotein hormone EPO and the recognition that the maturation of erythroid cells is critically dependent on this cytokine. Studies in hepatoma cell lines indicated that the synthesis of EPO was transcriptionally regulated by oxygen through the regulation of hypoxia inducible factor 1α (HIF1α) (Semenza and Wang 1992). Under normoxic conditions HIF1α is ubiquitinated and degraded after proline hydroxylation and binding the von Hippel Lindau protein (Maxwell et al. 1999; Jaakkola et al. 2001). This hydroxylase requires oxygen for its enzymatic function. In contrast, under hypoxic conditions, HIF1α is not hydroxylated on its proline residues and is therefore freed to bind an enhancer sequence of the EPO gene and upregulate EPO transcription. Thus, the ever-present signal to synthesize EPO is effectively down-modulated by normal oxygen levels in the tissue.

In the adult, EPO is synthesized primarily in the kidneys. Thus, hormone synthesis is at a site distant from its action. Studies in the fetuses of various organisms reveal the liver as the primary site of both erythroid cell maturation as well as EPO synthesis. The source of EPO for primitive erythroid cells in the yolk sac has not been delineated. Targeted disruption of EPO and its receptor (EPO-R) in mice produce the same phenotype of embryonic lethality at E13.5 indicating that the primary mode of EPO action in erythropoiesis is through engagement of EPO-R (Wu et al. 1995). Loss of EPO signaling leads to a significant decrease in primitive erythroid cell numbers and a complete block in the ability of definitive erythroid precursors to mature (Wu et al. 1995; Lin et al. 1996; Kieran et al. 1996). A similar differential effect of EPO function in primitive vs. definitive erythropoiesis has recently been noted in zebrafish embryos (Paffett-Lugassy et al. 2007).

In the adult, EPO-R is expressed primarily by CFU-E and immature erythroid precursors (Fig. 4.1). Both BFU-E and CFU-E require EPO to form colonies in vitro, however, only CFU-E are EPO-responsive in vivo (Stephenson et al. 1971; Iscove 1977). It is clear that EPO acts as a critical survival factor for CFU-E and this activity is thought to be the primary regulatory mechanism for maintaining the homeostasis of red cell mass (Koury and Bondurant 1990; Erslev and Besarab 1997). Thus, a drop in red cell numbers causes tissue hypoxia, leading to increased EPO synthesis and the survival of more CFU-E that can rapidly mature into erythrocytes.

EPO engages its receptor inducing EPO-R dimerization, autophosphorylation of the Janus kinase 2 (JAK2), and phosphorylation of several tyrosine residues of the intracytoplasmic portion of EPO-R (Remy et al. 1999). The critical role of JAK2 in EPO-R signaling was established by examination of the phenotype of JAK2-null mice, which phenocopy the severe fetal anemia and fetal death of the EPO and EPO-R knockouts (Neubauer et al. 1998). Phosphorylation of EPO-R leads to recruitment of multiple signaling molecules, including STAT5, Grb2-SOS, and p85. STAT5 translocates to the nucleus to regulate the transcription of the antiaoptotic gene Bcl-x(L) and enhance erythroid cell survival (Silva et al. 1999). Bcl-x(L) functions to protect both primitive and definitive erythroid cells from apoptosis (Motoyama et al. 1999). Mice lacking STAT5 signaling develop transient fetal anemia and have reduced ability to respond to acute anemia as adults (Socolovsky et al. 1999). These findings suggest that STAT5 signaling plays an important role in stress erythropoiesis. Grb2-SOS recruitment to the EPO-R stimulates the Ras/Raf and MAP kinase signaling pathways (Arcasoy and Jiang 2005). p85 binding to the EPO-R activates PI3 kinase, which in turn activates AKT leading to phosphorylation of GATA-1 and Foxo3a (Uddin et al. 2000). PI3 kinase activation in erythroid progenitors also leads to the upregulation of several cell cycle genes as well as increased transcription of c-kit (Sivertsen et al. 2006). Lyn, a Src family member tyrosine kinase, also associates with EPO-R, phosphorylating it and STAT5 (Chin et al. 1998). Erythroid cells lacking lyn have decreased levels of several key erythroid transcription factors including Stat5, GATA-1 and EKLF. Lyn-null mice develop progressive anemia due to ineffective erythropoiesis (Ingley et al. 2005). While the effects of EPO on CFU-E and immature erythroid progenitors have received considerable attention, EPO may also plays roles at late stages of erythroid maturation. It has recently been recognized that EPO alters the expression of adhesive molecules, such as podocalyxin in target cells and promotes rapid reticulocyte egress from the bone marrow into the bloodstream (Sathyanarayana et al. 2007). In summary, EPO signaling through its specific receptor stimulates multiple intracellular signaling cascades to protect erythroid cells from apoptosis, induce more rapid lineage differentiation, and stimulate reticulocyte release.

Several cytokines have been shown to synergize with EPO to enhance erythroid cell differentiation, including IL6, IL3, and stem cell factor (SCF, also called kit ligand as well as mast cell growth factor). SCF interacts with its receptor, c-kit, that is expressed by multiple hematopoietic progenitors including BFU-E and CFU-E (Olweus et al. 1996). Mice carrying mutations of SCF (Sl/Sl^d) or c-kit (W/W^v) have decreased numbers of erythroid progenitors in the marrow and variable degrees of anemia (reviewed by Broudy (1997)). SCF synergizes with EPO and other cytokines to expand the number and size of erythroid colonies formed in vitro from the bone marrow, in part by stimulating the proliferation of erythroid progenitors (McNiece et al. 1991; Muta et al. 1995). SCF is necessary for in vitro growth of BFU-E in the absence of serum and for the rapid in vivo erythroid "stress" response that occurs in the spleen of mice following the induction of acute anemia (Dai et al. 1991; Broudy et al. 1996). These actions may occur through physical interactions of c-kit and EPO-R in hematopoietic progenitors leading to the activation of EPO-R through tyrosine phosphorylation (Wu et al. 1997). Furthermore, SCF signaling through c-kit appears to maintain EPO-R and STAT5 protein expression, thus enhancing survival of erythroid progenitors in response to EPO (Kapur and Zhang 2001).

Transcription Factors in Erythroid Differentiation

Lineage differentiation in the hematopoietic system and terminal maturation of erythroid cells is regulated in part by the differential expression and the combinatorial action of transcription factors. The central erythroid-specific transcription factor is GATA-1, which forms complexes with multiple other proteins to upregulate the expression of erythroid-specific genes. Furthermore, several Kruppel-like transcription factors, in particular erythroid Kruppel-like factor (EKLF, KLF1), also play important roles in the regulation of erythroid-specific genes.

GATA-1 is the founding member of the GATA transcription factor family of dual zinc finger proteins that bind a WGATAR consensus motif present in essentially all erythroid-specific genes. Targeted disruption of GATA-1 in the mouse leads to a block of primitive erythroid cell maturation at the proerythroblast stage resulting in embryonic lethality at E9.5–10.5, prior to the shift of hematopoiesis from the yolk sac to the fetal liver (Fujiwara et al. 1996). GATA-1-null embryonic stem cells fail to contribute to definitive erythropoiesis in chimeric mice consistent with a role for GATA-1 in definitive erythropoiesis (Pevny et al. 1991). Conditional expression of GATA-1 in differentiating GATA-1-null embryonic stem cells revealed roles for GATA-1 both in the proliferation of immature erythroid precursors and in their terminal differentiation (Zheng et al. 2006). Different functional domains of GATA1 are required for activation of target genes in primitive vs. definitive erythroid cells suggesting that different transcriptional complexes may form in these lineages (Shimizu et al. 2001).

A screen for proteins that bind to GATA-1 led to the identification of Friend of GATA (FOG-1), a nuclear zinc finger protein that binds the amino zinc finger of GATA-1 (Tsang et al. 1997). Consistent with its predicted GATA-1 interaction, targeted disruption of FOG in mice results in severe anemia and embryonic lethality at E10.5–11 (Tsang et al. 1998). Like mice lacking GATA1, FOG-1-null primitive erythroid cells also arrest at the proerythroblast stage of maturation. The importance of GATA-1/FOG-1 interactions is exemplified by the finding of individuals with severe congenital dyserythropoietic anemia due to a GATA-1 mutation that disrupts its physical contact with FOG-1 (Nichols et al. 2000). FOG-1 mediates GATA-1 interactions with the MeCP1 complex that is implicated in mediating gene repressive actions (Rodriguez et al. 2005).

GATA-1 can also interact physically with the CREB-binding protein (CBP), a ubiquitously expressed nuclear protein with histone acetyltransferase activity (Blobel et al. 1998). Since acetylation of histones is associated with transcriptionally active regions of chromatin, this finding suggests that GATA-1 may also function to recruit histone modifiers to erythroid specific genes to enhance their expression (Letting et al. 2003).

The stem cell leukemia gene (SCL, TAL1) is another transcription factor found in a multiprotein complex with GATA-1. SCL is a helix–loop–helix transcription factor that heterodimerizes with the E2A gene product and binds E-box (CAGGTG) DNA sequences (Begley et al. 1989). SCL is expressed both in primitive and in definitive erythroid cells (Silver and Palis 1997; Zhang et al. 2005). Targeted disruption of SCL in mice leads to a complete block in the emergence of embryonic hematopoietic cells and lethality at E9.5–10.5 of gestation (Shivdasani et al. 1995; Robb et al. 1995). SCL-null embryonic stem cells fail to contribute to adult hematopoiesis, suggesting important roles for SCL in definitive hematopoiesis (Robb et al. 1996; Porcher et al. 1996). Consistent with this notion, conditional knockdown of SCL in adult mice causes a significant impairment of erythropoiesis and megakaryopoiesis (Hall et al. 2003).

Lbd1 is a member of the erythroid multimeric complex containing GATA-1–FOG–SCL (Wadman et al. 1997). A screen of Ldb-1 binding partners in erythroid cells led to the identification of ETO-2, Cdk9, and LMO4 (Meier et al. 2006). Morpholino-mediated downregulation of these latter transcription factors in zebrafish reveals that they function in definitive, but not primitive, hematopoiesis. These results support the notion that primitive and definitive erythropoiesis contain different transcriptional complexes. ETO2 appears to function in the GATA-1–FOG–SCL complex by repressing SCL target genes (Schuh et al. 2005).

The Lim domain-containing transcription factor LMO2 is also a member of the GATA1–FOG–SCL transcriptional complex (Wadman et al. 1997). Similar to SCL, targeted disruption of LMO2 in mice blocks the emergence of hematopoietic cells and causes embryonic lethality at ER9.5–10.5 (Warren et al. 1994). LMO2-null embryonic stem cells fail to contribute to adult erythropoiesis in chimeric mice, consistent with a functional role for LMO2 in definitive erythropoiesis (Yamada et al. 1998).

Several members of the Kruppel-like transcription factor family are known to activate erythroid-specific genes. The prototypical member, KLF1 (EKLF) was originally cloned from erythroid cells and is expressed both in primitive and in definitive erythroid precursors (Miller and Bieker 1993; Southwood et al. 1996). Targeted disruption of EKLF in the mouse leads to severe abnormalities of definitive erythroid cells resulting in fetal death at E15.5 (Perkins et al. 1995; Nuez et al. 1995). While it was initially thought that EKLF mainly regulated the adult beta-globin gene through interactions with its CACC motif, it is now recognized that EKLF also regulates the expression of multiple erythroid-specific genes, including cytoskeletal proteins and AHSP (Hodge et al. 2006; Pilon et al. 2006). EKLF associates with the SWI/SNF-related chromatin remodeling complex suggesting that its activity in erythroid cells may in part be related to altering the configuration of chromatin (Amstrong et al. 1998). Consensus binding sites for EKLF and GATA-1 are found near the cis regulatory regions of many erythroid-specific genes. Interestingly, GATA-1 binding motifs are present upstream of the EKLF gene and GATA-1 activates EKLF transcription (Anderson et al. 2000).

Two other KLF family members have been implicated in the regulation of globin genes. Studies of KLF2-null mouse embryos revealed significant decreases in βH1- and εy-globin transcripts in primitive erythroid cells (Basu et al. 2005).

Human ε-globin transgenes were also reduced in mice lacking KLF2, providing evidence that this transcription factor may play a similar role in humans. KLF4 knockdown in zebrafish embryos leads to decreased embryonic globin gene expression (Gardiner et al. 2007). Consistent with this finding, KLF4 preferentially binds the CACC sites in the promoters of the embryonic compared to the adult β-globin genes.

As hematopoietic progenitors become committed to specific lineages and undergo terminal maturation, different batteries of genes are up- and downregulated. The continued identification of interacting transcription factors and their downstream targets will provide a better understanding of erythroid cell maturation and will facilitate the construction of increasingly complex models of the genetic regulatory networks that exist in erythroid cells (Swiers et al. 2006; Nemenman et al. 2007). The relative simple composition of mature erythrocytes, lacking a nucleus and intracellular organelles and containing hemoglobin as the predominant protein, makes the erythroid lineage an attractive model for such modeling, as well as for studies of the cellular proteome (Pasini et al. 2006).

MicroRNAs and Erythropoiesis

MicroRNAs are endogenous small noncoding RNAs that target mRNAs in a sequence-specific manner to modulate their stability or their ability to undergo translation. To date, several hundred microRNAs have been identified in plants and animals. Recent screens have revealed more than 100 microRNAs in definitive erythroid cells. Of these, miR-451 is found to be most abundantly upregulated during erythroid cell maturation (Zhan et al. 2007; Choong et al. 2007). The forced overexpression of miR-451 induced the maturation of murine erythroleukemia cells in vitro, suggesting that it plays a functional role in erythroid cell maturation (Zhan et al. 2007). miR-451 and miR-144 were recently identified as GATA-1 regulated sequences (Dore et al. 2008). Morpholino-mediated knockdown of miR-451, but not miR-144, leads to a significant block in terminal erythroid differentiation in zebrafish embryos. These in vitro and in vivo findings indicate that miR-451 is necessary for terminal erythropoiesis and highlights the emerging importance of microRNAs in hematopoietic lineage maturation. While most studies of erythroid cell maturation have focused on the induction of erythroid-specific genes, these recent studies of microRNAs highlight the potential importance of gene downregulation during erythroid cell maturation.

Globin Gene Regulation

Hemoglobin molecules contain globin chains derived from both the α- and the β-globin gene loci. The β-globin gene cluster has served as a prototypical model system for the study of cell type-specific and developmental stage-specific gene regulation. The overall structures of the β-globin locus from multiple species have been elucidated and found to contain multiple adjacent β-like globin genes and an upstream locus control region (LCR) that contains several DNase I hypersensitive sites (Fig. 4.2). These hypersensitive sites each contain multiple transcription factor binding motifs, including those for GATA1 and NFE2. As shown in Fig. 4.2, the human β-globin locus contains five active genes (ε, Gγ, Aγ, δ, and β) located in tandem in the order of their expression during development. The ε-globin gene is expressed during early embryogenesis, followed by the γ-globin genes in the fetus, and the δ- and β-globin genes in the adult. The regulation of "globin switching" has been an area of active research for the past three decades (Stamatoyannopoulos 2005).

The LCR facilitates the high level expression of the β-globin genes in erythroid cells. Removal of the LCR in mice caused a 100-fold reduction in globin gene expression (Bender et al. 2000). It is thought that the LCR functions to bring transcriptional regulators in close proximity to the basal transcriptional machinery immediately upstream of the various globin genes. Individual hypersensitive sites in the LCR appear to have different roles in chromatin remodeling and control of globin gene switching (reviewed by Harju et al. 2002). Transgenic mice containing one or two tandem globin transgenes downstream of LCR elements have provided evidence in support of a competition model of globin gene regulation whereby globin genes vie for LCR interactions based on their proximity to the LCR (Enver et al. 1990). Besides competition, there is also evidence that the embryonic and the fetal globin genes are regulated by autonomous silencing (Raich et al. 1990). For example, the embryonic ε-globin gene is silenced in part within adult (definitive) erythroid cells by the TR2 and the TR4 orphan nuclear receptors (Tanabe et al. 2007).

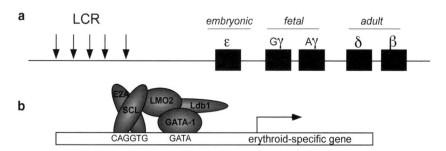

Fig. 4.2 (**a**) Simplified schema of the human beta-globin locus that consists of five active globin genes (ε, Gγ, Aγ, δ, and β) and an upstream locus control region (LCR) containing five DNase hypersensitive sites (*arrows*). The globin genes in the human are expressed sequentially with the ε-globin gene expressed in the embryo, the γ-globin genes expressed in the fetus and the β-globin gene expressed postnatally. (**b**) Model of a multiprotein transcription factor complex regulating erythroid-specific gene expression. The SCL-E2A dimer interacts with the E box motif (CAGGTG) and GATA-1 interacts with the nearby GATA motif in the cis regulatory region of an erythroid-specific gene. LMO2 and Ldb1 provide protein–protein interactions bridging these molecules (adapted from Wadman et al. 1997)

Fetal and adult definitive red cells in the mouse, unlike the human, express only adult (α1, α2, β1, β2) globins. In contrast, primitive red cells express both embryonic (βH1, ϵy, ζ) and adult globins. The differential expression of globin genes in primitive vs. definitive erythroid cells suggests that the regulatory mechanisms in these distinct lineages also differ. Analysis of globin gene expression in murine primitive erythroid cells indicates that the embryonic ζ- and βH1-globin genes are superseded by the α1-, α2- and ϵy-globin genes as these cells mature in the bloodstream (Kingsley et al. 2006). This process of "maturational" globin switching occurs in the same cells as they complete their terminal differentiation. These changes in globin transcript levels are associated with changes in RNA polymerase II density at their promoters. Furthermore, the βH1- and ϵy-globin genes in primitive erythroid cells reside in a single large hyperacetylated domain suggesting that maturational globin switching is regulated by altered transcription factor presence instead of chromatin accessibility (Kingsley et al. 2006). In contrast, embryonic globin genes in definitive erythroid cells lie outside of hyperacetylated domains and are not expressed in definitive erythroid cells (Bulger et al. 2003; Trimborn et al. 1999; Kingsley et al. 2006). Primitive erythroid cells in human embryos also appear to undergo maturational globin switching, since ζ- to α-globin and ϵ- to γ-globin gene switches have been described between 5 and 7 weeks gestation (Peschle et al. 1985). Definitive erythroid cells also undergo a more subtle γ- to β-globin maturational switch as they terminally differentiate (Papayannopoulou et al. 1979). A better understanding of the mechanisms regulating the activation and silencing of the embryonic, fetal, and adult globin genes has important clinical implications for the treatment and cure of patients with hemoglobinopathies, particularly sickle cell disease and the thalassemia syndromes.

References

Allen, T. D., and Dexter, T. M. 1984. The essential cells of the hematopoietic microenvironment. Exp Hematol 12:517–521.

Amstrong, J. A., Bieker, J. J., and Emerson, B. M. 1998. A SWI/SNF-related chromatin remodeling complex, E-RC1, is required for tissue-specific transcriptional regulation by EKLF in vitro. Cell 95:93–104.

Anderson, K. P., Crable, S. C., and Lingrel, J. B. 2000. The GATA-E box-GATA motif in the EKLF promoter is required for in vivo expression. Blood 95:1652–1655.

Arcasoy, M. O., and Jiang, X. 2005. Co-operative signalling mechanisms required for erythroid precursor expansion in response to erythropoietin and stem cell factor. Br J Haematol 130:121–129.

Basu, P., Morris P. E., Haar J. L., Wani, M. A., Lingrel, J. B., Gaensler, K. M., and Lloyd, J. A. 2005. KLF2 is essential for primitive erythropoiesis and regulates the human and murine embryonic beta-like globin genes in vivo. Blood 106:2566–2571.

Begley, C. G., Aplan, P. D., Denning, S. M., Haynes, B. F., Waldmann, T. A., and Kirsch, I. R. 1989. The gene SCL is expressed during early hematopoiesis and encodes a differentiation-related DNA-binding motif. Proc Natl Acad Sci USA 86:10128–10132.

Bender, M. A., Bulger, M., Close, J., and Groudine, M. 2000. Beta-globin gene switching and DNase I sensitivity of the endogenous beta-globin locus in mice do not require the locus control region. Mol Cell 5:387–393.

Bessis, M., Mize, C., and Prenant, M. 1978. Erythropoiesis: Comparison of in vivo and in vitro amplification. Blood Cells 4:155–174.

Bethlenfalvay, N. C., and Block, M. 1970. Fetal erythropoiesis. Maturation in megaloblastic (yolk sac) erythropoiesis in the C57Bl/6J mouse. Acta Haematol 44:240–245.

Blobel, G. A., Nakajima, T., Eckner, R., Montminy, M., and Orkin, S. H. 1998. CREB-binding protein cooperates with transcription factor GATA-1 and is required for erythroid differentiation. Proc Natl Acad Sci USA 95:2061–2066.

Boas, F. E., Forman, L., and Beutler, E. 1998. Phosphatidylserine exposure and red cell viability in red cell aging and in hemolytic anemia. Proc Natl Acad Sci USA 95:3077–3081.

Broudy, V. C. 1997. Stem cell factor and hematopoiesis. Blood 90:1345–1364.

Broudy, V. C., Lin, N. L., Priestley, G. V., Nocka, K., and Wolf, N. S. 1996. Interaction of stem cell factor and its receptor c-kit mediates lodgment and acute expansion of hematopoietic cells in the murine spleen. Blood 88:75–81.

Bulger, M., Schubeler, D., Bender, M. A., Hamilton, J., Farrell, C. M., Hardison, R. C., and Groudine, M. 2003. A complex chromatin landscape revealed by patterns of nuclease sensitivity and histone modification within the mouse beta-globin locus. Mol Cell Biol 23:5234–5244.

Chasis, J. A., Prenant, M., Leung, A., and Mohandas, N. 1989. Membrane assembly and remodeling during reticulocyte maturation. Blood 74:1112–1120.

Chin, H., Arai, A., Wakao, H., Kamiyama, R., Miyasaka, N., and Miura, O. 1998. Lyn physically associates with the erythropoietin receptor and may play a role in activation of the Stat5 pathway. Blood 91:3734–3745.

Choong, M. L., Yang, H. H., and McNiece, I. 2007. MicroRNA expression profiling during human cord blood-derived CD34 cell erythropoiesis. Exp Hematol 35:551–564.

Dai, C. H., Krantz, S. B., and Zsebo, K. M. 1991. Human burst-forming units – erythroid need direct interaction with stem cell factor for further development. Blood 78:2493–2497.

de la Chapelle, A., Fantoni, A., and Marks, P. 1969. Differentiation of mammalian somatic cells: DNA and hemoglobin synthesis in fetal mouse yolk sac erythroid cells. Proc Natl Acad Sci USA 63:812–819.

de Maria, R., Testa, U., Luchetti, L., Zeuner, A., Stassi, G., Pelosi, E., Riccioni, R., Felli, N., Samoggia, P., and Peschle, C. 1999. Apoptotic role of Fas/Fas ligand system in the regulation of erythropoiesis. Blood 93:796–803.

Dore, L. C., Amigo, J. D., Dos Santos, C. O., Zhang, Z., Gai, X., Tobias, J. W., Yu, D., Klein, A. M., Dorman, C., Wu, W., Hardison, R. C., Paw, B. H., and Weiss, M. J. 2008. A GATA-1-regulated microRNA locus essential for erythropoiesis. Proc Natl Acad Sci USA 105:3333–3338.

Emerson, S. G., Shanti, T., Ferrara, J. L., and Greenstein, J. L. 1989. Developmental regulation of erythropoiesis by hematopoietic growth factors: Analysis on populations of BFU-E from bone marrow, peripheral blood, and fetal liver. Blood 74:49–55.

Enver, T., Raich, N, Ebens, A. J., Papayannopoulou, T., Costantini, F., and Stamatoyannopoulos, G. 1990. Developmental regulation of human fetal-to-adult globin gene switching in transgenic mice. Nature 344:309–313.

Erslev, A. J. and Besarab, A. 1997. Erythropoietin in the pathogenesis and treatment of the anemia of chronic renal failure. Kidney Int 51:622–630.

Fader, C. M., and Colombo, M. I. 2006. Multivesicular bodies and autophagy in erythrocyte maturation. Autophagy 2:122–125.

Ferkowicz, M. J., and Yoder, M. C. 2005. Blood island formation: Longstanding observations and modern interpretations. Exp Hematol 33:1041–1047.

Fraser, S., Isern, J., and Baron, M. 2007. Maturation and enucleation of primitive erythroblasts during mouse embryogenesis is accompanied by changes in cell surface antigen expression. Blood 109:343–352.

Fujiwara, Y., Browne, C. P., Cunniff, K., Goff, S. C., and Orkin, S. H. 1996. Arrested development of embryonic red cell precursors in mouse embryos lacking transcription factor GATA-1. Proc Natl Acad Sci USA 93:12355–12358.

Gaiduschek J. B., and Singer, S. J. 1979. Molecular changes in the membranes of mouse erythroid cells accompanying differentiation. Cell 16:149–163.

Gardiner, M. R., Gongora, M. M., Grimmond, S. M., and Perkins, A. C. A global role for zebrafish klf4 in embryonic erythropoiesis. Mech Dev 2007; 124:762–774.

Gifford, S. C., Derganc, J., Shevkoplyas, S. S., Yoshida, T., Bitensky, M. W. 2006. A detailed study of time-dependent changes in human red blood cells: From reticulocyte maturation to erythrocyte senescence. Br J Haematol 135: 395–404.

Gulliver, G. 1875. Observations on the sizes and shapes of red corpuscles of the blood of vertebrates, with drawings of them to a uniform scale, and extended and revised tables of measurements. Proc Zool Soc London 474–495.

Haar, J. L., and Ackerman, G. A. 1971. A phase and electron microscopic study of vasculogenesis and erythropoiesis in the yolk sac of the mouse. Anat Rec 170:199–223.

Hall, M. A., Curtis, D. J., Metcalf D., Elefanty, A. G., Sourris, K., Robb, L., Gothert, J. R., Jane, S. M., and Begley, C. G.2003. The critical regulator of embryonic hematopoiesis, SCL, is vital in the adult for megakaryopoiesis, erythropoiesis, and lineage choice in CFU-S12. Proc Natl Acad Sci USA 100:992–997.

Hanspal, M., and Hanspal, J. S. 1994. The association of erythroblasts with macrophages promotes erythroid proliferation and maturation: A 30-kD heparin-binding protein is involved in this contact. Blood 84:3494–3504.

Hanspal, M., Smockova, Y., and Uong, Q. 1998. Molecular identification and functional characterization of a novel protein that mediates the attachment of erythroblasts to macrophages. Blood 92:2940–2950.

Harju, S., McQueen, K. J., and Peterson, K. R. 2002. Chromatin structure and control of beta-like globin gene switching. Exp Biol Med 227:683–700.

Heath, D. S., Axelrod, A. A., McLeod, D. L., and Shreeve, M. M. 1976. Separation of the erythropoietin-responsive BFU-E and CFU-E in mouse bone marrow by unit gravity separation. Blood 47:777–792.

Hodge, D., Coghill, E., Keys, J., Maguire, T., Hartmann, B., McDowall, A., Weiss, M., Grimmond, S., and Perkins, A. 2006. A global role for EKLF in definitive and primitive erythropoiesis. Blood 107:3359–3370.

Ingley, E., McCarthy, D. J., Pore, J. R., Sarna, M. K., Adenan, A. S., Wright, M. J., Erber, W., Tilbrook, P. A., and Klinken, S. P. 2005. Lyn deficiency reduces GATA-1, EKLF and STAT5, and induces extramedullary stress erythropoiesis. Oncogene 24:336–343.

Iscove, N. N. 1977. The role of erythropoietin in regulation of population size and cell cycling of early and late erythroid precursors in mouse bone marrow. Cell Tissue Kinet 10:323–334.

Jaakkola, P., Mole, D. R., Tian, Y. M., Wilson, M. I., Gielbert, J., Gaskell, S. J., Kriegsheim, Av., Hebestreit, H. F., Mukherji, M., Schofield, C. J., Maxwell, P. H., Pugh, C. W., and Ratcliffe, P. J.2001. Targeting of HIF-alpha to the von Hippel-Lindau ubiquitylation complex by O2-regulated prolyl hydroxylation. Science 292:468–472.

Ji, R. P., Phoon, C. K. L., Aristizábal, O., McGrath, K. E., Palis, J., and Turnbull, D. H. 2003. Onset of cardiac function during early mouse embryogenesis coincides with entry of primitive erythroblasts into the embryo proper. Circ Res 92:133–135.

Ji, P., Jayapal, S. R., and Lodish, H. F. 2008. Enucleation of cultured mouse fetal erythroblasts requires Rac GTPases and mDia2. Nat Cell Biol 10:314–321.

Johnson, G. R., and Moore, M. A. S. 1975. Role of stem cell migration in initiation of mouse foetal liver haemopoiesis. Nature 258:726–728.

Johnstone, R. M., Mathew, A., Mason, A. B., and Teng, K. 1991. Exosome formation during maturation of mammalian and avian reticulocytes: Evidence that exosome release is a major route for externalization of obsolete membrane proteins. J Cell Physiol 147:27–36.

Kapur, R., and Zhang, L. 2001. A novel mechanism of cooperation between c-Kit and erythropoietin receptor. Stem cell factor induces the expression of Stat5 and erythropoietin receptor, resulting in efficient proliferation and survival by erythropoietin. J Biol Chem 276:1099–1106.

Kawane, K., Fukuyama, H., Kondoh, G., Takeda, J., Ohsawa, Y., Uchiyama, Y., and Nagata, S. 2001. Requirement of DNase II for definitive erythropoiesis in the mouse fetal liver. Science 292:1546–1549.

Color Plates

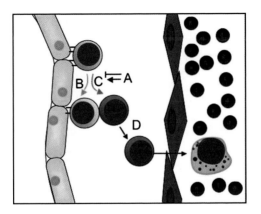

Fig. 1.1 Model of hematopoiesis with three tiers of lineage restriction. (See Complete Caption on Page 3)

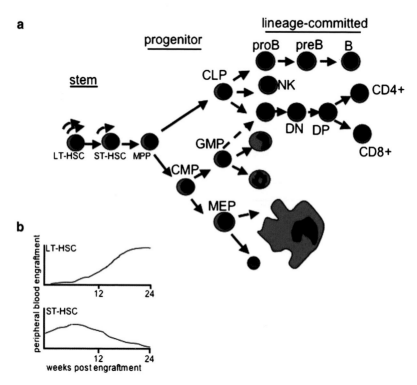

Fig. 1.2 Model of HSC cell division modes in the bone marrow environment. (See Complete Caption on Page 10)

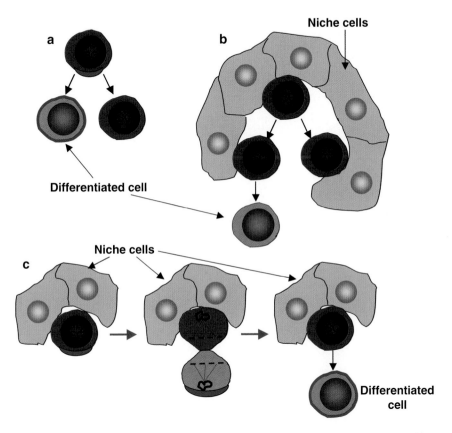

Fig. 3.1 Three ways to achieve asymmetric cell division. (See Complete Caption on Page 50)

a

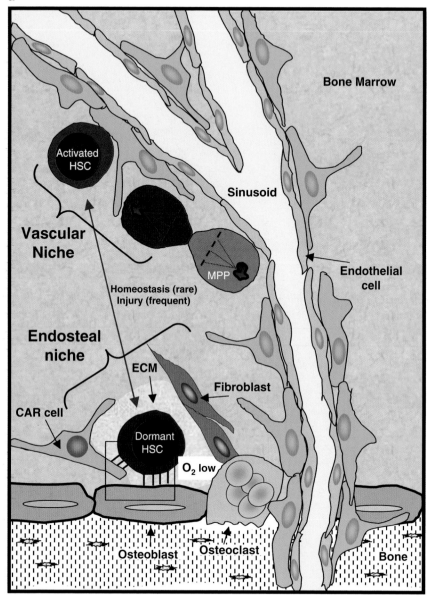

Fig. 3.2 (**a**) Two types of bone marrow niche. (See Complete Caption on Pages 52 & 53)

b

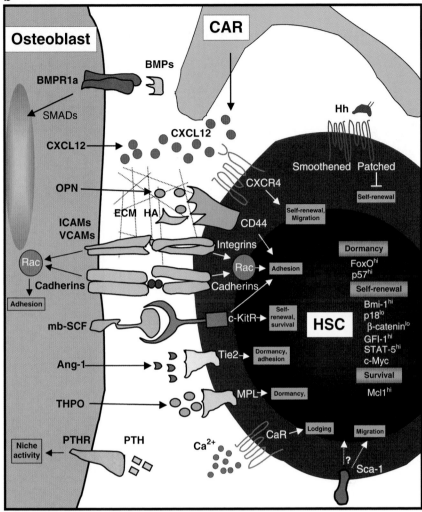

Fig. 3.2 (continued)

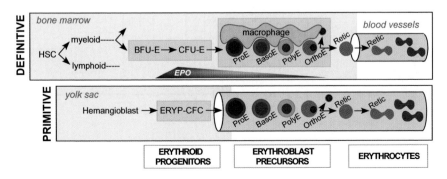

Fig. 4.1 Schema of definitive and primitive erythropoiesis. (See Complete Caption on Page 75)

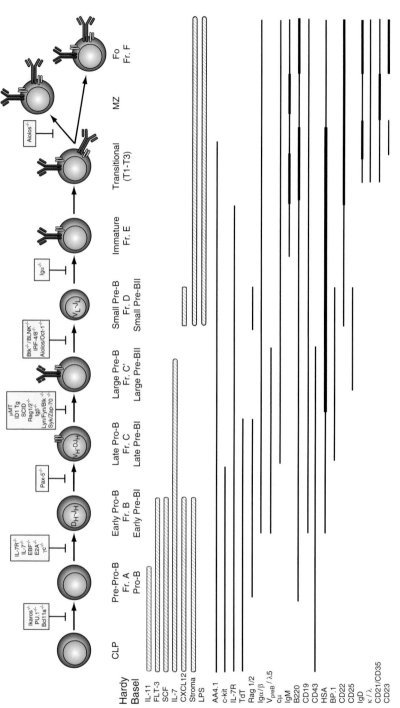

Fig. 8.1 B cell developmental scheme: Developmental progression of B lineage cells from CLPs to mature B cells in mice. *Hatched bars* denote growth factor and stroma-dependent stages. *Solid lines* represent expression of cell phenotype markers, with line thickness indicating relative expression levels. Developmental blocks arising from mutant or transgenic gene expression are denoted at appropriate stages

Khandelwal, S., van Rooijen, N., and Saxena, R. K. 2007. Reduced expression of CD47 during murine red blood cell (RBC) senescence and its role in RBC clearance from the circulation. Transfusion 7:1725–1732.

Kieran, M. W., Perkins, A. C., Orkin, S. H., and Zon, L. Z. 1996. Thrombopoietin rescues in vitro erythroid colony formation from mouse embryos lacking the erythropoietin receptor. Proc Natl Acad Sci USA 93:9126–9131.

Kingsley, P. D., Malik, J., Fantauzzo, K., and Palis, J. 2004. Yolk sac-derived primitive erythroblasts enucleate during mammalian embryogenesis. Blood 104:19–25.

Kingsley, P. D., Malik, J., Emerson, R. L., Bushnell, T. P., McGrath, K. E., Bloedorn, L. A., Bulger, M., Palis, J. 2006. "Maturational" globin switching in primary primitive erythroid cells. Blood 104:1665–1672.

Knoll, W. 1927. Blut und blutbildende organe menschlicher embryonen. Schriften der Schweizerischen Naturforschenden Gesellschaft 64:1–81.

Koury, M. J., and Bondurant, M. C. 1990. Erythropoietin retards DNA breakdown and prevents programmed cell death in erythroid progenitor cells. Science 248:378–381.

Koury, S. T., Koury, M. J., and Bondurant, M. C. 1989. Cytoskeletal distribution and function during the maturation and enucleation of mammalian erythroblasts. J Cell Biol 109:3005–3013.

Kunisaki, Y., Masuko, S., Noda, M., Inayoshi, A., Sanui, T., Harada, M., Sasazuki, T., and Fukui, Y. 2004. Defective fetal liver erythropoiesis and T lymphopoiesis in mice lacking the phosphatidylserine receptor. Blood 103:3362–3364.

Kurata, H., Mancini, G. C., Alespieti, G., Migliaccio, A. R., and Migliaccio, G. 1998. Stem cell factor induces proliferation and differentiation of fetal progenitor cells in the mouse. Br J Haematol 101:676–687.

Lee, J. C., Gimm, J. A., Lo, A. J., Koury, M. J., Krauss, S. W., Mohandas, N., and Chasis, J. A. 2004. Mechanism of protein sorting during erythroblast enucleation: Role of cytoskeletal connectivity. Blood 103:1912–1919.

Lee, G., Lo, A., Short, S. A., Mankelow, T. J., Spring, F., Parsons, S. F., Yazdanbakhsh, K., Mohandas, N., Anstee, D. J., and Chasis, J. A. 2006. Targeted gene deletion demonstrates that the cell adhesion molecule ICAM-4 is critical for erythroblastic island formation. Blood 108:2064–2071.

Letting, D. L., Rakowski, C., Weiss, M. J., Blobel, G. A. 2003. Formation of a tissue-specific histone acetylation pattern by the hematopoietic transcription factor GATA-1. Mol Cell Biol 23:1334–1340.

Lin, C.-S., Lim, S.-K., D'Agati, V., and Constantini, F. 1996. Differential effects of an erythropoietin receptor gene disruption on primitive and definitive erythropoiesis. Genes Dev 10:154–164.

Lux, C. T., Yoshimoto, M., McGrath, K., Conway, S. J., Palis, J., and Yoder, M. C. 2008. All primitive and definitive hematopoietic progenitor cells emerging prior to E10 in the mouse embryo are products of the yolk sac. Blood, in press.

Maximow, A. A. 1909. Untersuchungen uber blut und bindegewebe 1. Die fruhesten entwicklungsstadien der blut- und binde- gewebszellan bein saugetierembryo, bis zum anfang der blutbildung unden leber. Arch Mikroskop Anat. 73:444–561.

Maxwell, P. H., Wiesener, M. S., Chang, G. W., Clifford, S. C., Vaux, E. C., Cockman, M. E., Wykoff, C. C., Pugh, C. W., Maher, E. R., and Ratcliffe, P. J. 1999. The tumour suppressor protein VHL targets hypoxia-inducible factors for oxygen-dependent proteolysis. Nature 399:271–275.

McGrath, K. E., Koniski, A. D., Malik, J., and Palis, J. 2003. Circulation is established in a stepwise pattern in the mammalian embryo. Blood 101:1669–1676.

McGrath, K. E., Kingsley, P. D., Koniski, A. D., Porter, R. L., Bushnell, T. P., and Palis, J. 2008. Enucleation of primitive erythroid cells generates a transient population of "pyrenocytes" in the mammalian fetus. Blood 111:2409–2417.

McNiece, I. K., Langley, K. E., and Zsebo, K. M. 1991. Recombinant human stem cell factor synergises with GM-CSF, G-CSF, IL-3 and EPO to stimulate human progenitor cells of the myeloid and erythroid lineages. Exp Hematol 19:226–231.

Meier, N., Krpic, S., Rodriguez, P., Strouboulis, J., Monti, M., Krijgsveld, J., Gering, M., Patient, R., Hostert, A., and Grosveld, F. 2006. Novel binding partners of Ldb1 are required for haematopoietic development. Development 133:4913–4924.

Meyron-Holz, E. G., Fibach, E., Gelvan, D., and Konijn, A. M. 1994. Binding and uptake of exogenous isoferritins by cultured human erythroid precursor cells. Br J Haematol 86:635–641.

Migliaccio, A. R. and Migliaccio, G. 1988. Human embryonic hemopoiesis: Control mechanisms underlying progenitor differentiation in vitro. Dev Biol 125:127–134.

Miller, I. J., and Bieker, J. J. 1993. A novel, erythroid cell-specific murine transcription factor that binds to the CACCC element and is related to the Krüppel family of nuclear proteins. Mol Cell Biol 13:2776–2786.

Mohandas, N. and Prenant M. 1978. Three-dimensional model of bone marrow. Blood 51:633–643.

Motoyama, N., Kimura, T., Takahashi, T., Watanabe, T., and Nakano, T. 1999. bcl-x prevents apoptotic cell death of both primitive and definitive erythrocytes at the end of maturation. J Exp Med 11:1691–1698.

Muta, K., Krantz, S. B., Bondurant, M. C., and Dai, C. H. 1995. Stem cell factor retards differentiation of normal human erythroid progenitor cells while stimulating proliferation. Blood 86:572–580.

Nemenman, I., Escola, G. S., Hlavacek, W. S., Unkefer, P. J., Unkefer, C. J., and Wall, M. E. 2007. Reconstruction of metabolic networks from high-throughput metabolite profiling data: In silico analysis of red blood cell metabolism. 2007. Ann N Y Acad Sci 1115:102–115.

Neubauer, H., Cumano, A., Müller, M., Wu, H., Huffstadt, U., and Pfeffer, K. 1998. Jak2 deficiency defines an essential developmental checkpoint in definitive hematopoiesis. Cell 93:397–409.

Nichols, K. E., Crispino, J. D., Poncz, M., White, J. G., Orkin, S. H., Maris, J. M., and Weiss, M. J.2000. Familial dyserythropoietic anaemia and thrombocytopenia due to an inherited mutation in GATA1. Nat Genet 24:266–270.

Nuez, B., Michalovich, D., Bygrave, A., Ploemacher, R., and Grosveld, F. 1995. Defective haematopoiesis in fetal liver resulting from inactivation of the EKLF gene. Nature 375:316–318.

Olweus, J., Terstappen, L. W., Thompson, P. A., and Lund-Johansen, F. 1996. Expression and function of receptors for stem cell factor and erythropoietin during lineage commitment of human hematopoietic progenitor cells. Blood 88:1594–1607.

Paffett-Lugassy, N., Hsia, N., Fraenke, P. G., Paw, B., Leshinsky, I., Barut, B., Bahary, N., Caro, J., Handin, R., and Zon, L. I. 2007. Functional conservation of erythropoietin signaling in zebrafish. Blood 110:2718–2726

Palis, J., Robertson, S., Kennedy, M., Wall, C. and Keller, G. 1999. Development of erythroid and myeloid progenitors in the yolk sac and embryo proper of the mouse. Development 126, 5073–5084.

Papayannopoulou, T., Kalmantis, T., and Stamatoyannopoulos, G. 1979. Cellular regulation of hemoglobin switching: Evidence for inverse relationship between fetal hemoglobin synthesis and degree of maturity of human erythroid cells. Proc Natl Acad Sci USA 76:6420–6424.

Pasini, E. M., Kirkegaard, M., Mortensen, P., Lutz, H. U., Thomas, A. W., and Mann, M. 2006. In-depth analysis of the membrane and cytosolic proteome of red blood cells. Blood 108:791–801.

Perkins, A. C., Sharpe, A. H., and Orkin, S. H. 1995. Lethal beta-thalassaemia in mice lacking the erythroid CACCC-transcription factor EKLF. Nature 375:318–322.

Peschle, C., Mavilio, F, Care, A., Migliaccio, G., Migliaccio, A. R., Salvo, G., Samoggia, P., Petti, S., Guerriero, R., Marinucci, M., Lazzaro, D., Russo, G., and Mastroberardino, G.1985. Haemoglobin switching in human embryos: Asynchrony of the ζ - > α and ε - > γ-globin switches in primitive and definitive erythropoietic lineage. Nature 313:235–238.

Pevny, L., Simon, M. C., Robertson, E., Klein, W. H., Tsai, S. F., D'Agati, V., Orkin, S. H., and Costantini, F. 1991. Erythroid differentiation in chimaeric mice blocked by a targeted mutation in the gene for transcription factor GATA-1. Nature 349:257–260.

Pilon, A. M., Nilson, D. G., Zhou, D., Sangerman, J., Townes, T. M., Bodine, D. M., and Gallagher, P. G. 2006. Alterations in expression and chromatin configuration of the alpha hemoglobin-stabilizing protein gene in erythroid Kruppel-like factor-deficient mice. Mol Cell Biol 26:4368–4377.

Porcher, C., Swat, W., Rockwell, K., Fujiwara, Y., Alt, F. W., and Orkin, S. H. 1996. The T cell leukemia oncoprotein SCL/tal-1 is essential for development of all hematopoietic lineages. Cell 86:47–57.

Raich, N., Enver, T., Nakamoto, B., Josephson, B., Papayannopoulou, T., and Stamatoyannopoulos, G. 1990. Autonomous developmental control of human embryonic globin gene switching in transgenic mice. Science 250:1147–1149.

Remy, I., Wilson, I. A., and Michnick, S. W. 1999. Erythropoietin receptor activation by a ligand-induced conformation change. Science 283:990–993.

Rhodes, M. M., Kopsombut, P., Bondurant, M. C., Price, J. O., and Koury, M. J. 2008. Adherence to macrophages in erythroblastic islands enhances erythroblast proliferation and increases erythrocyte production by a different mechanism than erythropoietin. Blood 111:1700–1708.

Rich, I. N. 1986. A role for the macrophage in normal hematopoiesis. Exp Hematol 14:746–751.

Rich, I. N. and Kubanek, B. 1976. Erythroid colony formation in foetal liver and adult bone marrow and spleen from the mouse. Blood 33:171–180.

Rich, I. N. and Kubanek, B. 1979. The ontogeny of erythropoiesis in the mouse detected by the erythroid colony-forming technique. J Embryol Exp Morphol 50:57–74.

Robb, L., Lyons, I., Li, R., Hartley, L., Köntgen, F., Harvey, R. P., Metcalf, D., Begley, C. G. 1995. Absence of yolk sac hematopoiesis from mice with a targeted disruption of the scl gene. Proc Natl Acad Sci USA 92:7075–7079.

Robb, L., Elwood, N. J., Elefanty, A. G., Kontgen, F., Li, R., Barnett, L. D., and Begley, C. G. 1996. The scl gene product is required for the generation of all hematopoietic lineages in the adult mouse. EMBO J 15:4123–4129.

Rodriguez, P., Bonte, E., Krijgsveld, J., Kolodziej, K. E., Guyot, B., Heck, A. J., Vyas, P., de Boer, E., Grosveld, F., and Strouboulis, J. 2005. GATA-1 forms distinct activating and repressive complexes in erythroid cells. EMBO J 24:2354–2366.

Sadahira, Y., Yoshino, T., and Monobe, Y. 1995. Very late activation antigen 4-vascular cell adhesion molecule 1 interaction is involved in the formation of erythroblast islands. J Exp Med 181:411–415.

Sangiorgi, F., Woods, C. M. and Lazarides, E. 1990. Vimentin downregulation is an inherent feature of murine erythropoiesis and occurs independently of lineage. Development 110:85–96.

Sasaki, K., and Matsumura, G. 1986. Haemopoietic cells of yolk sac and liver in the mouse embryo: A light and electron microscopical study. J Anat 148:87–97.

Sathyanarayana, P., Menon, M. P., Bogacheva, O., Bogachev, O., Niss, K., Kapelle, W. S., Houde, E., Fang, J., and Wojchowski, D. M. 2007. Erythropoietin modulation of podocalyxin and a proposed erythroblast niche. Blood 110:509–518.

Schuh, A. H., Tipping, A. J., Clark, A. J., Hamlett, I., Guyot, B., Iborra, F. J., Rodriguez, P., Strouboulis, J., Enver, T., Vyas, P., and Porcher, C. 2005. ETO-2 associates with SCL in erythroid cells and megakaryocytes and provides repressor functions in erythropoiesis. Mol Cell Biol 25:10235–10250.

Schweers, R. L., Zhang, J., Randall, M. S., Loyd, M. R., Li, W., Dorsey, F. C., Kundu, M., Opferman, J. T., Cleveland, J. L., Miller, J. L., and Ney, P. A.2007. NIX is required for programmed mitochondrial clearance during reticulocyte maturation. Proc Natl Acad Sci USA 104:19500–19505.

Semenza, G. L. and Wang, G. L. 1992. A nuclear factor induced by hypoxia via de novo protein synthesis binds to the human erythropoietin gene enhancer at a site required for transcriptional activation. Mol Cell Biol 12:5447–5454.

Shimizu, R., Takahashi, S., Ohneda, K., Engel, J. D. and Yamamoto, M. 2001. In vivo requirements for GATA-1 functional domains during primitive and definitive erythropoiesis. EMBO J 20:5250–5260.

Shivdasani, R. A., Mayer, E. L., and Orkin, S. H. 1995. Absence of blood formation in mice lacking T-cell leukemia oncoprotein tal-1/SCL. Nature 373:432–434.

Silva, M., Benito, A., Sanz, C., Prosper, F., Ekhterae, D., Nuñez, G., and Fernandez-Luna, J. L. 1999. Erythropoietin can induce the expression of bcl-x(L) through Stat5 in erythropoietin-dependent progenitor cell lines. J Biol Chem 274:22165–9.

Silver, L, and Palis, J. 1997. Initiation of murine embryonic erythropoiesis: A spatial analysis. Blood 89:1154–1164.

Simpson, C. F. and Kling, J. M. 1967. The mechanism of denucleation in circulating erythroblasts. J Cell Biol 35:237–245.

Sivertsen, E. A., Hystad, M. E., Gutzkow, K. B., Dosen, G., Smeland, E. B., Blomhoff, H. K., and Myklebust, J. H. 2006. PI3K/Akt-dependent Epo-induced signalling and target genes in human early erythroid progenitor cells. Br J Haematol 135:117–128.

Skutelsky, E., and Danon, D. 1970. Comparative study of nuclear expulsion from the late erythroblast and cytokinesis. Exp Cell Res 60:427–436.

Socolovsky, M., Fallon, A. E., Wang, S., Brugnara, C., and Lodish, H. F. 1999. Fetal anemia and apoptosis of red cell progenitors in Stat5a–/–5b–/– mice: A direct role for Stat5 in Bcl-X(L) induction. Cell 98:181–191.

Soni, S., Bala, S., Gwynn, B., Sahr, K. E., Peters, L. L., and Hanspal, M. 2006. Absence of erythroblast macrophage protein (Emp) leads to failure of erythroblast nuclear extrusion. J Biol Chem 281:20181–20189.

Southwood, C. M., Downs, K. M., and Bieker, J. J. 1996. Erythroid Krüppel-like factor exhibits an early and sequentially localized pattern of expression during mammalian erythroid ontogeny. Dev Dyn 206:248–259.

Spring, F. A., Parsons, S. F., Ortlepp, S., Olsson, M. L., Sessions, R., Brady, R. L., and Anstee D. J. 2001. Intercellular adhesion molecule-4 binds alpha(4)beta(1) and alpha(V)-family integrins through novel integrin-binding mechanisms. Blood 98:458–466.

Stamatoyannopoulos, G. 2005. Control of globin gene expression during development and erythroid differentiation. Exp Hematol 33:259–271.

Steiner, R., and Vogel, H. 1973. On the kinetics of erythroid cell differentiation in fetal mice: I. Microspectrophotometric determination of the hemoglobin content in erythroid cells during gestation. J Cell Physiol 81:323–338.

Stephenson, J. R., Axelrad, A., McLeod, D., and Shreeve, M. 1971. Induction of colonies of hemoglobin-synthesizing cells by erythropoietin in vitro. Proc Natl Acad Sci USA 68:1542–1546.

Swiers, G., Patient, R., and Loose, M. 2006. Genetic regulatory networks programming hematopoietic stem cells and erythroid lineage specification. Dev Biol 294:525–540.

Tanabe, O., McPhee, D., Kobayashi, S., Shen, Y., Brandt, W., Jiang, X., Campbell, A. D., Chen, Y. T., Chang, C., Yamamoto, M., Tanimoto, K., and Engel, J. D. 2007. Embryonic and fetal beta-globin gene repression by the orphan nuclear receptors, TR2 and TR4. EMBO J 26:2295–2306.

Tavassoli, M. 1991. Embryonic and fetal hemopoiesis: An overview. Blood Cells 1:269–281.

Tinsley, J. C., Jr., Moore, C. V., Dubach, R., Minnich, V., and Grinstein, M. 1949. The role of oxygen in the regulation of erythropoiesis; depression of the rate of delivery of new red cells to the blood by high concentrations of inspired oxygen. J Clin Invest 28:1544–1564.

Trimborn, T., Bribnau, J., Grosveld, F., and Fraser, P. 1999. Mechanisms of developmental control of transcription in the murine α- and β-globin loci. Genes Dev 13:112–124.

Tsang, A. P., Visvader, J. E., Turner, C. A., Fujiwara, Y., Yu, C., Weiss, M. J., Crossley, M., and Orkin, S. H. 1997. FOG, a multitype zinc finger protein, acts as a cofactor for transcription factor GATA-1 in erythroid and megakaryocytic differentiation. Cell 90:109–119.

Tsang, A. P., Fujiwara, Y., Hom, D. B. and Orkin, S. H. 1998. Failure of megakaryopoiesis and arrested erythropoiesis in mice lacking the GATA-1 transcriptional cofactor FOG. Genes Dev 15:1176–1188.

Uddin, S., Kottegoda, S., Stigger, D., Platanias, L. C., and Wickrema, A. 2000. Activation of the Akt/FKHRL1 pathway mediates the antiapoptotic effects of erythropoietin in primary human erythroid progenitors. Biochem Biophys Res Commun 275:16–19.

Wadman, I. A., Osada, H., Grütz, G. G., Agulnick, A. D., Westphal, H., Forster, A., and Rabbitts, T. H. 1997. The LIM-only protein Lmo2 is a bridging molecule assembling an erythroid, DNA-binding complex which includes the TAL1, E47, GATA-1 and Ldb1/NLI proteins. EMBO J 16:3145–3157.

Warren, A. J., Colledge, W. H., Carlton, M. B. L., Evans, M., Smith, A. J. H., and Rabbitts, T. H. 1994. The oncogenic cysteine-rich LIM domain protein is essential for erythroid development. Cell 78:45–57.

Wong, P. M. C., Chung, S. W., Chui, D. H. K., and Eaves, C. J. 1986. Properties of the earliest clonogenic hematopoietic precursors to appear in the developing murine yolk sac. Proc Natl Acad Sci USA 83:3851–3854.

Wu, H., Liu, X., Jaenisch, R., and Lodish, H. F. 1995. Generation of committed erythroid BFU-E and CFU-E progenitors does not require erythropoietin or the erythropoietin receptor. Cell 83:59–67.

Wu, H., Klingmüller, U., Acurio, A., Hsiao, J. G., and Lodish, H. F. 1997. Functional interaction of erythropoietin and stem cell factor receptors is essential for erythroid colony formation. Proc Natl Acad Sci USA 94:1806–1810.

Yamada, Y., Warren, A. J., Dobson, C., Forster, A., Pannell, R., and Rabbitts, T. H. 1998. The T cell leukemia LIM protein Lmo2 is necessary for adult mouse hematopoiesis. Proc Natl Acad Sci USA 95:3890–3895.

Yoshida, H., Kawane, K., Koike, M., Mori, Y., Uchiyama, Y. and Nagata, S. 2005. Phosphatidylserine-dependent engulfment by macrophages of nuclei from erythroid precursor cells. Nature 437:754–758.

Zhan, M., Miller, C., and Papayannopoulou, T., Stamatoyannopoulos, G., and Song, C.-Z. 2007. MicroRNA expression dynamics during murine and human erythroid differentiation. Exp Hematol 35:1015–1025.

Zhang, Y., Payne, K. J., Zhu, Y., Price, M. A., Parrish, Y. K., Zielinska, E., Barsky, L. W., and Crooks, G. M. 2005. SCL expression at critical points in human hematopoietic lineage commitment. Stem Cells 23:852–860.

Zheng, J., Kitajima, K., Sakai, E., Kimura, T., Minegishi, N., Yamamoto, M., and Nakano, T. 2006. Differential effects of GATA-1 on proliferation and differentiation of erythroid lineage cells. Blood 107:520–527.

Chapter 5
Development of Megakaryocytes

Nicholas Papadantonakis and Katya Ravid

Abstract Megakaryocytes (MKs) comprise a rare population of bone marrow cells, responsible for the production of platelets. MKs are derived from hematopoietic stem cells and share some common progenitors with the erythroid lineage. Through a partially elucidated interplay of transcription and growth factors, cells committed to the MK lineage are formed. Diploid MKs undergo multiple rounds of endomitosis, including aborted mitosis and cytokinesis. The mediators of endomitosis include cyclins, proteins involved in mitosis and cytokinesis, and other yet unrecognized proteins. Several signaling pathways are activated during endomitosis but their precise role remains largely uncharacterized. Endomitosis leads to high states of ploidy, which are accompanied by a cytoplasmatic volume increase. During the final stages of the MK life cycle biogenesis of platelets occurs. The precise mechanism of this aspect remained controversial for many years, but the implementation of sophisticated imaging modalities has gradually elucidated the process of proplatelet formation. Several disorders have been described affecting MK and platelet physiology. For some of them, the molecular pathology has been elucidated. Translational research has led to the development of thrombopoietic agents that are engineered to overcome changes in platelet levels associated with these states. In this chapter, we discuss key aspects of MK physiology and structure and we explore the molecular pathways governing these fascinating cells under normal and some pathological conditions.

K. Ravid (✉)
Department of Biochemistry, K225
Boston University School of Medicine,
Boston, MA 02118, USA
e-mail: ravid@biochem.bumc.bu.edu

A. Wickrema and B. Kee (eds.), *Molecular Basis of Hematopoiesis*,
DOI: 10.1007/978-0-387-85816-6_1, © Springer Science+Business Media, LLC 2009

Overview

Megakaryocytes (MKs) constitute only a small fraction of cells residing in bone marrow and are releasing daily 1×10^{11} platelets in human adults (Branehog et al. 1975), which can be increased severalfold under conditions of stress (Deutsch and Tomer 2006). Platelets, one of the most numerous elements of blood, attach in areas where blood vessel integrity is compromised, preventing hemorrhage. However, recently MKs and platelets have expanded their role in the context of immunity, cancer biology and cardiovascular disorders. This chapter will survey processes that control MK development, as well as related pathologies.

The Origin of Megakaryocytes

In invertebrates, hemolymph has a key role in the clotting cascade, which also acts as an immune defense mechanism against pathogen invaders (Muta and Iwanaga 1996). Fish were the first vertebrates to evolve approximately half a billion years ago, and their hemostasis is carried out by thrombocytes (translated from Greek: *Clotting cells*). Their thrombocytes are nucleated cells and their progenitor remains elusive. Zebrafish thrombocytes express several transcription factors involved in megakaryopoiesis and hence, have attracted attention as a model system of hemopoiesis (de Jong and Zon 2005). Chicken thrombocytes have a more established role in immunity and can respond to LPS stimulation through Toll-like receptor 4 (TLR4) signaling (Scott and Owens 2008). Evidence for the presence of functional TLR4 was also recently reported in human platelets and MKs (Andonegui et al. 2005). The reason why these thrombocytes, with a dual role in hemostasis and immunity, were not retained in mammalian species remains enigmatic (Brass 2005).

In mammals, megakaryopoiesis occurs in two waves: the primitive and definite. However, ontogeny studies are hampered by the lack of markers to differentiate between them. MK progenitors and MK erythroid progenitors (MEP) of primitive hemopoiesis are first detected in primitive streaks and persist to embryonic day 9 (E9) in mouse (Tober et al. 2008). MK progenitors are detected at the yolk sac and fetal liver approximately at E7.5-10.5 and E11.5, respectively. It is unknown whether MK progenitors detected in the fetal liver at that period belong to primitive or definite hemopoiesis. Platelets are detected between E10.5 and E11.5 (Tober et al. 2007), and evidence is mounting that they are derived from primitive MK progenitors (Tober et al. 2008). In humans, platelets can be detected at 5 weeks of gestation and their level reaches a plateau at 18 weeks of gestation (Forestier et al. 1991). Fetal liver is the main site of hemopoiesis for the most part of embryonic life, with bone marrow exhibiting a key role in late stages of gestation.

In adult megakaryopoiesis, MKs are generated by hematopoietic stem cells (HSC) that reside in specialized areas of the bone marrow called niches (Jones and Wagers 2008). The most well studied is in the vicinity of endosteum, where osteoblasts as well as bone marrow stromal cells (BMSC), matrix proteins, and minerals constitute a unique environment for HSC homeostasis (Adams and Scadden 2006). Importantly, Notch signaling affects murine niches (Duncan et al. 2005), and data derived from an erythroleukemic cell line support a role of notch signaling in megakaryopoiesis (Lam et al. 2000). Recently, evidence is mounting for the existence and role of the vascular niche that is placed around the bone marrow sinusoids (Kiel et al. 2005; Li and Li 2006). These are highly permeable blood vessels, and their wall is composed by a single layer of endothelial cells. The great majority of mature MKs are found in the vicinity of sinusoids (Tavassoli and Aoki 1989) of the vascular niche, and these can through sinusoids (a) shed platelets in the blood stream, or (b) more rarely gain entrance to blood circulation (Tavassoli and Aoki 1981; Junt et al. 2007). The MKs that gain entrance to blood circulation are trapped in the capillary network of the lungs with increased numbers reported in biopsies of patients suffering from diffuse alveolar damage (Mandal et al. 2007). A knockout model of CD31 (PECAM-1) indicated that immature MKs could not migrate to sinusoids, but the platelet levels remained within normal range by increased extramedullary megakaryopoiesis (Wu et al. 2007). The localization of MKs in the vicinity of sinusoids may play a role in the phenomenon of emperipolesis (Samii and Pasteur 1998), where bone marrow blood cells such as neutrophils enter MK cytoplasm.

During megakaryopoiesis HSC produce MKs through the generation of progressively more differentiated progenitors. In the most established, *canonical*, pathway, HSC generate common myeloid progenitors (CMP) cells (Akashi et al. 2000) from which the megakaryocyte/erythrocyte progenitor (MEP) population is formed (Debili et al. 1996). In the recently proposed noncanonical pathway, HSC can directly give rise to MEP population without an intermediate progenitor (Adolfsson 2005). MEP gives rise both to the megakaryocytic and erythroid lineages; however, in MEP colonies the number of MKs is much smaller than erythroblasts. The most definite primitive progenitors of MKs are the high proliferation potential-colony-forming unit-megakaryocyte (HPP-CFU-MK) (Jackson et al. 1994; Bruno et al. 1996). A more differentiated progenitor is the burst-forming unit-megakaryocyte (BFU-MK) that gives rise to colony-forming unit-megakaryocyte (CFU-MK) (Briddell et al. 1989). Both human and murine CFU-MK-derived colonies are formed faster than those of BFU-MK, but contain fewer cells.

The CFU-MK then gives rise to megakaryoblasts, (alternatively designated group I MKs), the most primitive cell of thrombopoiesis that can be recognized in histological preparations of bone marrow. It is characterized by a large, nucleus with several nucleoli, which is surrounded by a basophilic cytoplasm that lacks granules.

Immature MKs (alternatively designated stage II MKs or promegakaryocytes) have a larger size and cytoplasm-to-nucleus ratio than megakaryoblasts. Their single, multiple lobes containing nucleus have a horseshoe shape. In addition, a scarce amount of azurophilic granules is dispersed in their cytoplasm, which contains numerous mitochondria. Eventually, cells with a large nucleus are generated and are called mature MKs (alternatively designated stage III MKs, or granular megakaryocytes). Their cytoplasm is acidophilic and harbors a sizeable amount of azurophilic granules. The hallmark of Stage IV MKs is giving birth to platelets.

MK developmental staging is based, besides the aforementioned morphological characteristics (Weinstein et al. 1981; Levine et al. 1982), predominantly on studies with flow cytometry that detect characteristic clusters of differentiation (CD) patterns (reviewed in Tomer 2004; Deutsch and Tomer 2006). The most widely recognized surface markers for the study of megakaryopoiesis are integrin GPIIIa (CD61), receptor of von Willebrand factor (CD42b), GPIIb/IIIa, alternatively designated as integrin $a_{IIb} \beta_3$ (CD41a), GPIIb (CD41b), platelet factor 4 (PF4), von Willebrand factor coupled sometimes with CD34 and CD38.

Namely, human BFU-MK expresses CD34 but only at low levels of HLA-DR compared with CFU-MK that expresses both CD34 and HLA-DR. By contrast, isolation of murine CFU-MK requires a more elaborate panel of surface antigens (Nakorn et al. 2003).

MK Abundance, Structure, and Physiology

The MK population is rare, namely, in human bone marrow only 0.02–0.05% of nucleated cells are MKs (Levine et al. 1982), precluding large-scale isolation experiments. Alternative sources (Fig. 5.1) of MKs, apart from murine bone marrow, are murine fetal liver cultures obtained at days 12–15 post coitum (Lecine et al. 1998), or CD34+ cells isolated from cord blood and cultured with a combination of

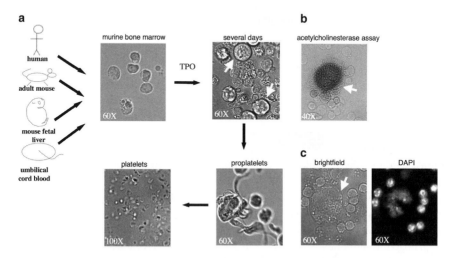

Fig. 5.1 From MKs to platelets (**a**) overview of MK sources and platelet biogenesis. (**b**) A murine MK stained for acetylcholinesterase assay. (**c**) Brightfield microscopy and DAPI staining of the nucleus in a polyploid MK surrounded by diploid cells

growth factors that commonly include thrombopoietin (TPO) (Schipper et al. 2003; van Hensbergen et al. 2006).

Murine bone marrow and spleen together harbor 18×10^4 MKs, and this number can be increased twofold after TPO stimulation (Junt et al. 2007). MKs are large cells with a median diameter size of 30–70 µm (range 20–100 µm), and a single large polyploid nucleus that is multisegmented with coarse-grained chromatin. The cytoplasm of MKs can be divided into three zones. The perinuclear territory remains attached to the nucleus after release of platelets and contains several organelles, such as ER, Golgi apparatus, and centrioles. The intermediate zone contains the demarcation membrane system (DMS), which plays a role in the biogenesis of platelets. The DMS consists of an extensive network of vesicles and tubules and serves as the reservoir of proplatelet formation (Schulze et al. 2006). It is estimated that a single MK cell can shed approximately 10^9 platelets (Junt et al. 2007). Finally, the marginal zone involves the megakaryocytic cell membrane and is composed mainly of cytoskeletal filaments.

Although the mean ploidy of MKs in the bone marrow of humans and almost all mice strains is 16N (32N for CH3 strain) (Jackson et al. 1990b), some MKs can acquire a DNA content of up to 128N, where 2N is the DNA content of a diploid cell. The increase in MK ploidy is also followed by increase in MK size (Levine et al. 1982; Tomer 2004) with the diameter of human 2N MK being 21 ± 4 µm compared with 55.80 ± 8.1 µm of 64N MK (Tomer 2004).

Mouse models have been indispensable for the study of hemopoiesis and thrombopoiesis. However, platelets and MKs of mouse and human origin are not identical (Schmitt et al. 2001). For example, in humans, platelet levels range from 150–400 $\times 10^9$/L while in mice these levels are in the range of 100–150 $\times 10^{10}$/L (Levin and Ebbe 1994). Human platelets have a mean diameter of 1–2 µm and platelet volume of 7–11 fl, and their average lifespan is 9 days (Drachman 2004), compared with murine platelets that have a mean diameter of 0.5 µm, platelet volume of 8–12 fl, and a lifespan of only 4 days. Furthermore, the murine population's α-granules are not as uniform with respect to morphology as in humans and also the number of dense granules is slightly lower in mouse platelets (Schmitt et al. 2001).

Differences also exist between MKs derived from human and mouse sources. Acetylcholinesterase assay (Jackson 1973) can be used to identify MKs of murine origin, but is not useful for human MKs. In humans, megakaryopoiesis occurs in bone marrow but in rodents spleen may also participate (Fig. 5.2). Importantly, cultured human MKs typically attain ploidy levels up to 16–32N (Mattia et al. 2002) compared with murine cultures that can reach higher ploidy levels (Tajika et al. 2000).

Cellular prion protein (PrPc) can be converted to scarpie prion protein that is instrumental in the pathogenesis of Creutzfieldt-Jacob disease (Kretzschmar 1993). PrPc can be detected in whole blood, and it was recently confirmed that primary MKs and platelets express PrPc (MacGregor et al. 1999; Starke et al. 2005). The functional role of PrPc in MKs and platelets as well as their potential link to the development of Creutzfieldt-Jacob disease after blood transfusions from donors harboring scarpie prion protein remain elusive.

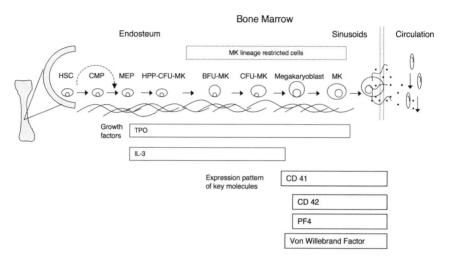

Fig. 5.2 Synopsis of megakaryopoiesis. Overview of the intermediate stages from HSC to platelets. The *dashed arrow* indicates the noncanonical pathway. The pattern of expression of some key CD molecules are depicted, as well as the TPO and IL-3 frame of stimulation

MK Endomitosis: In vitro and In vivo Studies

The State of Polyploidy

Mature MKs have 4N–128N content and achieve this state of high ploidy through endomitosis (Ravid et al. 2002). Polyploidy is not unique to MKs and has been documented in cardiomyocytes, hepatocytes, vascular smooth muscle cells of ageing, or hypertensive animals (Yang et al. 2007; McCrann et al. 2008a), and also in the giant cells of trophoblast (Ravid et al. 2002). Polyploidy, especially tetraploidy, also may serve as an intermediate step toward aneyploidy, a common theme of many neoplasias (Nguyen and Ravid 2006; Ganem et al. 2007).

The significance of MK polyploidy in platelet production has not been clearly elucidated (Ravid et al. 2002; Raslova et al. 2007). The relation between MK ploidy level and platelets remains controversial, with some studies linking platelet quality to MK ploidy level (reviewed in Zimmet and Ravid 2000), while other studies demonstrated production of human platelets by 2N or 4N MKs (Chang et al. 2007), in neonatal and myelodyplastic conditions.

The mechanisms that lead to polyploidy in nonmalignant cells are not uniform in all cell types (Zimmet and Ravid 2000). In the field of MK research, the terms endomitosis and endoreplication are commonly used. During endoreplication (alternatively known as endoreduplication) cells enter rounds of DNA synthesis that are interrupted by a gap phase without entering mitosis. Drosophila polytene chromosomes are the most studied examples of this process (Sauer et al. 1995). The term endomitosis was originally coined to describe polyploidization during which mitosis is blocked at prophase and the nuclear membrane is intact. However, it does not reflect precisely the process occurring in MKs.

The Endomitotic Cell Cycle

During MK endomitosis S-phase is followed by mitosis, during which anaphase and telophase are skipped and a gap phase precedes the subsequent round of DNA synthesis (Zhang et al. 1996). The MK endomitotic process is characterized by nuclear breakdown and the presence of multiple spindle poles (Roy et al. 2001).

The duration of an endomitotic MK cell cycle is estimated to be 7 h for S phase, and 9.3 hr in total (Jackson et al. 1990a; Wang et al. 1995). The majority (greater than 85%) of murine polyploid MKs isolated from bone marrow are cell cycle arrested (Wang et al. 1995). Moreover, only 60% are positive for cyclin D3 or p21, which are markers of G1 phase. Taken together, these data suggest that the majority of polyploid MKs are arrested at G1, while the rest at Go. However, cell cycle-arrested MKs may be recruited into endomitotic cycles by TPO treatment. This is suggested by the fact that 15% of freshly isolated bone marrow MKs are in S phase and this percentage is doubled after 12 h of TPO treatment (Wang et al. 1995 our unpuplished Data). This relatively fast increase of MK population in S phase may arise from the recruitment of polyploid MK cells that are cell cycle-arrested in parallel with diploid MKs entering endomitotic cycles.

Approximately 50% of diploid murine MKs cultured in the presence of TPO are in G1 phase. However, in the presence of TPO, as the MKs increase their ploidy state up to 32N the percentage of cells in G1 phase is reduced with concomitant increase in the number of cells found in G2/M and S phase. By contrast, in the MK subpopulation going from 32N to 64N the percentage of cells in G1 phase increases with a parallel decrease of cells in the G2/M phase (Crow et al. 2001). Taken together, these data indicate that MKs that reach ploidy states greater than 32N tend to exit endomitotic cycles.

Visualization studies of chromosomal dynamics and endomitotic machinery during MK endomitosis are predominantly based on immunofluorescence microscopy (Nagata et al. 1997; Roy et al. 2001), with only one study in the literature pursuing ex-vivo time lapse microscopy (Geddis et al., 2007). The latter study involves transduced cells and focuses on the microtubule network of MKs. On the basis of these studies, it has been concluded that during early anaphase the sister chromatids separate and depart toward the poles of the spindle network, consistent with anaphase A. An integral part of anaphase is the formation of spindle midzone. The spindle midzone is formed by antiparallel microtubules that are not associated with kinetochores and are placed between segregating chromosomes (Straight and Field 2000). Anaphase B, which involves spindle elongation and further movement of chromatids toward the poles, is not observed in MKs. The formation of spindle midzone is instrumental for the concentration of an array of proteins involved in cytokinesis (Straight and Field 2000; Geddis and Kaushansky 2006).

During furrowing of diploid cells, a contractile actomyosin network under the regulation of RhoA protein is formed that partitions the cell into two territories. However, many aspects of RhoA activity regulation remain speculative (Barr and Gruneberg 2007). In the last steps of cytokinesis the contractile ring is constricted, generating two daughter cells, but the terminal steps of cytokinesis require the addition of membrane vesicles to cover the area between the edges of the contractile rings and to also surround the daughter cells with membrane surface area.

Recently, the Polo-like kinase (PLK) family of proteins (Glover et al. 1996) has attracted attention in the context of MK endomitosis. PLK proteins have a crucial role for the formation of mitotic spindle and completion of mitosis, and in their absence mitotic arrest occurs followed by apoptosis (van Vugt et al. 2004). However, if in cells lacking PLK the spindle checkpoint is bypassed, then mitosis is completed without anaphase, resulting in polyploidy (Sumara et al. 2004). A recent study (Yagi and Roth 2006) indicated that PLK-1 was absent in polyploid MKs, and forced expression of PLK-1 was correlated with attenuated polyploidization.

On the basis of data obtained from a study of fixed MKs (Roy et al. 2001) it was found that the late events of mitosis related to telophase such as formation of nucleus around each group of chromatids, generation of two daughter cells, and furrowing are aborted. However, the single ex-vivo study of MK endomitosis indicated that significant furrowing occurs at least in MKs of low ploidy (Geddis et al. 2007). High-ploidy MKs (designated as those with four or more spindle poles) exhibited transient and attenuated furrowing. However, studies of spermatogenesis and oogenesis indicate that aborted cytokinesis does not necessarily promote poly-ploidy (Li 2007). Namely, the testicular specific Tex14 protein recruits MKLP1 and other key proteins that localize in midzone to stabilize a transient furrowing during male gametogenesis (Greenbaum et al. 2007). Despite the fact that during endomi-tosis the mechanisms that refrain diploid cells from attaining polyploidy status (Malumbres and Barbacid 2001; Margolis et al. 2003) are circumvented, the whole process remains tightly controlled (Nagata et al. 1997). Several layers of regulation have been described involving cyclins/Cdk, chromosomal passenger complex (CPC), and microtubule organizing proteins.

The Regulation of G1 Phase

Cyclins interact with Cdks and regulate progression through the cell cycle (Bloom and Cross 2007). In the regulation of the mitotic cycle in MKs, cyclins of type D (D3 and to a lesser extent D1) and type E (E1 and E2) have a significant role. Type D cyclins interact with Cdk 4 and 6 and phosphorylate pRB and pRB-related proteins (p107 and p130) priming progression to S-phase (Muntean et al. 2007). Overexpression of cyclin D3 in MKs increases MK ploidy level (Zimmet et al. 1997). Type E cyclins (Moroy and Geisen 2004) associate with Cdk 2 and 3 and hyperphosphorylate pRB, also promoting progression to S-phase. Importantly, knockout models of type E cyclins indicate that these proteins are absolutely required for progression of cells from Go to S phase and that they have a significant role in promoting polyploidy in MKs and trophoblast giant cells (Geng et al. 2003). Although the significance of cyclins has been documented, their mediators or mechanisms that result in high-ploidy MKs are poorly described.

The Regulation of Mitosis

Microtubule organization and spindle assembly checkpoints have an important role in preventing cells from attaining a polyploidy/aneuploidy state (Musacchio and Salmon 2007). MKs must overcome these regulatory circuits when switching from normal mitosis to endomitosis. Chromosomal passenger proteins have attracted attention as important mediators of endomitosis because they are involved in the association of microtubules with chromosomes, in the spindle checkpoint and during late events of telophase (Lens et al. 2006).

During early mitosis CPC relocates from chromosomal arms to centromeres while at telophase CPC translocates to midbody (Ruchaud et al. 2007).

Aurora-B, INCENP, Borealin/Desra, and survivin are members of the chromosomal passenger complex (Nguyen and Ravid 2006) (CPC). It has been hypothesized that the mechanisms leading to MK polyploidy may be linked to mislocalization or absence of one or more components of CPC during endomitosis.

Survivin is also a member of the IAP (inhibitor of apoptosis) family of proteins, and its role in the context of apoptosis has been recently reviewed (Altieri 2006), with several cancer clinical trials targeting survivin (Fukuda and Pelus 2006). Survivin has a firmly established role in cell cycle regulation in the context of chromosomal passenger proteins (Uren 2000). A survivin murine knockout model was embryonic lethal at day 5.5 post coitum. Gross anomalies included a small number of cells with abnormal and large nuclei, coupled with defects in microtubule formation and cytokinesis (Uren 2000). A study adopting a siRNA approach (Gurbuxani et al. 2005) indicated that survivin is essential for the proliferation of megakaryocyte–erythroid progenitors, and is also required for the differentiation of the more committed progenitors of the erythroid lineage. Furthermore, reduction of survivin levels through RNA interference in vitro augmented megakaryopoiesis. However, two transgenic mouse models in which the PF4 or GATA-1 promoter drives survivin expression indicated that survivin does not affect the MK ploidy profile in vivo, or the commitment to megakaryocytic lineage (McCrann et al. 2008b). In addition, although localization of survivin in human (low ploidy) MKs is normal (Geddis and Kaushansky 2004), its localization in murine models during all phases of endomitosis has not been precisely defined (Zhang et al. 2004; Gurbuxani et al. 2005).

Another member of the chromosomal passenger complex is Aurora B/AIM-1 (Nguyen et al. 2005). Aurora B is implicated in cytokinesis and segregation of chromosomes (Kotwaliwale 2006). It interacts with survivin and INCENP and by phosphorylating vimentin exerts control on vimentin-mediated cytokinesis. In a transgenic model of Aurora B, small MKs have higher levels of Aurora B compared with large MKs (Zhang et al. 2004). Excess Aurora B was removed with an internal mechanism, and high levels had an inhibitory effect on endomitosis. Furthermore, knockout models of Aurora B had mitotic defects, leading to the formation of polyploid cells (Murata-Hori et al. 2002). However, Aurora B localization was reported to be similar with that of mitotic diploid cells in MKs of human (Geddis and Kaushansky 2004) and murine (Geddis 2006) origins.

Recently, new players were implicated in endomitosis, adding another layer of complexity. Stathmin (Rubin et al. 2003) is a protein involved in microtubule depolymerization, and high levels are necessary in early steps of megakaryocyte maturation. However, downregulation of stathmin protein levels may be required for the switch from mitotic to endomitotic cycle (Iancu-Rubin et al. 2005).

The potential development of fluorescently tagged proteins with a key role in MK physiology, coupled with imaging of living MKs could address many unresolved issues. This is exemplified not only by the ex vivo study of MK endomitosis, but also by the generation of a CD41-YFP knock-in mouse model (Zhang et al. 2007). MKs of this model are intensely stained, while HSC and MK progenitors are stained only weakly. This model was used to visualize in vivo, murine BM through multiphoton intravital microscopy (Junt et al. 2007). This approach validated the proplatelet model and offered significant information regarding platelet biogenesis. Taken together these data clearly demonstrate that the application of time lapse microscopy in the field of MK physiology will significantly enhance our understanding of the control of MK polyploidy.

Growth Factors and MKs

Effects of Growth Factors on MKs

An inhibitory effect to megakaryopoiesis is exerted by TGF-β (Kuter et al. 1992), IL-4 (Sonoda et al. 1993), rapamycin (Raslova et al. 2006), and platelet factor 4 (PF4) (Lambert et al. 2007). The PF4 is a 7.8-kDa protein (with a half life of 25 min) that is produced by MKs and is packed in α-granules (von Hundelshausen et al. 2007). Autocrine production of PF4 attenuates megakaryopoiesis both in vitro (Miller et al. 2005) and in vivo (Lambert et al. 2007), and neutralizing antibodies against PF4 ameliorate thrombocytopenia induced by 5-FU in vivo. On the other hand, several cytokines exert a positive effect on MKs, including EPO, SCF, GM-CSF, IL-3, IL-6, IL-12, and IL-11 (Kaushansky 2005). IL-11 is produced by stromal cells and can directly stimulate megakaryopoiesis in vitro through gp130 signaling, and it can also act in synergy with TPO (Weich et al. 1997). Moreover, IL-11 administration to bone marrow-transplanted mice expedited platelet recovery levels (Du et al. 1993). IL-11 is approved for treatment of thrombocytopenia associated with myelosupressive treatments (Tepler et al. 1996). Similarly, IL-6 signals through gp130 and clinical trials had encouraging results (Nagler et al. 1995) but were accompanied with severe side effects (Ciurea and Hoffman 2007). IL-3 can also affect early stages of megakarypoiesis (Quesenberry et al. 1985). In addition, the water-soluble vitamin B3 (nicotamide) has been reported to increase ploidy levels of human primary MKs and megakaryocytic cell lines up to 64N, and augmented proplatelet formation without affecting other aspects of MK physiology (Giammona et al. 2006).

Effect of TPO on MKs

The major regulator of MK physiology is TPO (or mpl ligand) (reviewed in Kato et al. 1998; Kaushansky and Drachman 2002). Under normal conditions TPO levels in the human plasma are in the range of 12–61 pg/mL (Kato et al. 1998) and can increase several fold under conditions of stress, (<1 ng/mL). The human *TPO* gene is located on chromosome 3q26.3-3q27 and encodes a 353 amino acid precursor protein that contains a 21 amino acid signal peptide; a splice variant lacking four aminoacids is also described but it can not activate c-mpl (Gurney et al. 1995a). Modification of the precursor leads to production of the acidic and highly glycosylated 95-kD TPO protein. TPO protein can be divided into the N- and C-terminal domains. The N-terminal domain is sufficient for signaling (Linden and Kaushansky 2000) and consists of 153 amino acids sharing 23% homology with erythropoietin. Importantly, the N-terminal domain has two interaction sites with the TPO receptor, one conferring high- and the other low-affinity binding (Feese et al. 2004). On the other hand, the C-terminal domain has 179 amino acids, and is predominantly involved in TPO stability and secretion (Kaushansky and Drachman 2002). Murine TPO shares extensive homology (93%) with human TPO in the N-terminal domain, although in the C-terminal domain homology is less pronounced (74%). In addition, several residues of TPO with a key functional role have been identified (Jagerschmidt et al. 1998). TPO acts on HSC and is very important for their homeostasis (Carver-Moore et al. 1996; Solar et al. 1998). TPO exerts its effect on HSC by upregulating gene expression or by facilitating nuclear localization of transcription factors, such as homeobox proteins (Kirito et al. 2004). However, TPO does not alter the proportion of blood cell progenitors formed by HSC (Goncalves et al. 1997), but rather influences cells committed to megakaryocytic lineage (Kaushansky et al. 1995), and has a drastic effect on MK maturation by significantly increasing their size and ploidy status (Debili et al. 1995). On the other hand, terminal stages of MK life cycle that include formation of proplatelets are TPO-independent (Ito et al. 1996).

TPO is produced constitutively by two pathways and its effect on megakaryopoiesis is largely regulated by the level of circulating platelets (Kuter and Rosenberg 1995). In accordance, c-mpl knockout mice have hepatic TPO mRNA levels similar to wild-type mice despite severe thrombocytopenia (Fielder et al. 1996). Under steady state conditions, TPO is produced predominantly in the liver (Peck-Radosavljevic et al. 2000) and also in kidneys (Kuter et al. 1994). It is captured by circulating platelets through their c-mpl receptors, followed by internalization and proteolytic digestion. The free (unbound) TPO that reaches bone marrow exerts its effect on MKs. Production of TPO can be significantly increased in septic conditions (Zakynthinos et al. 2004). Furthermore, TPO is upregulated in inflammatory conditions, such as inflammatory bowel disease (Heits et al. 1999), although the precise mechanism has not been completely deciphered.

In the second pathway TPO is produced by the stromal cells of bone marrow and is released in the vicinity of MKs (Kuter et al. 1994). Under nonpathological conditions, TPO production via this pathway is minimal, but it can increase significantly in states of thrombocytopenia (Sungaran et al. 1997). It has been reported that PDGF and FGF-2 have a positive effect, while TGF-β has a negative effect on the production of TPO by stromal cells of human bone marrow (Sungaran et al. 2000). However, a different study indicated that TGF-β increased TPO mRNA levels in bone marrow (Sakamaki et al. 1999). Intriguingly, the regulation of megakaryopoiesis by TPO is not straightforward (Kaushansky 2005). For example, the number of mpl receptors in endothelial cells surpasses those found on platelets and MKs, but endothelial cells are not implicated in the regulation of TPO levels (Geddis et al. 2006). Furthermore, in patients suffering from idiopathic thrombocytopenic purpura, characterized by massive consumption of platelets, the levels of TPO are not increased proportionately to platelet destruction (Kosugi et al. 1996). In addition, the NF-E2 knockout mice do not exhibit increased levels of TPO even though platelets are absent (Shivdasani et al. 1997a).

A key component of the TPO signaling cascade is the TPO receptor. The TPO receptor (c-mpl) (Vigon et al. 1992), a type I transmbrane protein, exhibits the typical structure of the hemopoietic receptor superfamily with an extracellular domain, a 25 amino acid transmembrane domain, an approximately 120 amino acid intracellular domain (Kuter and Begley 2002). However, c-mpl does not interact with other hemopoietic receptor subunits or undergoes autophosphorylation, and its intracellular part does not share homology with any known protein. The extracellular domain has a proximal and a distal cytokine receptor homology module (CRM). TPO interacts only with the distal module (CRM1) abolishing suppression of c-mpl activation exerted by the CRM1 (Kaushansky and Drachman 2002). The intracellular domain consists of 122 amino acids in humans and 121 amino acids in mice. The intracellular domain of the c-mpl contains regions crucial for the association with JAK2 and other signaling transducers. The Box1 region consists of Pro-Ser-Iso-Pro (aminoacids 17–20), while the Box2 region is characterized by high content of proline and acidic amino acid residues (Gurney et al. 1995b). Moreover, although the cytoplasmatic region contains five tyrosine residues, only Tyr 112 and to a lesser extent Tyr 117 are the main targets of phosphorylation and the c-mpl emanating signaling cascades (Drachman and Kaushansky 1997; Rouyez et al. 1997). Furthermore, mutations affecting Tyr 112 decrease internalization of c-mpl–TPO complex by 50% (Dahlen et al. 2003). An α-helical amphipathic motif in the vicinity of the transmembrane region keeps the receptor inactive during unstimulated state (Staerk et al. 2006). It remains elusive whether TPO binds preformed c-mpl dimers or is responsible for c-mpl dimerization.

TPO exerts its action through three different signaling pathways: PI3K-AKT (phosphoinositol-3-Kinase/AKT), MAPK (mitogen activated protein kinase), and JAK-STAT (Janus kinase-signal transducers and activators of transcription). Seminal studies indicated that MAPK pathway is activated by TPO (Rouyez et al. 1997) but consensus for its TPO modulation was lacking (Guerriero et al. 2006). A recent study indicated that small amounts of MEK inhibitor increased ploidy levels in vitro (Guerriero et al. 2006). PI3K-AKT signaling pathway has a key role in cell cycle progression and one of its targets is the mammalian target of rapamycin (mTOR) (Inoki et al. 2005). TPO stimulation leads to mTOR activation, and addition of its inhibitor, rapamycin, has as a result decreased MK size and ploidy levels (Raslova et al. 2006). Another target of PI3K-AKT pathway is the transcription factor family of FoxO proteins. FoxO proteins regulate transcription of genes, such as *Fas Ligand*, *TRAIL*, *Bim*, and *p27 kip1* (Huang and Tindall 2007). PI3K-AKT signaling mediates their transport from the nucleus to the cytoplasm, allowing cells to progress through the cell cycle. FoxO3 was demonstrated to modulate the activity of p27 Kip1, and upon TPO stimulation to downregulate the expression of p27 Kip 1 (Nakao et al. 2007).

TPO-mediated changes in AKT and/or STAT have also been implicated in the modulation of hypoxia inducible factor (HIF) (Kirito and Kaushansky 2006), a transcription factor with a key role in oxygen tension sensing and VEGF production (Poellinger and Johnson 2004).

Interaction of TPO with its receptor activates the JAK-STAT pathway (Gurney et al. 1995b). Activation of c-mpl promotes JAK2 association and phosphorylation of key Tyr residues in the cytoplasmatic domain of the TPO receptor, enabling the docking and subsequent activation of STAT family of proteins (Baker et al. 2007). In addition, JAK2 can increase c-mpl surface localization and protect c-mpl from degradation (Royer et al. 2005) in vitro. Furthermore, the JAK2 V617F gain of function mutation is frequently detected in essential thrombocythemia and polycythemia vera (PV) patients (Lippert et al. 2006) and inhibitors are under development (Gaikwad and Prchal 2007). STAT 3, STAT 5α, and STAT 5β are involved in megakaryopoiesis. Upon dimerization, they translocate into the nucleus and activate transcription of an array of genes. A dominant negative STAT3 form was shown not to affect baseline MK or platelet levels in vivo, but it has a significant effect on platelet recovery after 5-fluoracil/myelosuppresion treatment (Kirito et al. 2002). An intriguing recent finding indicates that STAT3 can directly interact with GATA-1 and GATA-2 transcription factors. Even though the double knockout STAT5α/β mouse model exhibits reduced platelet levels and attenuated megakaryopoiesis (coupled with cytopenias), the precise role of STAT5α/β in this process has not been completely delineated (Snow et al. 2002). Namely, STAT5α/β activation has a key role in HSC homeostasis, which is mediated by an array of growth factors, apart from TPO (Kirito and Kaushansky 2006).

Effect of TPO on Platelets

Importantly, platelets express c-mpl and can respond to TPO stimulation but their direct activation requires an excessive amount of TPO. However, TPO was demonstrated to induce JAK2 Tyr phosphorylation and activation in platelets (Miyakawa et al. 1995). On the other hand, although TPO can activate RAS in human platelets, ERK-1/2 that are RAS downstream targets are not phosphorylated (Tulasne et al. 2002).

TPO can render platelets more susceptible to stimulation by a second agonist. Namely, TPO primed platelet activation upon stimulation with ADP, epinephrine, or shear stress (Oda et al. 1996). A similar effect has also been reported for thrombin (Akkerman 2006). Moreover, TPO enhances, through PI3K pathway, the signaling of collagen receptor in platelets, promoting a more vigorous platelet aggregation and secretion response (Pasquet et al. 2000).

Disruption of the production of TPO or its receptor c-mpl reduces platelet production by 85% (Kaushansky 1995), the knockout mice are viable and do not exhibit spontaneous bleeding. Importantly, double knockout mouse models of c-mpl and one of IL-3, IL-6, IL-11, LIF, or GM-CSF do not exhibit further platelet impairment (Gainsford et al. 1998; Gainsford et al. 2000). These findings indicate that other redundant pathways exist that can promote thrombopoiesis in the absence of TPO/mpl.

Effects of Other Receptors on MKs

MKs express N-methyl-D-aspartate (NMDA) and NMDA-type glutamate receptors that can affect many aspects of MK physiology, including maturation (Zheng et al. 2008). Importantly, antagonists of NMDA receptors severely affected MK maturation in vitro. Namely, MKs did not exhibit lobulation of nucleus and formation of DMS or proplatelets, even in the presence of TPO (Hitchcock et al. 2003).

Bone marrow endothelial cells (BMECs) can interact with MK progenitors and through the action of stromal derived factor -1 (SDF-1) and fibroblast growth factor 4 (FGF-4) rescue thrombopoiesis of TPO or its receptor knock out mice (Avecilla et al. 2004). The evolving concept is that SDF-1 and FGF-4 promote localization of MKs in the vascular niche, where the microenvironment can sustain their proliferation, maturation, and ultimately their release of platelets (Avecilla et al. 2004).

TPO in the Clinic

In translational research, several clinical conditions are manifested with low levels of platelets, such as ITP-, HIV- induced thrombocytopenia and liver disease, rendering TPO a potential therapeutic agent. Initial efforts to produce TPO led to the development of recombinant human thrombopoietin (rhTPO) and pegylated recombinant human megakaryocyte growth and development factor (PEG-rHuMGDF) (Begley and Basser 2000). Clinical trials utilizing these first generation agents indicated a beneficial increase of nadir platelet levels and shortening of thrombocytopenia duration in myelosuppresive conditions. However, in a clinical trial of PEG-rHuMGDF, formation of antibodies against endogenous TPO was reported, resulting in thrombocytopenia and other complications (Kuter and Begley 2002).

The issue of cross-reactive antibodies led to search of novel molecules. These can be grouped into three groups: TPO agonist antibodies, peptides mimicking TPO, and nonpeptide molecules mimicking TPO (Kuter 2007). The molecule AMG 531 has entered clinical trials and it was engineered by linking two peptides bridged by polylysine to two human IgG heavy constant regions held together by disulfide bonds (Wang et al. 2004). The AMG 531 has an extended half life and was not associated with formation of cross-reacting antibodies (Kuter 2007). An orally administrated molecule mimicking TPO, the Eltrombopag (Erickson-Miller et al. 2005), has also entered clinical trials with encouraging results (Newland 2007).

Transcription Factors and MKs

GATA Factors

The GATA family of transcription factors has a key role in dictating megakaryocytic fate (Crispino 2005). This family of transcription factors recognizes the (A/T) GATA(A/G) sequence motif, and interaction with DNA is mediated through a Zn-binding motif (Orkin 1992). The GATA family currently has six members but in megakaryopoiesis only GATA-1 and GATA-2 are involved.

The X-linked GATA-1 was initially studied in the context of erythropoiesis, but later its expression was verified also on mast cells, eosinophils, and in megakaryocytic lineage (Ravid et al. 1991; Zon et al. 1993; Hirasawa et al. 2002). GATA-1 null mice are embryonic lethal at day E10.5-11.5 due to anemia. The role of GATA-1 in megakaryopoiesis was elucidated through the development of GATA-1^{low} and GATA-1.05 mouse models. The GATA-1^{low} mice harbor a deletion of a DNAse I hypersensitivity site flanked by the two GATA promoters (Shivdasani et al. 1997b). The expression levels of GATA-1 are only 20% compared to wild type, and 99% of GATA-1^{low} mice die between late gestation and within the first 48 h of life due to anemia. However, the remaining 1% survive to adulthood but eventually develop myelofibrosis and extramedullary hemopoiesis. In the bone marrow, the number of primitive MKs is significantly increased and they exhibit a large nucleus that is segmented and surrounded by scarce amount of cytoplasm, which contains scant amounts of DMS. In addition, MK ploidy levels are attenuated and proplatelet formation is significantly reduced.

Only female heterozygous are alive in the GATA-1.05 mouse model that closely resembles the MDS phenotype. Hallmarks of this mouse model are severe thrombocytopenia and accumulation of primitive MKs in tissues (Takahashi et al. 1998).

GATA-1 missense mutations have been reported in families suffering from inherited X-linked thrombocytopenias coupled with varying degrees of anemia (Crispino 2005). The mutations are located in the N finger of GATA-1 and their clinical manifestation is macrothrombocytopenia coupled in the case of (a) D218G with dyserythropoiesis and severe anemia, (b) D218Y with dyserythropoiesis and mild anemia, (c) R216Q with thalasemia, and (d) V205M with anemia. An exception is G208S, which exhibits only thrombocytopenia. These mutations affect interaction of GATA-1 with the Friend of GATA-1 (FOG-1) apart from R216Q that affects GATA-1 binding to DNA.

Approximately 10% of patients with Down's syndrome develop during their childhood transient myeloproliferative disorder (TMD), and in the overwhelming majority of cases GATA-1 mutations are implicated. These mutations had, as result, the generation of a short, truncated GATA-1 protein (GATA-1s) that lacks the N terminus (Wechsler et al. 2002). The hallmarks of TMD are expansion of abnormal megakaryoblasts and their detection in peripheral circulation. The majority of cases are subclinical, but 20–30% will eventually progress to acute megakaryocytic leukemia (AMKL). A mouse model of GATA-1s indicated that although MKs progressed through the terminal stages of their life cycle, MK progenitors from fetal liver and yolk sac had hyperproliferating capacity up to late gestation (Li et al. 2005). The molecular mechanisms linking GATA-1s with the development of AMKL are not well understood.

The GATA-2 expression is highest in the level of HSC and progenitors and has overlapping, but distinct, effects compared with GATA-1 (Cheng et al. 1996). Although, GATA-2 knockout mice are embryonic lethal before E11.5 (Tsai et al. 1994), data derived from erythroleukemia cell lines indicate that GATA-2 overexpression favors megakaryocytic differentiation (Ikonomi et al. 2000).

A yeast two-hybrid assay indicated that FOG-1, a 998 amino acid protein with 9 zinc fingers interacts with GATA-1 (Tsang et al. 1997). FOG-1 expression pattern closely resembles that of GATA-1. FOG-1 knockout mice do not form any MK progenitors and are embryonic lethal at E11.5 (Tsang et al. 1998). A GATA-1 mutant that can not associate with FOG-1 indicated that FOG-1 may also interact with GATA-2 to exert GATA-1-independent actions (Chang et al. 2002). Furthermore, a study utilizing conditional expression of FOG-1 indicated a stage-specific modulation of megakarypoiesis (Tanaka et al. 2004). Currently, FOG-1 mutations have not been linked to any human pathological condition.

NF-E2 Transcription Factor

The basic leucine zipper transcription factor p45 NF-E2 is a heterodimer with a key role in late stages of the MK life cycle. The heterodimer consists of a 45-kDa subunit restricted to hemopoietic lineages, associated with p18 MafG or MafK subunits (Ney et al. 1993). NF-E2 knockout mice had undetectable levels of platelets, even though the MK population had significantly expanded (Shivdasani et al. 1995).

Although MKs exhibited very low amount of granules coupled with a disorganized DMS, endomitosis was not affected. The same level of thrombocytopenia was also exhibited by MafG–MafK double knockout mice (Onodera et al. 2000). These data, taken together and coupled with in vitro studies demonstrating absence of proplatelet formations indicate that the molecular defect lies at the terminal stages of MK maturation. The genes modulated by NF-E2 have not been precisely defined, but they include β1 tubulin, Rab27, and caspase 12, indicating a broad repertoire of functions (Lecine et al. 2000; Tiwari et al. 2003; Kerrigan et al. 2004). A study involving transduced cells overexpressing NF-E2 indicated that biogenesis of platelets was considerably enhanced (Fock et al. 2008). On the other hand, NF-E2 overexpression is implicated in the development of polycythemia vera (Goerttler et al. 2005).

CBF Transcription Factor

The core binding factor (CBF) is composed of two subunits. Only the α subunit (CBFA) can bind DNA, an attribute that is further enhanced by the β subunit (CBFβ) (Wang et al. 1993). RUNX1 associates with CBFβ and its expression is increased along the MK lineage. Although RUNX1 knockout mice are embryonic lethal, conditional knock out exhibited low-ploidy micro-MKs with rudimentary DMS (Ichikawa et al. 2004). Genetic lesions that cause RUNX1 haplo insufficiency were linked to the disorder of familial thrombocytopenia with propensity to AML (Song et al. 1999).

ETS Transcription Factors

Fli-1 is a member of the ETS family of transcription factors (Watson et al. 1992). Fli-1 is detected on platelets and its expression in erythroleukemic lines favors differentiation toward the MK lineage Fli-1(Athanasiou et al. 1996). Knockout mice exhibit aberrant megakaryopoiesis and die due to hemorrhage at E12.5 (Hart et al. 2000). Fli-1 is implicated in the development of Paris Trusseau syndrome.

From MK to Platelets

The terminal steps of MK maturation involve the production of platelets. The mechanism underlying platelet production has been controversial for several years. One hypothesis, based predominantly on electron microscopy studies (Mori et al. 1993) and supported by phase contrast observations (Kosaki 2005), proposed that platelets are produced in the cytoplasm of MKs and subsequently are released. The second model proposed that platelets are generated by proplatelets (Tablin et al. 1990;

Italiano et al. 1999), an array of cytoplasmatic processes that extend from the surface of MKs with a highly variable pattern and can reach several hundred μm in length. Proplatelets are formed through an unknown triggering mechanism, in one pole of MK, and their number usually range from 1 to 10 (Patel et al. 2005a). The proplatelet model was further strengthened with pioneering ex-vivo real time studies pursued by Italiano and coworkers, which demonstrated that microtubules are the main component of proplatelets, and their sliding action is responsible for the proplatelet elongation, which has a duration of 4–10 h (Patel et al. 2005b; Hartwig and Italiano 2006). Proplatelets have a diameter of 2–4 μm and have numerous bulges, which are interconnected with a thin bridge of cytoplasm. Furthermore, as the proplatelets extend from the MK surface the thick bundle of microtubules proximal to MK cell membrane gradually thins out to approximately 20 microtubules at the tip of proplatelet.

Platelets are released only from nascent tips of proplatelets. An elegant mechanism that increases the number of nascent tips has been documented that implicates rearrangement of actin cytoskeleton (Rojnuckarin and Kaushansky 2001) through Protein kinase C (PLK-C) action although details for the precise mechanism are lacking (Patel et al. 2005a).

The generation and elongation of proplatelets continues until almost all cytoplasm is consumed. Ultimately, proplatelets are extruded in the bone marrow sinusoids (Junt et al. 2007).

Proplatelet release from mature MKs heralds the formation of senescent (apoptotic) MKs, which are removed by macrophages. The apoptosis of MKs is very complex and involves tight regulation through Bcl-2, Bcl-xL, and caspase 3 and 9 proteins. Namely, activation of caspase 3 during proplatelet formation appears localized but in the senescent MK the pattern is diffuse (De Botton et al. 2002). The precise timing of the apoptotic events in nucleus in relation to proplatelet formation may require time lapse imaging in order to be elucidated.

A unique characteristic of platelets is that they carry MK mRNA and a functional splicesome machinery that is active in the absence of nucleus (reviewed in Gnatenko et al. 2006). Platelets carry approximately 1,600–3,000 mRNA gene transcripts, and a proteomic analysis indicated that platelets carry 641 proteins in resting state. Approximately 300 proteins are secreted during platelet activation including chemokines such as CXCL7, PF4, CCL5, and VEGF (von Hundelshausen et al. 2007). Intriguingly, some of the secreted proteins are only found as atherosclerotic lesions rendering them therapeutic targets.

Megakaryocytes and Bone Formation

Several studies provided evidence for the interaction of MKs with the milieu of bone niche and their potential role in bone homeostasis (Kacena et al. 2006a; Kacena and Horowitz 2006). Namely, MKs harbor receptors for receptor activator of NFKβ (RANKL) (Kartsogiannis et al. 1999) and bone morphogenetic proteins

(BMP) 2,4, and 6 (Sipe et al. 2004), which affect bone remodeling. In addition, MKs in vitro can inhibit osteoclast formation through an osteoprotegerin-independent mechanism that remains elusive (Beeton et al. 2006; Kacena et al. 2006b).

Transplantation mouse models in which TPO was overexpressed exhibited osteosclerosis and myelofibrosis before the first year of life (Villeval et al. 1997), and similar results were obtained by using retroviral constructs expressing TPO (Yan et al. 1995). Intriguingly, c-mpl knockout mice do not exhibit abnormal bone homeostasis (Perry et al. 2007).

In accordance, GATA-1^{low} mice exhibited increased bone formation and mass that was followed by myelofibrosis after the first year of life. It is proposed that myelofibrosis occurred through a TGF-β1 mechanism (Vannucchi et al. 2005) while osteosclerosis by elevated levels of BMPs produced by the expanded MK population (Garimella et al. 2007). Also, deletion of the 45-kDa subunit of NF-E2 in a mouse model was accompanied by increased MK number as well as elevated bone mass and bone formation index (Kacena et al. 2005). A common characteristic of the MK lineage GATA-1 and NF-E2 null mouse models is the increased number of osteoblasts.

On the other hand, MKs can attenuate octeoclast activity in vitro (Beeton et al. 2006). A recently engineered mouse model (Suva et al. 2008) of platelet-type von Willebrand disease (Miller and Castella 1982) highlighted a potential link between MKs and osteoclasts in vivo. The mouse model was engineered with a single amino acid change of G233V affecting *GP Ia* gene. Hallmarks of this mouse model are an increased spleen size with a very high number of MKs coupled with increased bone mass. The levels of osteoblasts were not increased but formation of osteoclasts was significantly attenuated.

Clinical Conditions

Acute Megakaryoblastic Leukemia

Acute megakaryoblastic leukemia (AMKL) is designated as M7 subtype of acute myeloid leukemia according to the French American British (FAB) classification (Bennett et al. 1985). Hallmarks of AMKL are the presence of an excess number of blasts in the bone marrow with more than 50% of them of megakaryocytic lineage, coupled with myelofibrosis, organomegaly, and pancytopenia (Ribeiro et al. 1993). AMKL blasts express at least two of the platelet/MK-related markers (CD41, CD42b, CD61b, vWF), as well as CD33 and/or CD13 (Arber 2001; Orazi 2007). AMKL exhibits a bimodal distribution with the majority of cases in adulthood and early childhood. AMKL in childhood cases can be grouped into those related to Down's syndrome (DS) and those occurring outside this context. The incidence of pediatric AMKL on a DS background is 1:500 with median age of presentation at 2 years, and the usual age range between 1 and 3 years of life. Essentially, in all cases, GATA-1 mutations are found in the blasts. The blasts express reduced amounts of cytidine deaminase, an enzyme that inactivates Ara-C, rendering these cells sensitive to chemotherapy (Ge et al. 2004). Pediatric AMKL not related to DS, includes cases affecting infants, in which 1;22 translocation leads to the fusion of RNA binding motif protein 15 (RBM15, alternatively known as OTT) with megakaryoblastic leukemia 1 gene (*MKL1*) (Bernstein et al. 2000; Ma et al. 2001). The chimeric protein contains all the functional domains of RBM15 and MKL1 (Mercher et al. 2001). The SAP domain (Aravind and Koonin 2000) of MKL1 has a putative role of linking nuclear scaffolds with DNA, and may translocate RBM15 in eucromatin regions. The median age of pediatric AMKL outwidth the context of DS is 2.2 years and the prognosis is dismal. The cytogenetic lesions occasionaly detected are t(9;11), and (10;11) that leads to generation of CALM-AF10 chimeric protein. On the other hand, adult AMKL is characterized by a much broader repertoire of molecular lesions and is estimated to account for 1–10% of all adult AML cases (Tallman et al. 2000). Prognosis is dismal with few patients achieving remission. Recently, it has been reported that mutations of JAK2 and JAK3 were detected in patients suffering from AMKL (Jelinek et al. 2005; Walters et al. 2006).

Other Mutations and Pathologies

Several other disorders have been described affecting MKs, but in many cases their molecular pathology has not been completely elucidated. A rare autosomal recessive disorder that affects mpl is the congenital amegakaryocytic thrombocytopenia (CAMT) (Nurden and Nurden 2007b). It is manifested in early years of life with high levels of circulating TPO, very low levels of platelets, and MKs. Patients fall into two groups: those who harbor nonsense mutations in the *mpl* gene (Ballmaier et al. 2001) and are predicted to have minimal amounts of mpl protein (CAMT type 1), and those who have misense mutations affecting the extracellular domain of the receptor (CAMT type 2). Patients suffering from CAMT type 2 have increased levels of platelets in comparison with patients suffering from CAMT type 1 during the first year of life, albeit significantly lower than healthy controls.

Mutations affecting the promoter of the *TPO* gene and the *mpl* gene are implicated in the development of hereditary essential thrombocytosis (ET) (Steensma and Tefferi 2002). Hallmarks of ET include high levels of platelets, hyperplastic MKs, and episodes of hemorrhage (predominantly from gastrointestinal tract) and thrombosis. Approximately 7% of African Americans harbor a single nucleotide polymorphism (1238 G-T) of *mpl* gene that leads to the substitution of lysine 39 by asparagine (K39N) in the c-mpl protein (Moliterno et al. 2004).

The autosomal dominant May-Hegglin anomaly closely resembles Bernard-Soulier Syndrome and is the most common hereditary cause of macrothrombocytopenia (Seri et al. 2000). The gene responsible for this disorder lies in 22q12-13 and is encoding nonmuscle myosin heavy chain IIA – the only isoform of myosin found in platelets. An intriguing finding is that MKs are not affected and platelets have normal life span, suggesting that the molecular defect may lie in proplatelet formations (Hartwig and Italiano 2003).

The hallmark of Gray platelet syndrome is the very low number of α granules in platelets that appear gray in common platelet staining techniques (Balduini and Savoia 2004). The molecular defect and the exact mode of inheritance remains elusive, but inability of MKs to properly package α granules has been implicated (Nurden and Nurden 2007a, b). The phenotype is variable, and patients exhibit moderate thrombocytopenia and some degree of myelofibrosis linked to the action of α-granule content leaked to bone marrow (Falik-Zaccai et al. 2001).

Paris trousseau syndrome (PTS) is caused by deletions in the vicinity of *Fli-1* gene (11q23.3-11q24.2) location and is characterized by mild thrombocytopenia with a fraction of platelets having large α-granules (Favier et al. 2003). Platelet life span is normal, and the bone marrow harbors a large population of MKs with a significant proportion of them being micro-MKs. It was proposed that in TPS patients harboring the deletion in the stage of CD41+/CD42− progenitor cells a population without Fli-1 expression is generated and the produced MK is structurally defective (Raslova et al. 2004). A similar phenotype is manifested by Jacobsen syndrome, which also includes mental retardation and cardiac defects (Wenger et al. 2006).

Thrombocytopenia with absent radius (TAR) syndrome is characterized by thrombocytopenia, absent radius with presence of thumbs and also defects in cardiac and genitourinary development (Geddis 2006). Thrombocytopenia is usually severe, but is ameliorated toward adulthood. The mode of inheritance is complex and a deletion of 11 genes in a 200-kb region of chromosome 11 is postulated to be perquisite for the development of TAR syndrome in the presence of an unknown modifier (Klopocki et al. 2007) gene(s).

Perspectives

Recent studies revealed that MKs and platelets also have intriguing roles in immunity, in the context of neoplasias (Folkman 2007), skeletal homeostasis (Kacena et al. 2006a), and as carriers of prognostic markers related to cardiovascular diseases (Healy et al. 2006). Further developments in these emerging fields of research are likely to outline yet newer roles for megakaryocytes and their progenies.

Acknowledgments We apologize to those whose work was not cited, owing to limited space. KR is funded by NHLBI (HL 80442) and is an established investigator with the American Heart Association. We would like to thank Robin McDonald and Shannon Carroll for their valuable help in preparing the manuscript.

References

Adams, G. B., D. T. Scadden (2006). "The hematopoietic stem cell in its place." *Nat Immunol* **7**(4): 333–337.

Adolfsson, J., R. Mansson, et al. (2005). "Identification of Flt3 + lympho-myeloid stem cells lacking erythro-megakaryocytic potential a revised road map for adult blood lineage commitment." *Cell* **121**(2): 295–306.

Akashi, K., D. Traver, et al. (2000). "A clonogenic common myeloid progenitor that gives rise to all myeloid lineages." *Nature* **404**(6774): 193–197.

Akkerman, J. W. (2006). "Thrombopoietin and platelet function." *Semin Thromb Hemost* **32**(3): 295–304.

Altieri, D. C. (2006). "The case for survivin as a regulator of microtubule dynamics and cell-death decisions." *Current Opinion in Cell Biology* **18**(6): 609–615.

Andonegui, G., S. M. Kerfoot, et al. (2005). "Platelets express functional Toll-like receptor-4." *Blood* **106**(7): 2417–2423.

Uren, A. G., L. Wong, M. Pakusch, K. J. Fowler, F. J. Burrows, D. L. Vaux, K. H. Choo (2000). "Survivin and the inner centromere protein INCENP show similar cell-cycle localization and gene knockout phenotype." *Current Biology* **10**(21): 1319–1328.

Aravind, L., E. V. Koonin (2000). "SAP – a putative DNA-binding motif involved in chromosomal organization." *Trends Biochem Sci* **25**(3): 112–114.

Arber, D. A. (2001). "Realistic pathologic classification of acute myeloid leukemias." *Am J Clin Pathol* **115**(4): 552–560.

Athanasiou, M., P. A. Clausen, et al. (1996). "Increased expression of the ETS-related transcription factor FLI-1/ERGB correlates with and can induce the megakaryocytic phenotype." *Cell Growth Differ* **7**(11): 1525–1534.

Avecilla, S. T., K. Hattori, et al. (2004). "Chemokine-mediated interaction of hematopoietic progenitors with the bone marrow vascular niche is required for thrombopoiesis." *Nat Med* **10**(1): 64–71.

Baker, S. J., S. G. Rane, et al. (2007). "Hematopoietic cytokine receptor signaling." *Oncogene* **26**(47): 6724–6737.

Balduini, C. L., A. Savoia (2004). "Inherited thrombocytopenias: molecular mechanisms." *Semin Thromb Hemost* **30**(5): 513–523.

Ballmaier, M., M. Germeshausen, et al. (2001). "c-mpl mutations are the cause of congenital amegakaryocytic thrombocytopenia." *Blood* **97**(1): 139–146.

Barr, F. A., U. Gruneberg (2007). "Cytokinesis: placing and making the final cut." *Cell* **131**(5): 847–860.

Beeton, C. A., S. Bord, et al. (2006). "Osteoclast formation and bone resorption are inhibited by megakaryocytes." *Bone* **39**(5): 985–990.

Begley, C. G., R. L. Basser (2000). "Biologic and structural differences of thrombopoietic growth factors." *Semin Hematol* **37**(2 Suppl 4): 19–27.

Bennett, J. M., D. Catovsky, et al. (1985). "Criteria for the diagnosis of acute leukemia of megakaryocyte lineage (M7). A report of the French-American-British Cooperative Group." *Ann Intern Med* **103**(3): 460–462.

Bernstein, J., N. Dastugue, et al. (2000). "Nineteen cases of the t(1;22)(p13;q13) acute megakaryblastic leukaemia of infants/children and a review of 39 cases: report from a t(1;22) study group." *Leukemia* **14**(1): 216–218.

Bloom, J., F. R. Cross (2007). "Multiple levels of cyclin specificity in cell-cycle control." *Nat Rev Mol Cell Biol* **8**(2): 149–160.

Branehog, I., B. Ridell, et al. (1975). "Megakaryocyte quantifications in relation to thrombokinetics in primary thrombocythaemia and allied diseases." *Scand. J. Haematol.* **15**(5): 321–332.

Brass, L. F. (2005). "Did dinosaurs have megakaryocytes? New ideas about platelets and their progenitors." *J Clin Invest* **115**(12): 3329–3331.

Briddell, R. A., J. E. Brandt, et al. (1989). "Characterization of the human burst-forming unit-megakaryocyte." *Blood* **74**(1): 145–151.

Bruno, E., L. J. Murray, et al. (1996). "Detection of a primitive megakaryocyte progenitor cell in human fetal bone marrow." *Exp Hematol* **24**(4): 552–558.

Carver-Moore, K., H. E. Broxmeyer, et al. (1996). "Low levels of erythroid and myeloid progenitors in thrombopoietin-and c-mpl-deficient mice." *Blood* **88**(3): 803–808.

Chang, A. N., A. B. Cantor, et al. (2002). "GATA-factor dependence of the multitype zinc-finger protein FOG-1 for its essential role in megakaryopoiesis." *Proc Natl Acad Sci USA* **99**(14): 9237–9242.

Chang, Y., D. Bluteau, et al. (2007). "From hematopoietic stem cells to platelets." *J Thromb Haemost* 5 Suppl **1**: 318–327.

Cheng, T., H. Shen, et al. (1996). "Temporal mapping of gene expression levels during the differentiation of individual primary hematopoietic cells." *Proc Natl Acad Sci USA* **93**(23): 13158–13163.

Ciurea, S. O., R. Hoffman (2007). "Cytokines for the treatment of thrombocytopenia." *Semin Hematol* **44**(3): 166–182.

Crispino, J. D. (2005). "GATA1 in normal and malignant hematopoiesis." *Semin Cell Dev Biol* **16**(1): 137–147.

Crow, C. E., N. E. Fox, et al. (2001). "Kinetics of endomitosis in primary murine megakaryocytes." *J Cell Physiol* **188**(3): 291–303.

Dahlen, D. D., V. C. Broudy, et al. (2003). "Internalization of the thrombopoietin receptor is regulated by 2 cytoplasmic motifs." *Blood* **102**(1): 102–108.

Debili, N., F. Wendling, et al. (1995). "The Mpl-ligand or thrombopoietin or megakaryocyte growth and differentiative factor has both direct proliferative and differentiative activities on human megakaryocyte progenitors." *Blood* **86**(7): 2516–2525.

Debili, N., L. Coulombel, et al. (1996). "Characterization of a bipotent erythro-megakaryocytic progenitor in human bone marrow." *Blood* **88**(4): 1284–1296.

De Botton, S., S. Sabri, et al. (2002). "Platelet formation is the consequence of caspase activation within megakaryocytes." *Blood* **100**(4): 1310–1317.

de Jong, J. L., L. I. Zon (2005). "Use of the zebrafish system to study primitive and definitive hematopoiesis." *Annu Rev Genet* **39**: 481–501.

Deutsch, V. R., A. Tomer (2006). "Megakaryocyte development and platelet production." *Br J Haematol* **134**(5): 453–466.

Drachman, J. G. (2004). "Inherited thrombocytopenia: when a low platelet count does not mean ITP." *Blood* **103**(2): 390–398.

Drachman, J. G., K. Kaushansky (1997). "Dissecting the thrombopoietin receptor: functional elements of the Mpl cytoplasmic domain." *Proc Natl Acad Sci USA* **94**(6): 2350–2355.

Du, X. X., T. Neben, et al. (1993). "Effects of recombinant human interleukin-11 on hematopoietic reconstitution in transplant mice: acceleration of recovery of peripheral blood neutrophils and platelets." *Blood* **81**(1): 27–34.

Duncan, A. W., F. M. Rattis, et al. (2005). "Integration of Notch and Wnt signaling in hematopoietic stem cell maintenance." *Nat Immunol* **6**(3): 314–322.

Erickson-Miller, C. L., E. DeLorme, et al. (2005). "Discovery and characterization of a selective, nonpeptidyl thrombopoietin receptor agonist." *Exp Hematol* **33**(1): 85–93.

Falik-Zaccai, T. C., Y. Anikster, et al. (2001). "A new genetic isolate of gray platelet syndrome (GPS): clinical, cellular, and hematologic characteristics." *Mol Genet Metab* **74**(3): 303–313.

Favier, R., K. Jondeau, et al. (2003). "Paris-Trousseau syndrome: clinical, hematological, molecular data of ten new cases." *Thromb Haemost* **90**(5): 893–897.

Feese, M. D., T. Tamada, et al. (2004). "Structure of the receptor-binding domain of human thrombopoietin determined by complexation with a neutralizing antibody fragment." *Proc Natl Acad Sci USA* **101**(7): 1816–1821.

Fielder, P. J., A. L. Gurney, et al. (1996). "Regulation of thrombopoietin levels by c-mpl-mediated binding to platelets." *Blood* **87**(6): 2154–2161.

Fock, E. L., F. Yan, et al. (2008). "NF-E2-mediated enhancement of megakaryocytic differentiation and platelet production in vitro and in vivo." *Exp Hematol* **36**(1): 78–92.

Folkman, J. (2007). "Angiogenesis: an organizing principle for drug discovery?" *Nat Rev Drug Discov* **6**(4): 273–286.

Forestier, F., F. Daffos, et al. (1991). "Developmental hematopoiesis in normal human fetal blood." *Blood* **77**(11): 2360–2363.

Fukuda, S., L. M. Pelus (2006). "Survivin, a cancer target with an emerging role in normal adult tissues." *Mol Cancer Ther* **5**(5): 1087–1098.

Gaikwad, A., J. T. Prchal (2007). "Study of two tyrosine kinase inhibitors on growth and signal transduction in polycythemia vera." *Exp Hematol* **35**(11): 1647–1656.

Gainsford, T., A. W. Roberts, et al. (1998). "Cytokine production and function in c-mpl-deficient mice: no physiologic role for interleukin-3 in residual megakaryocyte and platelet production." *Blood* **91**(8): 2745–2752.

Gainsford, T., H. Nandurkar, et al. (2000). "The residual megakaryocyte and platelet production in c-mpl-deficient mice is not dependent on the actions of interleukin-6, interleukin-11, or leukemia inhibitory factor." *Blood* **95**(2): 528–534.

Ganem, N. J., Z. Storchova, et al. (2007). "Tetraploidy, aneuploidy and cancer." *Curr Opin Genet Dev* **17**(2): 157–162.

Garimella, R., M. A. Kacena, et al. (2007). "Expression of bone morphogenetic proteins and their receptors in the bone marrow megakaryocytes of GATA-1(low) mice: a possible role in osteosclerosis." *J Histochem Cytochem* **55**(7): 745–752.

Ge, Y., T. L. Jensen, et al. (2004). "The role of cytidine deaminase and GATA1 mutations in the increased cytosine arabinoside sensitivity of Down syndrome myeloblasts and leukemia cell lines." *Cancer Res* **64**(2): 728–735.

Geddis, A. E. (2006). "Inherited thrombocytopenia: congenital amegakaryocytic thrombocytopenia and thrombocytopenia with absent radii." *Semin Hematol* **43**(3): 196–203.

Geddis, A. E., K. Kaushansky (2004). "Megakaryocytes express functional Aurora-B kinase in endomitosis." *Blood* **104**(4): 1017–1024.

Geddis, A. E., K. Kaushansky (2006). "Endomitotic megakaryocytes form a midzone in anaphase but have a deficiency in cleavage furrow formation." *Cell Cycle* **5**(5): 538–545.

Geddis, A. E., N. E. Fox, et al. (2006). "The Mpl receptor expressed on endothelial cells does not contribute significantly to the regulation of circulating thrombopoietin levels." *Exp Hematol* **34**(1): 82–86.

Geddis, A. E., N. E. Fox, et al. (2007). "Endomitotic megakaryocytes that form a bipolar spindle exhibit cleavage furrow ingression followed by furrow regression." *Cell Cycle* **6**(4): 455–460.

Geng, Y., Q. Yu, et al. (2003). "Cyclin E ablation in the mouse." *Cell* **114**(4): 431–443.

Giammona, L. M., P. G. Fuhrken, et al. (2006). "Nicotinamide (vitamin B3) increases the polyploidisation and proplatelet formation of cultured primary human megakaryocytes." *Br J Haematol* **135**(4): 554–566.

Glover, D. M., H. Ohkura, et al. (1996). "Polo kinase: the choreographer of the mitotic stage?" *J Cell Biol* **135**(6, Part 2): 1681–1684.

Gnatenko, D. V., P. L. Perrotta, et al. (2006). "Proteomic approaches to dissect platelet function: half the story." *Blood* **108**(13): 3983–3991.

Goerttler, P. S., C. Kreutz, et al. (2005). "Gene expression profiling in polycythaemia vera: over-expression of transcription factor NF-E2." *Br J Haematol* **129**(1): 138–150.

Goncalves, F., C. Lacout, et al. (1997). "Thrombopoietin does not induce lineage-restricted commitment of Mpl-R expressing pluripotent progenitors but permits their complete erythroid and megakaryocytic differentiation." *Blood* **89**(10): 3544–3553.

Greenbaum, M. P., L. Ma, et al. (2007). "Conversion of midbodies into germ cell intercellular bridges." *Dev Biol* **305**(2): 389–396.

Guerriero, R., I. Parolini, et al. (2006). "Inhibition of TPO-induced MEK or mTOR activity induces opposite effects on the ploidy of human differentiating megakaryocytes." *J Cell Sci* **119** (Part 4): 744–752.

Gurbuxani, S., Y. Xu, et al. (2005). "Differential requirements for survivin in hematopoietic cell development." *Proc Natl Acad Sci USA* **102**(32): 11480–11485.

Gurney, A. L., W. J. Kuang, et al. (1995a). "Genomic structure, chromosomal localization, and conserved alternative splice forms of thrombopoietin." *Blood* **85**(4): 981–988.

Gurney, A. L., S. C. Wong, et al. (1995b). "Distinct regions of c-Mpl cytoplasmic domain are coupled to the JAK-STAT signal transduction pathway and Shc phosphorylation." *Proc Natl Acad Sci USA* **92**(12): 5292–5296.

Hart, A., F. Melet, et al. (2000). "Fli-1 is required for murine vascular and megakaryocytic development and is hemizygously deleted in patients with thrombocytopenia." *Immunity* **13**(2): 167–177.

Hartwig, J. H., J. E. Italiano, Jr. (2003). "The birth of the platelet." *J Thromb Haemost* **1**(7): 1580–1586.

Hartwig, J. H., J. E. Italiano, Jr. (2006). "Cytoskeletal mechanisms for platelet production." *Blood Cells Mol Dis* **36**(2): 99–103.

Healy, A. M., M. D. Pickard, et al. (2006). "Platelet expression profiling and clinical validation of myeloid-related protein-14 as a novel determinant of cardiovascular events." *Circulation* **113**(19): 2278–2284.

Heits, F., M. Stahl, et al. (1999). "Elevated serum thrombopoietin and interleukin-6 concentrations in thrombocytosis associated with inflammatory bowel disease." *J Interferon Cytokine Res* **19**(7): 757–760.

Hirasawa, R., R. Shimizu, et al. (2002). "Essential and instructive roles of GATA factors in eosinophil development." *J Exp Med* **195**(11): 1379–1386.

Hitchcock, I. S., T. M. Skerry, et al. (2003). "NMDA receptor-mediated regulation of human megakaryocytopoiesis." *Blood* **102**(4): 1254–1259.

Huang, H., D. J. Tindall (2007). "Dynamic FoxO transcription factors." *J Cell Sci* **120** (Part 15): 2479–2487.

Iancu-Rubin, C., C. A. Nasrallah, et al. (2005). "Stathmin prevents the transition from a normal to an endomitotic cell cycle during megakaryocytic differentiation." *Cell Cycle* **4**(12): 1774–1782.

Ichikawa, M., T. Asai, et al. (2004). "AML-1 is required for megakaryocytic maturation and lymphocytic differentiation, but not for maintenance of hematopoietic stem cells in adult hematopoiesis." *Nat Med* **10**(3): 299–304.

Ikonomi, P., C. E. Rivera, et al. (2000). "Overexpression of GATA-2 inhibits erythroid and promotes megakaryocyte differentiation." *Exp Hematol* **28**(12): 1423–1431.

Inoki, K., H. Ouyang, et al. (2005). "Signaling by target of rapamycin proteins in cell growth control." *Microbiol Mol Biol Rev* **69**(1): 79–100.

Italiano, J. E., Jr., P. Lecine, et al. (1999). "Blood platelets are assembled principally at the ends of proplatelet processes produced by differentiated megakaryocytes." *J Cell Biol* **147**(6): 1299–1312.

Ito, T., Y. Ishida, et al. (1996). "Recombinant human c-Mpl ligand is not a direct stimulator of proplatelet formation in mature human megakaryocytes." *Br J Haematol* **94**(2): 387–390.

Jackson, C. W. (1973). "Cholinesterase as a possible marker for early cells of the megakaryocytic series." *Blood* **42**(3): 413–421.

Jackson, C. W., N. K. Hutson, et al. (1990a). "Megakaryocytopoiesis in man and laboratory animals. Conclusions derived from comparative studies and recently discovered animal models with megakaryocyte anomalies." *Prog Clin Biol Res* **356**: 11–23.

Jackson, C. W., S. A. Steward, et al. (1990b). "An analysis of megakaryocytopoiesis in the C3H mouse: an animal model whose megakaryocytes have 32N as the modal DNA class." *Blood* **76**(4): 690–696.

Jackson, H., N. Williams, et al. (1994). "Classes of primitive murine megakaryocytic progenitor cells." *Exp Hematol* **22**(10): 954–958.

Jagerschmidt, A., V. Fleury, et al. (1998). "Human thrombopoietin structure–function relationships: identification of functionally important residues." *Biochem J* **333** (Part 3): 729–734.

Jelinek, J., Y. Oki, et al. (2005). "JAK2 mutation 1849G > T is rare in acute leukemias but can be found in CMML, Philadelphia chromosome-negative CML, and megakaryocytic leukemia." *Blood* **106**(10): 3370–3373.

Jones, D. L., A. J. Wagers (2008). "No place like home: anatomy and function of the stem cell niche." *Nat Rev Mol Cell Biol* **9**(1): 11–21.

Junt, T., H. Schulze, et al. (2007). "Dynamic visualization of thrombopoiesis within bone marrow." *Science* **317**(5845): 1767–1770.

Kacena, M. A., M. C. Horowitz (2006). "The role of megakaryocytes in skeletal homeostasis and rheumatoid arthritis." *Curr Opin Rheumatol* **18**(4): 405–410.

Kacena, M. A., C. M. Gundberg, et al. (2005). "Loss of the transcription factor p45 NF-E2 results in a developmental arrest of megakaryocyte differentiation and the onset of a high bone mass phenotype." *Bone* **36**(2): 215–223.

Kacena, M. A., C. M. Gundberg, et al. (2006a). "A reciprocal regulatory interaction between megakaryocytes, bone cells, and hematopoietic stem cells." *Bone* **39**(5): 978–984.

Kacena, M. A., T. Nelson, et al. (2006b). "Megakaryocyte-mediated inhibition of osteoclast development." *Bone* **39**(5): 991–999.

Kartsogiannis, V., H. Zhou, et al. (1999). "Localization of RANKL (receptor activator of NF kappa B ligand) mRNA and protein in skeletal and extraskeletal tissues." *Bone* **25**(5): 525–534.

Kato, T., A. Matsumoto, et al. (1998). "Native thrombopoietin: structure and function." *Stem Cells* **16**(5): 322–328.

Kaushansky, K. (1995). "Thrombopoietin: basic biology, clinical promise." *Int J Hematol* **62**(1): 7–15.

Kaushansky, K. (2005). "The molecular mechanisms that control thrombopoiesis." *J Clin Invest* **115**(12): 3339–3347.

Kaushansky, K., J. G. Drachman (2002). "The molecular and cellular biology of thrombopoietin: the primary regulator of platelet production." *Oncogene* **21**(21): 3359–3367.

Kaushansky, K., V. C. Broudy, et al. (1995). "Thrombopoietin, the Mp1 ligand, is essential for full megakaryocyte development." *Proc Natl Acad Sci USA* **92**(8): 3234–3238.

Kerrigan, S. W., M. Gaur, et al. (2004). "Caspase-12: a developmental link between G-protein-coupled receptors and integrin alphaIIbbeta3 activation." *Blood* **104**(5): 1327–1334.

Kiel, M. J., O. H. Yilmaz, et al. (2005). "SLAM family receptors distinguish hematopoietic stem and progenitor cells and reveal endothelial niches for stem cells." *Cell* **121**(7): 1109–1121.

Kirito, K., K. Kaushansky (2006). "Transcriptional regulation of megakaryopoiesis: thrombopoietin signaling and nuclear factors." *Curr Opin Hematol* **13**(3): 151–156.

Kirito, K., M. Osawa, et al. (2002). "A functional role of Stat3 in in vivo megakaryopoiesis." *Blood* **99**(9): 3220–3227.

Kirito, K., N. Fox, et al. (2004). "Thrombopoietin induces HOXA9 nuclear transport in immature hematopoietic cells: potential mechanism by which the hormone favorably affects hematopoietic stem cells." *Mol Cell Biol* **24**(15): 6751–6762.

Klopocki, E., H. Schulze, et al. (2007). "Complex inheritance pattern resembling autosomal recessive inheritance involving a microdeletion in thrombocytopenia-absent radius syndrome." *Am J Hum Genet* **80**(2): 232–240.

Kosaki, G. (2005). "In vivo platelet production from mature megakaryocytes: does platelet release occur via proplatelets?" *Int J Hematol* **81**(3): 208–219.

Kosugi, S., Y. Kurata, et al. (1996). "Circulating thrombopoietin level in chronic immune thrombocytopenic purpura." *Br J Haematol* **93**(3): 704–706.

Kotwaliwale, C., S. Biggins (2006). "Microtubule capture: a concerted effort." *Cell* **127**(6): 1105–1108.

Kretzschmar, H. A. (1993). "Human prion diseases (spongiform encephalopathies)." *Arch Virol Suppl* **7**: 261–293.

Kuter, D. J. (2007). "New thrombopoietic growth factors." *Blood* **109**(11): 4607–4616.

Kuter, D. J., C. G. Begley (2002). "Recombinant human thrombopoietin: basic biology and evaluation of clinical studies." *Blood* **100**(10): 3457–3469.

Kuter, D. J., R. D. Rosenberg (1995). "The reciprocal relationship of thrombopoietin (c-Mpl ligand) to changes in the platelet mass during busulfan-induced thrombocytopenia in the rabbit." *Blood* **85**(10): 2720–2730.

Kuter, D. J., D. M. Gminski, et al. (1992). "Transforming growth factor beta inhibits megakaryocyte growth and endomitosis." *Blood* **79**(3): 619–626.

Kuter, D. J., D. L. Beeler, et al. (1994). "The purification of megapoietin: a physiological regulator of megakaryocyte growth and platelet production." *Proc Natl Acad Sci USA* **91**(23): 11104–11108.

Lam, L. T., C. Ronchini, et al. (2000). "Suppression of erythroid but not megakaryocytic differentiation of human K562 erythroleukemic cells by notch-1." *J Biol Chem* **275**(26): 19676–19684.

Lambert, M. P., L. Rauova, et al. (2007). "Platelet factor 4 is a negative autocrine in vivo regulator of megakaryopoiesis: clinical and therapeutic implications." *Blood* **110**(4): 1153–1160.

Lecine, P., V. Blank, et al. (1998). "Characterization of the hematopoietic transcription factor NF-E2 in primary murine megakaryocytes." *J Biol Chem* **273**(13): 7572–7578.

Lecine, P., J. E. Italiano, Jr., et al. (2000). "Hematopoietic-specific beta 1 tubulin participates in a pathway of platelet biogenesis dependent on the transcription factor NF-E2." *Blood* **96**(4): 1366–1373.

Lens, S. M. A., G. Vader, et al. (2006). "The case for survivin as mitotic regulator." *Curr Opin Cell Biol* **18**(6): 616–622.

Levin, J., S. Ebbe (1994). "Why are recently published platelet counts in normal mice so low?" *Blood* **83**(12): 3829–3831.

Levine, R. F., K. C. Hazzard, et al. (1982). "The significance of megakaryocyte size." *Blood* **60**(5): 1122–1131.

Li, R. (2007). "Cytokinesis in development and disease: variations on a common theme." *Cell Mol Life Sci* **64**(23): 3044–3058.

Li, Z., L. Li (2006). "Understanding hematopoietic stem-cell microenvironments." *Trends Biochem Sci* **31**(10): 589–595.

Li, Z., F. J. Godinho, et al. (2005). "Developmental stage-selective effect of somatically mutated leukemogenic transcription factor GATA1." *Nat Genet* **37**(6): 613–619.

Linden, H. M., K. Kaushansky (2000). "The glycan domain of thrombopoietin enhances its secretion." *Biochemistry* **39**(11): 3044–3051.

Lippert, E., M. Boissinot, et al. (2006). "The JAK2-V617F mutation is frequently present at diagnosis in patients with essential thrombocythemia and polycythemia vera." *Blood* **108**(6): 1865–1867.

Ma, Z., S. W. Morris, et al. (2001). "Fusion of two novel genes, RBM15 and MKL1, in the t(1;22) (p13;q13) of acute megakaryoblastic leukemia." *Nat Genet* **28**(3): 220–221.

MacGregor, I., J. Hope, et al. (1999). "Application of a time-resolved fluoroimmunoassay for the analysis of normal prion protein in human blood and its components." *Vox Sang* **77**(2): 88–96.

Malumbres, M., M. Barbacid (2001). "To cycle or not to cycle: a critical decision in cancer." *Nat Rev Cancer* **1**(3): 222–231.

Mandal, R. V., E. J. Mark, et al. (2007). "Megakaryocytes and platelet homeostasis in diffuse alveolar damage." *Exp Mol Pathol* **83**: 327–331.

Margolis, R. L., O. D. Lohez, et al. (2003). "G1 tetraploidy checkpoint and the suppression of tumorigenesis." *J Cell Biochem* **88**(4): 673–683.

Mattia, G., F. Vulcano, et al. (2002). "Different ploidy levels of megakaryocytes generated from peripheral or cord blood CD34+ cells are correlated with different levels of platelet release." *Blood* **99**(3): 888–897.

McCrann, D. J., H. G. Nguyen, et al. (2008a). "Vascular smooth muscle cell polyploidy: an adaptive or maladaptive response?" *J Cell Physiol* **215**: 588–592

McCrann, D. J., T. Yezefski, et al. (2008b). "Survivin overexpression alone does not alter megakaryocyte ploidy nor interfere with erythroid/megakaryocytic lineage development in transgenic mice." *Blood* **111**: 4092–4095

Mercher, T., M. B. Coniat, et al. (2001). "Involvement of a human gene related to the Drosophila *spen* gene in the recurrent t(1;22) translocation of acute megakaryocytic leukemia." *Proc Natl Acad Sci USA* **98**(10): 5776–5779.

Miller, J. L., A. Castella (1982). "Platelet-type von Willebrand's disease: characterization of a new bleeding disorder." *Blood* **60**(3): 790–794.

Miller, J. S., Y. Soignier, et al. (2005). "Successful adoptive transfer and in vivo expansion of human haploidentical NK cells in patients with cancer." *Blood* **105**(8): 3051–3057.

Miyakawa, Y., A. Oda, et al. (1995). "Recombinant thrombopoietin induces rapid protein tyrosine phosphorylation of Janus kinase 2 and Shc in human blood platelets." *Blood* **86**(1): 23–27.

Moliterno, A. R., D. M. Williams, et al. (2004). "Mpl Baltimore: a thrombopoietin receptor polymorphism associated with thrombocytosis." *Proc Natl Acad Sci USA* **101**(31): 11444–11447.

Mori, M., J. Tsuchiyama, et al. (1993). "Proliferation, migration and platelet release by megakaryocytes in long-term bone marrow culture in collagen gel." *Cell Struct Funct* **18**(6): 409–417.

Moroy, T., C. Geisen (2004). "Cyclin E." *Int J Biochem Cell Biol* **36**(8): 1424–1439.

Muntean, A. G., L. Pang, et al. (2007). "Cyclin D-Cdk4 is regulated by GATA-1 and required for megakaryocyte growth and polyploidization." *Blood* **109**: 5199–5207.

Murata-Hori, M., M. Tatsuka, et al. (2002). "Probing the dynamics and functions of Aurora B kinase in living cells during mitosis and cytokinesis." *Mol. Biol. Cell* **13**(4): 1099–1108.

Musacchio, A., E. D. Salmon (2007). "The spindle-assembly checkpoint in space and time." *Nat Rev Mol Cell Biol* **8**(5): 379–393.

Muta, T., S. Iwanaga (1996). "The role of hemolymph coagulation in innate immunity." *Curr Opin Immunol* **8**(1): 41–47.

Nagata, Y., Y. Muro, et al. (1997). "Thrombopoietin-induced polyploidization of bone marrow megakaryocytes is due to a unique regulatory mechanism in late mitosis." *J. Cell Biol.* **139**(2): 449–457.

Nagler, A., V. R. Deutsch, et al. (1995). "Recombinant human interleukin-6 accelerates in-vitro megakaryocytopoiesis and platelet recovery post autologous peripheral blood stem cell transplantation." *Leuk Lymphoma* **19**(3–4): 343–349.

Nakao, T., A. E. Geddis, et al. (2007). "PI3K/Akt/FOXO3a pathway contributes to thrombopoie-tin-induced proliferation of primary megakaryocytes in vitro and in vivo via modulation of p27(Kip1)." *Cell Cycle* **7**(2): 257–266.

Nakorn, T. N., T. Miyamoto, et al. (2003). "Characterization of mouse clonogenic megakaryocyte progenitors." *Proc Natl Acad Sci* **100**(1): 205–210.

Newland, A. (2007). "Thrombopoietin mimetic agents in the management of immune thrombocy-topenic purpura." *Semin Hematol* **44**(4 Suppl 5): S35–S45.

Ney, P. A., N. C. Andrews, et al. (1993). "Purification of the human NF-E2 complex: cDNA cloning of the hematopoietic cell-specific subunit and evidence for an associated partner." *Mol Cell Biol* **13**(9): 5604–5612.

Nguyen, H. G., K. Ravid (2006). "Tetraploidy/aneuploidy and stem cells in cancer promotion: the role of chromosome passenger proteins." *J Cell Physiol* **208**(1): 12–22.

Nguyen, H. G., D. Chinnappan, et al. (2005). "Mechanism of Aurora-B degradation and its dependency on intact KEN and A-boxes: identification of an aneuploidy-promoting property." *Mol Cell Biol* **25**(12): 4977–4992.

Nurden, A. T., P. Nurden (2007a). "The gray platelet syndrome: clinical spectrum of the disease." *Blood Rev* **21**(1): 21–36.

Nurden, A. T., P. Nurden (2007b). "Inherited thrombocytopenias." *Haematologica* **92**(9): 1158–1164.

Oda, A., Y. Miyakawa, et al. (1996). "Thrombopoietin primes human platelet aggregation induced by shear stress and by multiple agonists." *Blood* **87**(11): 4664–4670.

Onodera, K., J. A. Shavit, et al. (2000). "Perinatal synthetic lethality and hematopoietic defects in compound mafG::mafK mutant mice." *EMBO J* **19**(6): 1335–1345.

Orazi, A. (2007). "Histopathology in the diagnosis and classification of acute myeloid leukemia, myelodysplastic syndromes, and myelodysplastic/myeloproliferative diseases." *Pathobiology* **74**(2): 97–114.

Orkin, S. H. (1992). "GATA-binding transcription factors in hematopoietic cells." *Blood* **80**(3): 575–581.

Pasquet, J. M., B. S. Gross, et al. (2000). "Thrombopoietin potentiates collagen receptor signaling in platelets through a phosphatidylinositol 3-kinase-dependent pathway." *Blood* **95**(11): 3429–3434.

Patel, S. R., J. H. Hartwig, et al. (2005a). "The biogenesis of platelets from megakaryocyte pro-platelets." *J. Clin. Invest.* **115**(12): 3348–3354.

Patel, S. R., J. L. Richardson, et al. (2005b). "*Differential roles of microtubule assembly and slid-ing in proplatelet formation by megakaryocytes.*" **106**: 4076–4085.

Peck-Radosavljevic, M., M. Wichlas, et al. (2000). "Thrombopoietin induces rapid resolution of thrombocytopenia after orthotopic liver transplantation through increased platelet production." *Blood* **95**(3): 795–801.

Perry, M. J., K. A. Redding, et al. (2007). "Mice rendered severely deficient in megakaryocytes through targeted gene deletion of the thrombopoietin receptor c-Mpl have a normal skeletal phenotype." *Calcif Tissue Int* **81**(3): 224–231.

Poellinger, L., R. S. Johnson (2004). "HIF-1 and hypoxic response: the plot thickens." *Curr Opin Genet Dev* **14**(1): 81–85.

Quesenberry, P. J., J. N. Ihle, et al. (1985). "The effect of interleukin 3 and GM-CSA-2 on megakaryocyte and myeloid clonal colony formation." *Blood* **65**(1): 214–217.

Raslova, H., E. Komura, et al. (2004). "FLI1 monoallelic expression combined with its hemizygous loss underlies Paris-Trousseau/Jacobsen thrombopenia." *J Clin Invest* **114**(1): 77–84.

Raslova, H., V. Baccini, et al. (2006). "Mammalian target of rapamycin (mTOR) regulates both proliferation of megakaryocyte progenitors and late stages of megakaryocyte differentiation." *Blood* **107**(6): 2303–2310.

Raslova, H., A. Kauffmann, et al. (2007). "Interrelation between polyploidization and megakaryocyte differentiation: a gene profiling approach." *Blood* **109**(8): 3225–3234.

Ravid, K., T. Doi, et al. (1991). "Transcriptional regulation of the rat platelet factor 4 gene: interaction between an enhancer/silencer domain and the GATA site." *Mol Cell Biol* **11**(12): 6116–6127.

Ravid, K., J. Lu, et al. (2002). "Roads to polyploidy: the megakaryocyte example." *J Cell Physiol* **190**(1): 7–20.

Ribeiro, R. C., M. S. Oliveira, et al. (1993). "Acute megakaryoblastic leukemia in children and adolescents: a retrospective analysis of 24 cases." *Leuk Lymphoma* **10**(4–5): 299–306.

Rojnuckarin, P., K. Kaushansky (2001). "Actin reorganization and proplatelet formation in murine megakaryocytes: the role of protein kinase calpha." *Blood* **97**(1): 154–161.

Rouyez, M. C., C. Boucheron, et al. (1997). "Control of thrombopoietin-induced megakaryocytic differentiation by the mitogen-activated protein kinase pathway." *Mol Cell Biol* **17**(9): 4991–5000.

Roy, L., P. Coullin, et al. (2001). "Asymmetrical segregation of chromosomes with a normal metaphase/anaphase checkpoint in polyploid megakaryocytes." *Blood* **97**(8): 2238–2247.

Royer, Y., J. Staerk, et al. (2005). "Janus kinases affect thrombopoietin receptor cell surface localization and stability." *J Biol Chem* **280**(29): 27251–27261.

Rubin, C. I., D. L. French, et al. (2003). "Stathmin expression and megakaryocyte differentiation: a potential role in polyploidy." *Exp Hematol* **31**(5): 389–397.

Ruchaud, S., M. Carmena, et al. (2007). "Chromosomal passengers: conducting cell division." *Nat Rev Mol Cell Biol* **8**(10): 798–812.

Sakamaki, S., Y. Hirayama, et al. (1999). "Transforming growth factor-beta1 (TGF-beta1) induces thrombopoietin from bone marrow stromal cells, which stimulates the expression of TGF-beta receptor on megakaryocytes and, in turn, renders them susceptible to suppression by TGF-beta itself with high specificity." *Blood* **94**(6): 1961–1970.

Samii, K., E. Pasteur (1998). "Images in hematology. Emperipolesis." *Am J Hematol* **59**(1): 64.

Sauer, K., J. A. Knoblich, et al. (1995). "Distinct modes of cyclin E/cdc2c kinase regulation and S-phase control in mitotic and endoreduplication cycles of Drosophila embryogenesis." *Genes Dev* **9**(11): 1327–1339.

Schipper, L. F., A. Brand, et al. (2003). "Differential maturation of megakaryocyte progenitor cells from cord blood and mobilized peripheral blood." *Exp Hematol* **31**(4): 324–330.

Schmitt, A., J. Guichard, et al. (2001). "Of mice and men: comparison of the ultrastructure of megakaryocytes and platelets." *Exp Hematol* **29**(11): 1295–1302.

Schulze, H., M. Korpal, et al. (2006). "Characterization of the megakaryocyte demarcation membrane system and its role in thrombopoiesis." *Blood* **107**(10): 3868–3875.

Scott, T., M. D. Owens (2008). "Thrombocytes respond to lipopolysaccharide through Toll-like receptor-4, and MAP kinase and NF-κB pathways leading to expression of interleukin-6 and cyclooxygenase-2 with production of prostaglandin E2." *Mol Immunol* **45**(4): 1001–1008.

Seri, M., R. Cusano, et al. (2000). "Mutations in MYH9 result in the May-Hegglin anomaly, and Fechtner and Sebastian syndromes. The May-Heggllin/Fechtner Syndrome Consortium." *Nat Genet* **26**(1): 103–105.

Shivdasani, R. A., M. F. Rosenblatt, et al. (1995). "Transcription factor NF-E2 is required for platelet formation independent of the actions of thrombopoietin/MGDF in megakaryocyte development." *Cell* **81**(5): 695–704.

Shivdasani, R. A., P. Fielder, et al. (1997a). "Regulation of the serum concentration of thrombopoietin in thrombocytopenic NF-E2 knockout mice." *Blood* **90**(5): 1821–1827.

Shivdasani, R. A., Y. Fujiwara, et al. (1997b). "A lineage-selective knockout establishes the critical role of transcription factor GATA-1 in megakaryocyte growth and platelet development." *EMBO J* **16**(13): 3965–3973.

Sipe, J. B., J. Zhang, et al. (2004). "Localization of bone morphogenetic proteins (BMPs)-2, -4, and -6 within megakaryocytes and platelets." *Bone* **35**(6): 1316–1322.

Snow, J. W., N. Abraham, et al. (2002). "STAT5 promotes multilineage hematolymphoid development in vivo through effects on early hematopoietic progenitor cells." *Blood* **99**(1): 95–101.

Solar, G. P., W. G. Kerr, et al. (1998). "Role of c-mpl in early hematopoiesis." *Blood* **92**(1): 4–10.

Song, W. J., M. G. Sullivan, et al. (1999). "Haploinsufficiency of CBFA2 causes familial thrombocytopenia with propensity to develop acute myelogenous leukaemia." *Nat Genet* **23**(2): 166–175.

Sonoda, Y., Y. Kuzuyama, et al. (1993). "Human interleukin-4 inhibits proliferation of megakaryocyte progenitor cells in culture." *Blood* **81**(3): 624–630.

Staerk, J., C. Lacout, et al. (2006). "An amphipathic motif at the transmembrane-cytoplasmic junction prevents autonomous activation of the thrombopoietin receptor." *Blood* **107**(5): 1864–1871.

Starke, R., P. Harrison, et al. (2005). "The expression of prion protein (PrP(C)) in the megakaryocyte lineage." *J Thromb Haemost* **3**(6): 1266–1273.

Steensma, D. P., A. Tefferi (2002). "Cytogenetic and molecular genetic aspects of essential thrombocythemia." *Acta Haematol* **108**(2): 55–65.

Straight, A. F., C. M. Field (2000). "Microtubules, membranes and cytokinesis." *Curr Biol* **10**(20): R760–R770.

Sumara, I., J. F. Gimenez-Abian, et al. (2004). "Roles of polo-like kinase 1 in the assembly of functional mitotic spindles." *Curr Biol* **14**(19): 1712–1722.

Sungaran, R., B. Markovic, et al. (1997). "Localization and regulation of thrombopoietin mRNa expression in human kidney, liver, bone marrow, and spleen using in situ hybridization." *Blood* **89**(1): 101–107.

Sungaran, R., O. T. Chisholm, et al. (2000). "The role of platelet alpha-granular proteins in the regulation of thrombopoietin messenger RNA expression in human bone marrow stromal cells." *Blood* **95**(10): 3094–3101.

Suva, L. J., E. Hartman, et al. (2008). "Platelet dysfunction and a high bone mass phenotype in a murine model of platelet-type von Willebrand disease." *Am J Pathol* **172**: 430–439.

Tablin, F., M. Castro, et al. (1990). "Blood platelet formation in vitro. The role of the cytoskeleton in megakaryocyte fragmentation." *J Cell Sci* **97** (Part 1): 59–70.

Tajika, K., H. Nakamura, et al. (2000). "Thrombopoietin can influence mature megakaryocytes to undergo further nuclear and cytoplasmic maturation." *Exp Hematol* **28**(2): 203–209.

Takahashi, S., T. Komeno, et al. (1998). "Role of GATA-1 in proliferation and differentiation of definitive erythroid and megakaryocytic cells in vivo." *Blood* **92**(2): 434–442.

Tallman, M. S., D. Neuberg, et al. (2000). "Acute megakaryocytic leukemia: the Eastern Cooperative Oncology Group experience." *Blood* **96**(7): 2405–2411.

Tanaka, M., J. Zheng, et al. (2004). "Differentiation status dependent function of FOG-1." *Genes Cells* **9**(12): 1213–1226.

Tavassoli, M., M. Aoki (1981). "Migration of entire megakaryocytes through the marrow–blood barrier." *Br J Haematol* **48**(1): 25–29.

Tavassoli, M., M. Aoki (1989). "Localization of megakaryocytes in the bone marrow." *Blood Cells* **15**(1): 3–14.

Tepler, I., L. Elias, et al. (1996). "A randomized placebo-controlled trial of recombinant human interleukin-11 in cancer patients with severe thrombocytopenia due to chemotherapy." *Blood* **87**(9): 3607–3614.

Tiwari, S., J. E. Italiano, Jr., et al. (2003). "A role for Rab27b in NF-E2-dependent pathways of platelet formation." *Blood* **102**(12): 3970–3979.

Tober, J., A. Koniski, et al. (2007). "The megakaryocyte lineage originates from hemangioblast precursors and is an integral component both of primitive and of definitive hematopoiesis." *Blood* **109**(4): 1433–1441.

Tober, J., K. E. McGrath, et al. (2008). "Primitive erythropoiesis and megakaryopoiesis in the yolk sac are independent of c-myb." *Blood* **111**(5): 2636–2639.

Tomer, A. (2004). "Human marrow megakaryocyte differentiation: multiparameter correlative analysis identifies von Willebrand factor as a sensitive and distinctive marker for early (2N and 4N) megakaryocytes." *Blood* **104**(9): 2722–2727.

Tsai, F. Y., G. Keller, et al. (1994). "An early haematopoietic defect in mice lacking the transcription factor GATA-2." *Nature* **371**(6494): 221–226.

Tsang, A. P., J. E. Visvader, et al. (1997). "FOG, a multitype zinc finger protein, acts as a cofactor for transcription factor GATA-1 in erythroid and megakaryocytic differentiation." *Cell* **90**(1): 109–119.

Tsang, A. P., Y. Fujiwara, et al. (1998). "Failure of megakaryopoiesis and arrested erythropoiesis in mice lacking the GATA-1 transcriptional cofactor FOG." *Genes Dev* **12**(8): 1176–1188.

Tulasne, D., T. Bori, et al. (2002). "Regulation of RAS in human platelets. Evidence that activation of RAS is not sufficient to lead to ERK1-2 phosphorylation." *Eur J Biochem* **269**(5): 1511–1517.

van Hensbergen, Y., L. F. Schipper, et al. (2006). "Ex vivo culture of human CD34+ cord blood cells with thrombopoietin (TPO) accelerates platelet engraftment in a NOD/SCID mouse model." *Exp Hematol* **34**(7): 943–950.

van Vugt, M. A., B. C. van de Weerdt, et al. (2004). "Polo-like kinase-1 is required for bipolar spindle formation but is dispensable for anaphase promoting complex/Cdc20 activation and initiation of cytokinesis." *J Biol Chem* **279**(35): 36841–36854.

Vannucchi, A. M., L. Bianchi, et al. (2005). "A pathobiologic pathway linking thrombopoietin, GATA-1, and TGF-beta1 in the development of myelofibrosis." *Blood* **105**(9): 3493–3501.

Vigon, I., J. P. Mornon, et al. (1992). "Molecular cloning and characterization of MPL, the human homolog of the v-mpl oncogene: identification of a member of the hematopoietic growth factor receptor superfamily." *Proc Natl Acad Sci USA* **89**(12): 5640–5644.

Villeval, J. L., K. Cohen-Solal, et al. (1997). "High thrombopoietin production by hematopoietic cells induces a fatal myeloproliferative syndrome in mice." *Blood* **90**(11): 4369–4383.

von Hundelshausen, P., F. Petersen, et al. (2007). "Platelet-derived chemokines in vascular biology." *Thromb Haemost* **97**(5): 704–713.

Walters, D. K., T. Mercher, et al. (2006). "Activating alleles of JAK3 in acute megakaryoblastic leukemia." *Cancer Cell* **10**(1): 65–75.

Wang, B., J. L. Nichol, et al. (2004). "Pharmacodynamics and pharmacokinetics of AMG 531, a novel thrombopoietin receptor ligand." *Clin Pharmacol Ther* **76**(6): 628–638.

Wang, S., Q. Wang, et al. (1993). "Cloning and characterization of subunits of the T-cell receptor and murine leukemia virus enhancer core-binding factor." *Mol Cell Biol* **13**(6): 3324–3339.

Wang, Z., Y. Zhang, et al. (1995). "Cyclin D3 is essential for megakaryocytopoiesis." *Blood* **86**(10): 3783–3788.

Watson, D. K., F. E. Smyth, et al. (1992). "The *ERGB/Fli-1* gene: isolation and characterization of a new member of the family of human ETS transcription factors." *Cell Growth Differ* **3**(10): 705–713.

Wechsler, J., M. Greene, et al. (2002). "Acquired mutations in GATA1 in the megakaryoblastic leukemia of Down syndrome." *Nat Genet* **32**(1): 148–152.

Weich, N. S., A. Wang, et al. (1997). "Recombinant human interleukin-11 directly promotes megakaryocytopoiesis in vitro." *Blood* **90**(10): 3893–3902.

Weinstein, R., M. B. Stemerman, et al. (1981). "The morphological and biochemical characterization of a line of rat promegakaryoblasts." *Blood* **58**(1): 110–121.

Wenger, S. L., P. D. Grossfeld, et al. (2006). "Molecular characterization of an 11q interstitial deletion in a patient with the clinical features of Jacobsen syndrome." *Am J Med Genet A* **140**(7): 704–708.

Wu, Y., T. Welte, et al. (2007). "PECAM-1: a multifaceted regulator of megakaryocytopoiesis." *Blood* **110**(3): 851–859.

Yagi, M., G. J. Roth (2006). "Megakaryocyte polyploidization is associated with decreased expression of polo-like kinase (PLK)." *J Thromb Haemost* **4**(9): 2028–2034.

Yan, X. Q., D. Lacey, et al. (1995). "Chronic exposure to retroviral vector encoded MGDF (mpl-ligand) induces lineage-specific growth and differentiation of megakaryocytes in mice." *Blood* **86**(11): 4025–4033.

Yang, D., D. J. McCrann, et al. (2007). "Increased polyploidy in aortic vascular smooth muscle cells during aging is marked by cellular senescence." *Aging Cell* **6**(2): 257–260.

Zakynthinos, S. G., S. Papanikolaou, et al. (2004). "Sepsis severity is the major determinant of circulating thrombopoietin levels in septic patients." *Crit Care Med* **32**(4): 1004–1010.

Zhang, J., F. Varas, et al. (2007). "CD41-YFP mice allow in vivo labeling of megakaryocytic cells and reveal a subset of platelets hyperreactive to thrombin stimulation." *Exp Hematol* **35**(3): 490–499.

Zhang, Y., Z. Wang, et al. (1996). "The cell cycle in polyploid megakaryocytes is associated with reduced activity of cyclin B1-dependent cdc2 kinase." *J Biol Chem* **271**(8): 4266–4272.

Zhang, Y., Y. Nagata, et al. (2004). "Aberrant quantity and localization of Aurora-B/AIM-1 and survivin during megakaryocyte polyploidization and the consequences of Aurora-B/AIM-1-deregulated expression." *Blood* **103**(10): 3717–3726.

Zheng, C., R. Yang, et al. (2008). "TPO-independent megakaryocytopoiesis." *Crit Rev Oncol Hematol* **65**(3): 212–222.

Zimmet, J., K. Ravid (2000). "Polyploidy: occurrence in nature, mechanisms, and significance for the megakaryocyte-platelet system." *Exp Hematol* **28**: 3–16.

Zimmet, J. M., D. Ladd, C. W. Jackson, P. E. Stenberg, K. Ravid. (1997). "A role for cyclin D3 in the endomitotic cell cycle." *Mol Cell Biol* **17**(12): 7248–7259.

Zon, L. I., Y. Yamaguchi, et al. (1993). "Expression of mRNA for the GATA-binding proteins in human eosinophils and basophils: potential role in gene transcription." *Blood* **81**(12): 3234–3241.

Chapter 6
Development of Macrophages and Granulocytes

Richard Dahl

Abstract Myeloid cells (monocytes/macrophages and granulocytes) are derived from a common progenitor, the granulocyte–monocyte progenitor (GMP). Combinatorial interactions between hematopoietic transcription factors program the differentiation of the GMP into monocytes and the four granulocytic lineages: neutrophils, eosinophils, basophils, and mast cells. Cytokines and growth factors mainly function to promote proliferation and survival of the myeloid lineages. However, there is evidence demonstrating that cytokines can affect transcription factor expression and influence lineage commitment. Gene targeting experiments in mice have demonstrated that several transcription factors are required to specify distinct myeloid lineages. An essential role for cytokines in steady state hematopoiesis has not been observed; however, several cytokines are required to mobilize proper immune responses to infectious pathogens. A newly recognized class of small RNAs termed micro-RNAs (miRNAs) may be critical in lineage specification. These small RNA molecules regulate the translation of specific mRNAs. Recent research is beginning to support the role of miRNAs in lineage commitment by regulating the protein accumulation of specific hematopoietic transcription factors.

Introduction

This chapter describes the molecular events that direct the development of monocytes/macrophages and granulocytes during hematopoiesis. These cells are commonly referred to as myeloid cells and are responsible for innate immunity and the inflammatory response. Innate immunity is the first line of defense against a variety

R. Dahl
University of New Mexico Health Sciences Center,
915 Camino de Salud, CRF 125a, MSC10 5550,
Albuquerque, NM 87131, USA
e-mail: Rdahl@salud.unm.edu

A. Wickrema and B. Kee (eds.), *Molecular Basis of Hematopoiesis*,
DOI: 10.1007/978-0-387-85816-6_1, © Springer Science+Business Media, LLC 2009

of pathogens. Myeloid cells are phagocytic ingesting foreign materials and organisms. They will also act as antigen presenting cells for lymphocytes and secrete cytokines involved in mediating an inflammatory response.

Macrophages are large cells, which reside in peripheral tissues (Faller and Mentzer 2000). They are derived from bone marrow-produced monocytes, which enter the blood stream and develop into macrophages once localized to specific tissues. The three main functions of the macrophage are tissue maintenance, immune regulation, and fighting pathogens. Macrophages ingest cellular debris and engulf apoptotic cells, which may result from tissue injury or remodeling. They are involved in immune regulation through antigen presentation, coordinating lymphocyte function through cytokine release, and the recognition and engulfment of cellular pathogens.

Granulocytes are divided into three classes of cells: neutrophils, eosinophils, and basophils. They have been termed granulocytes due to the presence of granules in their cytoplasm that contain bactericidal and/or fungicidal proteins. These cells have also been termed polymorphonuclear phagocytes due to their characteristic segmented nuclei. Neutrophils are the predominant granulocytes in the blood. They function in killing bacteria and fungi as well as acute inflammatory responses (Baehner 2000). Eosinophils are present in low numbers in the blood. They tend to localize specifically in tissues that are exposed to the external environment, which include skin, gastrointestinal tract, urogenital tract, and lungs (Gleich et al. 1989). They are implicated as the predominant proinflammatory cell in bronchial asthma. Basophils are the least common granulocytes in peripheral blood. Basophils mediate allergic reactions such as asthma and anaphylaxis (Arock et al. 2002). They express high-affinity receptors for IgE (FcεRI) on their cell surface. Lastly, mast cells are a special type of granulocytes (Austen and Boyce 2001). They contain cytoplasmic granules and carry out similar functions to basophils. Unlike basophils they do not have segmented nuclei and do not complete maturation until they localize to tissue.

As discussed in Chap. 2 granulocytes and monocytes arise from a common hematopoietic progenitor, the granulocyte–monocyte progenitor (GMP). In this chapter molecular regulation of GMP generation and the programming of GMP to become the specific mature myeloid lineages will be discussed. Three important classes of molecules will be discussed: transcription factors, cytokines, and miR-NAs. Gene targeting in mice has demonstrated that transcription factors are critical for lineage specification. Recent work demonstrating the importance of combinatorial interactions between transcription factors will be highlighted. Cytokines (and growth factors) have long been known to affect hematopoietic differentiation and many are now used clinically. Their roles in instructing myelopoiesis versus promoting proliferation and survival will be highlighted. Lastly, miRNAs are a new class of small RNA molecules that are implicated in controlling developmental processes. What is currently known about their function in hematopoiesis and myeloid development will be reviewed.

Transcription Factors Involved in Myeloid Development

In this section the mechanism by which transcription factors direct the development of the GMP from the hematopoietic stem cell and how different combinations and concentrations of these factors specify specific lineage acquisition will be addressed. As mentioned previously, each of the innate immune cells carries out specific functions, which are mediated by the expression of distinct protein repertoires. Researchers identified transcription factors that were important for the regulation of these myeloid genes. Several of these transcription factors were later shown to be critical for the specification of distinct myeloid lineages. Fig. 6.1 summarizes the transcription factors that are required for commitment to distinct myeloid lineages.

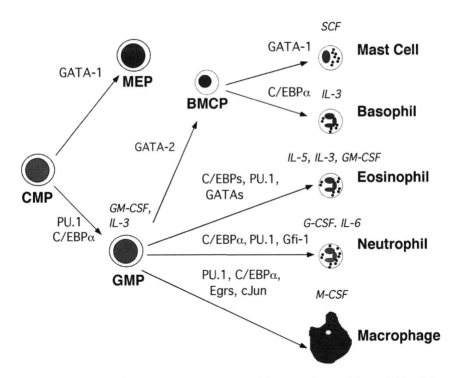

Fig. 6.1 Transcription factors and cytokines involved in promoting specific myeloid cell fate acquisition. Hematopoietic progenitors and mature myeloid cells are labeled in bold. Cytokines are listed in italics and are placed above the cell types they promote. Transcription factors are above the arrows indicating their importance in specifying an indicated cell type from a hematopoietic progenitor. CMP common myeloid progenitor, MEP megakaryocyte–erythroid progenitor, GMP granulocyte–monocyte progenitor, BMCP basophil mast cell progenitor

Specifying the GMP

Two transcription factors have been clearly shown to be essential for the generation of the GMP:PU.1 and C/EBPα (Dakic et al. 2005; Iwasaki et al. 2005; Zhang et al. 2004). PU.1 belongs to the large Ets family of transcription factors. It is expressed at low levels in early hematopoietic progenitors, and as cells mature, PU.1 expression is extinguished in erythrocytes, megakaryocytes, and T cells, but rises in monocytes, granulocytes, and B cells (Hromas et al. 1993). Gene targeting in mice demonstrated a critical role for PU.1 in hematopoietic differentiation. Mice lacking PU.1 do not make macrophages, granulocytes, or lymphoid cells (McKercher et al. 1996; Scott et al. 1994). Examination of bone marrow cells in PU.1 conditional knockout mice showed that $PU.1^{-/-}$ HSCs do not make detectable common lymphoid progenitors (CLPs), common myeloid progenitors (CMPs), or GMPs (Dakic et al. 2005; Iwasaki et al. 2005).

The other factor critical for GMP generation, C/EBPα, is a basic leucine zipper protein (bZIP), which functions either as a homodimer or heterodimer. It dimerizes with itself or with other bZIP proteins through its carboxy terminal leucine zipper. Within the hematopoietic system C/EBPα is predominantly expressed in granulocytes, monocytes, and the progenitors of these cells (Radomska et al. 1998; Scott et al. 1992). Ectopically expressing C/EBPα in cells capable of both monocyte and granulocyte differentiation induced granulocyte differentiation (Radomska et al. 1998). Lack of C/EBPα completely abrogates granulopoiesis and impairs monopoiesis (Heath et al. 2004; Zhang et al. 1997). Interestingly, no GMPs were identified from $CEBPA^{-/-}$ hematopoietic organs. Isolated CMPs were unable to generate granulocyte progenitors or monocyte progenitors in vitro suggesting that the monocytes observed in the germline mutants may be produced through an alternative progenitor pathway. Conditional mutants of CEBPA have demonstrated that once C/EBPα specifies the GMP it is no longer required for further myeloid differentiation. This is in contrast to PU.1, which when deleted in GMPs does not allow for further myeloid development (Iwasaki et al. 2005).

One of the mechanisms by which PU.1 is thought to specify the GMP from the CMP [and/or the lymphoid primed multipotential progenitor (LMPP) from the multipotential progenitors (MPP)] is to antagonize the function of the transcription factor GATA-1. Lack of GATA-1 in mice results in a block of both erythroid and megakaryocyte development (Pevny et al. 1991; Shivdasani et al. 1997). GATA-1 is a zinc finger protein expressed in erythrocytes, megakaryocytes, mast cells, eosinophils, and their progenitors (Zon et al. 1993). PU.1 and GATA-1 bind to each other and antagonize each other's ability to regulate lineage-specific genes (Cantor and Orkin 2001). GATA-1 binds to the Ets DNA binding domain of PU.1 through its carboxy proximal zinc finger. GATA-1 binding to DNA-bound PU.1 displaces the critical coactivator protein, c-Jun, decreasing PU.1's ability to promote transcription. Conversely, PU.1 recruits the retinoblastoma protein, Rb, to GATA1, which in turn recruits a transcriptional repression complex (Stopka et al. 2005). Both PU.1 and GATA-1 are expressed at low levels in the uncommitted CMP (Akashi et al. 2000). In these uncommitted cells the activities of these factors are thought to be in balance. Through stochastic changes in expression or cytokine signaling, the GATA-1:PU.1 ratio favors one of these factors and this factor activates its target genes while continuing to repress the alternative factor's target genes. This results in distinct lineage commitments – erythroid for GATA-1 and granulocyte/monocyte for PU.1 (Rekhtman et al. 1999; Rhodes et al. 2005).

Although antagonism between PU.1 and GATA-1 has become the dominant paradigm for how CMPs are directed toward megakaryocyte–erythroid progenitor (MEP) or GMP development, there is ample evidence that supports a role for C/EBPα. Erythroid development is significantly increased in fetal livers isolated from $CEBPA^{-/-}$ mice (Suh et al. 2006). It has also been shown that expression of C/EBPα in mouse myeloid primary progenitors and multipotential cell lines promotes myeloid development over erythroid development (Suh et al. 2006). Similar results were found when a hormone regulatable fusion protein between C/EBPα and the ligand binding domain of estrogen receptor, C/EBPα -ER, was expressed in primary human CD34+ hematopoietic cells (Cammenga et al. 2003). Activation of the fusion protein by hormone (estrogen) blocked the development of erythroid progenitors as measured by in vitro hematopoietic colony assays.

Specifying Monocyte Versus Granulocytes from the GMP

PU.1 is required for the development of monocytes, granulocytes, and B cells (McKercher et al. 1996; Scott et al. 1994). Besides being required for each of these lineages, distinct concentrations of PU.1 direct distinct lineage acquisition (Dahl et al. 2003; DeKoter and Singh 2000). In myeloid cell fate, high levels of PU.1 directed the GMP toward monocyte rather than granulocyte differentiation as shown by in vitro and in vivo studies (Dahl et al. 2003). It is not clear why the levels of PU.1 are deterministic for myeloid cell fate acquisition. One hypothesis is that the ratio of PU.1 to C/EBPα determines monocyte versus granulocyte cell fate similar to the GATA-1:PU.1 ratio determining erythroid versus myeloid cell fate (Dahl et al. 2003; Laslo et al. 2006; Reddy et al. 2002). In a $PU.1^{-/-}$ cell line in which PU.1 activity was restored by introduction of a 4-hydroxy-tamoxifen (OHT) regulatable PUER fusion protein, it was shown that high levels of PU.1 activity induced by high concentrations of OHT directed monocyte differentiation. Lower levels of OHT directed granulocytic differentiation. Interestingly, if the cells were pretreated with the granulocyte-promoting cytokine G-CSF the cells were primed to differentiate toward granulocytes even at the high dose of OHT. In the G-CSF primed cells it was shown that levels of C/EBPα increased approximately threefold. This suggested the hypothesis that when PU.1 activity exceeds that of C/EBPα, the expression of PU.1-responsive downstream target genes will direct monocytic differentiation. However, if C/EBPα activity is higher, then the GMP will commit to the granulocyte lineage. Consistent with this hypothesis if PU.1 was restored to the parental PU.1-null cell lines by retroviral transduction more monocytes than granulocytes developed from infected cells. However, if the cells were infected with both PU.1 and C/EBPα the percentage of granulocyte development increased compared to cells infected with PU.1 alone. Like GATA-1, C/EBPα binds PU.1 dissociating c-Jun, and blocking PU.1 transactivation (Reddy et al. 2002). A reciprocal inhibition of C/EBPα by PU.1 has not been reported. As discussed in detail in the next section mutual antagonism of transcription factors downstream of PU.1 and C/EBPα may be more important for cell fate specification.

Mutual Antagonism of Transcription Factors Downstream of PU.1 and C/EBPα

A model for how the PU.1:C/EBPα ratio directs cell fate came out of work with the PUER myeloid cell line. Gene array analysis demonstrated that within 24 h of activating PU.1, genes associated with both the monocytic and granulocytic lineages were induced in single cells (Laslo et al. 2006). The transcriptional regulators Egr-2 and Nab-2 were upregulated during the initial 24-h period of PUER activation and remained on during monocyte development. These two factors interact to form a repressive transcription complex. Potentially these two could be mediating the repression of granulocyte genes downstream of PU.1 activity. ShRNA-mediated knockdown of Egr-2 or Nab-2 inhibited the downregulation of granulocyte genes in OHT-treated PUER cells. *Egr2*-deficient mice did not demonstrate a defect in monocytes, however, monopoiesis was defective in $Egr1^{-/-}-Egr2^{+/-}$ mice demonstrating functional redundancy of these proteins in vivo. Previously Egr-1 was shown to direct monocyte development of mouse hematopoietic progenitors (Krishnaraju et al. 2001). In addition to functioning to inhibit the expression of granulocyte genes during monocyte development, it was also shown that Egrs are also needed for the activation of macrophage genes (Krysinska et al. 2006; Laslo et al. 2006).

The repression of granulocyte genes by Egr-1 and -2 during monocyte differentiation suggested that there are transcription factors that inhibit the expression of monocyte genes in developing granulocytes. A candidate for such a molecule was the zinc finger transcriptional repressor Gfi-1. Mice and humans with Gfi-1 mutations suffer from neutropenia (Hock et al. 2003; Karsunky et al. 2002; Person et al. 2003). Mutant mice appear to make granulocyte progenitors, but these progenitors fail to mature (Dahl et al. 2006; Hock et al. 2003; Karsunky et al. 2002). In addition, these granulocytes have aberrant expression of the monocyte genes *MCSFR, Mac3*, and *PU.1* (Hock et al. 2003; Karsunky et al. 2002). Opposite of Egr-1 expression, Gfi-1 retroviral infection of mouse bone marrow progenitors directs granulocyte differentiation (Dahl et al. 2006). Gfi-1 is downregulated in PUER cells during macrophage development (Laslo et al. 2006). Interestingly Egr-2 and Nab2 associate with the *Gfi1* promoter in PUER cells when they are differentiated into monocytes. These two also repress transcription from the *Gfi1* promoter. Conversely Gfi-1 is able to repress transcription mediated by the *Egr2* promoter. Similar to Egrs, Gfi-1 may also function in the activation of granulocyte genes as Gfi-1 cooperatively activates the neutrophil collagenase promoter with C/EBPε (Khanna-Gupta et al. 2007).

Other Transcription Factors Involved in Determining Myeloid Cell Fates

The identification of Egr-1, 2 and Gfi-1 functioning downstream of PU.1 and C/EBPα to commit cells to the monocyte and granulocyte cell fate is a compelling model. However, there is evidence of other factors functioning with PU.1 and/or C/EBPα to determine myeloid cell fate acquisition.

Macrophage

Other factors involved in monocyte/macrophage development include c-Jun, JunB, IRF-8, and KLF4. Each one will be discussed individually regarding its importance in monocyte development. As previously discussed c-Jun interacts with PU.1 but it may also have an important role in myelopoiesis through binding C/EBPα. C-Jun is a bZIP protein that is part of the AP-1 transcription factor complex dimerizing with c-Fos. However, c-Jun partners with many bZIP proteins including C/EBPα. Early studies implicated c-Jun in monocyte development as *Jun* transcripts were observed to increase during monocyte differentiation of cell lines (Hass et al. 1991; Sherman et al. 1990). It was also shown that Jun expression is downregulated by C/EBPα-induced differentiation of U937 cells. *Jun* downregulation is important as enforced c-Jun expression blocks C/EBPα-induced granulocytic differentiation of U937 cells. Recently, Alan Friedman's laboratory has engineered mutations in the leucine zipper domain of bZIP proteins that allows for the ability to control the partnering of bZIP proteins (Cai et al. 2007). Mutants of C/EBPα that have the ability to interact with specific bZIP family members had distinct effects on hematopoietic differentiation when expressed in hematopoietic progenitors (Cai et al. 2007). Heterodimers of C/EBPα and c-Jun were potent inducers of monocyte differentiation of primary hematopoietic progenitors. Homodimers of C/EBPα only modestly increased monopoiesis in liquid culture and did not significantly induce monocyte progenitor colonies when assayed by hematopoietic colony assay. This study intriguingly suggests that different bZIP protein interactions with C/EBPα may control monocyte or granulocyte cell fate decisions. It may also explain why recent studies expressing C/EBPα in mouse hematopoietic progenitors, pro-B and pro-T cells, result in macrophage differentiation as opposed to granulocyte differentiation previously observed with myeloid cell lines (Laiosa et al. 2006; Suh et al. 2006; Wang et al. 2006; Xie et al. 2004). The different fate outcomes could be due to the different pools of C/EBPα partner proteins available in target cells.

Jun$^{-/-}$ mice do not have a hematopoietic defect, which potentially could be due to compensation from AP-1 family members such as JunB (Eferl et al. 1999). JunB is expressed in hematopoietic tissue. *JunB*$^{-/-}$ mice die early due to placental vascular defects. When the placental defect is rescued by a JunB transgene (*JunB*$^{-/-}$*Ubi-junB*), *JunB*$^{-/-}$ mice develop a myeloproliferative disease (MPD) resembling chronic myeloid leukemia (CML) (Passegue et al. 2001). $JunB^{-/-}$ MPD is characterized by specific elevation of neutrophils in the blood and increased number of bone marrow granulocyte progenitors. Development of the disease correlated with loss of transgenic JunB expression. The requirement of losing JunB expression for MPD development was clearly demonstrated in mice carrying a conditional allele of *JunB* (Passegue et al. 2004). Mice induced to delete JunB at 4 weeks of age developed MPD 6 to 9 months later. Interestingly mice with JunB specifically deleted in granulocytic cells did not develop MPD. The development of a granulocytic disease versus a monocytic disease may indicate a role for JunB similar to c-Jun in directing monopoiesis versus granulopoiesis.

IRF8, also known as the interferon consensus sequence binding protein (ICSBP), is a member of the interferon regulatory factor family. It is most closely related to IRF4, which was originally identified as a protein that interacts with PU.1 (Eisenbeis et al. 1995). Both IRF8 and IRF4 bind to PU.1, however, interaction with PU.1 is dependent on phosphorylation of a conserved serine residue in the central PEST domain of PU.1 (Brass et al. 1999; Eisenbeis et al. 1995). The two bind to DNA together at composite ETS/IRF binding sites (Brass et al. 1999). Similar to *JunB* deficiency, mice deleted for *IRF8* develop a CML-like disease with 100% penetrance. The disease is more severe as approximately 30% of $IRF8^{-/-}$ mice develop a fatal granulocytic leukemia. As with JunB, this suggests that perhaps IRF-8 plays a role in inhibiting granulocyte and/or promoting monocyte development. Further evidence supporting a role in promoting monocyte differentiation came from experiments with an $IRF8^{-/-}$ cell line (Tamura et al. 2000). The cell line was GM-CSF-dependent and displayed immature blast morphology. If the cells were transferred to G-CSF-containing media approximately 70% of the cells would differentiate into granulocytes. In contrast if the cells were infected with an IRF-8 coding retrovirus the cells would become adherent macrophages. In addition, expression of IRF-8 in primary $IRF8^{-/-}$ progenitors would rescue the generation of monocyte colonies. This rescue is dependent on its DNA binding and PU.1 interaction domain. The data suggests that IRF-8 promotes monocyte/macrophage differentiation at the expense of granulocyte development.

KLF4 (formerly known as GKLF) is a kruppel-like zinc finger protein that is expressed in myeloid cells. In human hematopoietic cell lines KLF4 is expressed exclusively in the myeloid lineage (Feinberg et al. 2007). Expression was highest in primary human monocytes. It was not expressed in undifferentiated HL60 cells but expression was induced with TPA-directed monocyte differentiation. Retroviral expression of KLF4 in HL60 induced monocytic differentiation with upregulation of several monocytic genes including *CD11b, CD14, PU.1*, and *MCSFR* (Feinberg et al. 2007). Conversely, knockdown of KLF4 in TPA-treated HL60s blocked monocytic development. Interestingly expression of PU.1 was decreased in the knockdown cells, suggesting that KLF4 is involved in generating high levels of PU.1 necessary for directing monocytic differentiation. KLF4 transduction of murine HSCs results exclusively in monocytic differentiation whereas PU.1 transduction of HSCs results in granulocyte and monocyte differentiation. HSCs and CMPs from a mouse with a LoxP-flanked conditional *KLF4* allele generated decreased monocytes with compensatory increase in granulocytes when transduced with the CRE recombinase (Feinberg et al. 2007). Most intriguingly KLF4 expression in *PU.1⁻/⁻* fetal liver cells partially rescued monocytic differentiation. *PU.1⁻/⁻* KLF4 expressing cells had morphological features consistent with monocyte differentiation but did not appear to be as differentiated as cells rescued with a PU.1 retrovirus (Anderson et al. 1999; DeKoter et al. 1998; Feinberg et al. 2007). Chromatin immunoprecipitation and transient transfection assays demonstrate that *KLF4* is a direct transcriptional target of PU.1.

Neutrophil/Granulocyte Differentiation

In this section, function of C/EBPε, retinoic acid receptor alpha (RARα), and Lef1 in granulocyte development will be discussed. In addition, although early in development high PU.1 activity functions to repress granulocyte commitment, its activity is needed for terminal maturation of neutrophillic granulocytes.

C/EBPε is a basic leucine zipper transcription factor that is closely related to C/EBPα. It is highly expressed in granulocytic cells and their precursors (Koike et al. 1997; Yamanaka et al. 1997). Targeted deletion of *CEBPE* demonstrated an essential role in granulocyte differentiation (Lekstrom-Himes and Xanthopoulos 1999). Although they are viable and fertile, *CEBPE⁻/⁻* mice do not produce mature neutrophils or eosinophils. Because of this granulocytopenia, the mice die of opportunistic infection within a few months of birth. *CEBPE*-deficient granulocytes do not produce cytoplasmic secondary or tertiary granules. The mouse phenotype resembles the granulocytic disease-specific granule deficiency (SGD) (Breton-Gorius et al. 1980; Gombart and Koeffler 2002). Subsequent studies demonstrated that two of the five known SGD patients at the time had mutations that disrupted the activity of C/EBPε (Gombart et al. 2001; Lekstrom-Himes et al. 1999).

In a study of patients with severe congenital neutropenia (SCN) it was shown that expression of the HMG box transcription factor, Lef1, was downregulated approximately 20-fold in promyelocyte cells obtained from patients compared with healthy donors (Skokowa et al. 2006). Downregulation of Lef1 was specific to the granulocyte lineage as normal levels of Lef1 were detected in patient T cells and monocytes. Lentiviral expression of Lef1 rescued the granulocyte development of SCN CD34+ cells and induced the expression of C/EBPα. Lef1 functions in the Wnt signaling pathway by binding to nuclear translocated β-catenin, however, the upregulation of C/EBPα was shown to be independent of β-catenin. Lef1 shRNA in CD34+ cells blocked granulopoiesis.

RARα was first implicated in hematopoiesis since it is a frequent translocation partner in promyelocyte leukemia. It was hypothesized that the resultant RARα fusion proteins disrupted the normal function of RARα in granulocyte development. RARα is a member of the large nuclear hormone receptor family. These are heterodimeric transcription factors that are regulated by binding to retinoids. In the absence of retinoic acid (RA), this heterodimer binds corepressor complexes and represses transcription. In the presence of RA, corepressor complexes are exchanged for coactivator complexes and transcription is induced. RARα is predominantly expressed in myeloid cells (de The et al. 1989; Labrecque et al. 1998). A dominant negative RARα arrests granulocyte differentiation of a myeloid cell line at a promyelocyte stage (Tsai and Collins 1993). There are no defects seen in myeloid development of mice lacking RARα (Labrecque et al. 1998). However, mice lacking both RARα and γ have neutrophils that are blocked in development at a similar stage as *CEBPE* knockouts (Labrecque et al. 1998).

Lastly PU.1, unlike C/EBPα, is required for neutrophil maturation. Inactivation of PU.1 in adult mouse GMP cells is permissive for granulocyte lineage specification, but is still required for proper granulocyte maturation (Dakic et al. 2005). This agrees with previous data from the germline knockout mouse that demonstrated that early granulocyte cells were generated in *PU.1*$^{-/-}$ mice (Anderson et al. 1998; McKercher et al. 1996). Myeloid cells with the beginning of nuclear segmentation could be observed. These cells express early granulocytic genes but not genes for secondary and tertiary granule proteins that are associated with mature granulocytes. In the absence of PU.1, myeloid progenitors appear to be committing to the granulocyte lineage, but are unable to mature. Expression studies of developing granulocytes have shown that PU.1 levels increase during maturation (Yuan et al. 2007).

Specifying Eosinophils

C/EBP family members and PU.1 are also involved in the development of eosinophils. Both *CEBPA* and *PU.1* knockout mice do not generate eosinophils. Several eosinophil-specific genes are regulated by C/EBPα, β, and ε and PU.1. Additionally, exogenous expression of C/EBPα in a multipotential hematopoietic cell line or primary human CD34+ progenitors induces eosinophil differentiation (Iwama et al. 2002; Nerlov et al. 1998).

Another critical factor involved in eosinophil generation is GATA-1. This transcription factor is more associated with its critical role in erythroid and megakaryocyte development; however, it is also expressed in eosinophils, basophils, and mast cells (Harigae et al. 1998; Zon et al. 1993). The first evidence implicating GATA-1 in eosinophil development came from the study of avian hematopoietic cell lines. Retroviral expression of GATA1 in the multipotential E26-MEP cell line directed differentiation into erythrocytes, thrombocytes (avian megakaryocytes), and eosinophils (Kulessa et al. 1995). Lower expression directed eosinophil development and higher levels directed megakaryocyte and erythroid development. Similar results were obtained with human CD34+ hematopoietic progenitors (Hirasawa et al. 2002). Interestingly, a novel mutation that deleted a GATA binding site in the *GATA1* promoter demonstrated that this site was required for *GATA1* expression in eosinophils (Yu et al. 2002). Mice lacking this site were unable to produce eosinophils demonstrating an essential role for GATA-1 in eosinophil maturation.

The GATA-1 cofactor FOG-1 is a negative regulator of eosinophils. FOG-1 is expressed in erythroid and megakaryocytes but not in any of the myeloid lineages (Querfurth et al. 2000; Tsang et al. 1997). However, expression of FOG-1 in avian eosinophils causes the cells to revert to early progenitor cells as judged by morphology and expression of cell surface markers (Querfurth et al. 2000). In addition downregulation of FOG-1 was required for C/EBPβ to induce eosinophilic differentiation of E26-MEP cells. FOG-1 is able to repress the transcription of GATA-1 eosinophilic target genes (Querfurth et al. 2000). As discussed in more detail for mast cells/basophils, FOG-1 expression may be a determinant for whether GATA-1 functions as an inducer of erythroid-megakaryocyte or granulocyte development.

Specifying Mast Cells and Basophils

Mast cells and basophils both express the high-affinity IgE receptor, which mediates secretion of inflammatory proteins that defend against pathogens. Although the cells are similar, they do differ in that basophils will exist fully mature in the circulation whereas mast cells do not mature until they migrate into peripheral tissue. Additionally, basophils resemble traditional granulocytes with segmented nuclei whereas mast cells have non-segmented nuclei. Because of these similarities and differences the developmental relationship between these cells has been controversial. Recently, however, Akashi and colleagues identified a bipotential basophil/mast cell precursor (BMCP), which demonstrates that these lineages are developmentally related (Arinobu et al. 2005).

Mast cells and basophils require the activities of GATA family members GATA-1 and GATA-2. $GATA1^{-/-}$ mice have defects in mast cell maturation (Harigae et al. 1998). The absence of GATA-2 is more severe resulting in a complete loss of mast cells (Tsai and Orkin 1997). Enforced expression of GATA-2 in purified murine CLPs redirects their differentiation from lymphoid to basophils and mast cells (Iwasaki et al. 2006).

The development of mast cells and basophils was not examined in *CEBPA* mice. However, Akashi and colleagues observed that *CEBPA* is expressed at low levels in bipotent basophil mast cell progenitors (BMCPs) and its expression increases in committed basophil progenitors (BaPs) (Iwasaki et al. 2006). In contrast, *CEBPA* expression was reduced approximately 50-fold in mast cell progenitors (MCPs) compared with BMCPs. Deleting *CEBPA* from BMCPs resulted in only mast cell differentiation. Conversely retroviral expression of C/EBPα in BMCPs resulted in basophil development. In addition, expression of C/EBPα in mature intestinal mast cells resulted in their reprogramming into basophils. However, deletion of *CEBPA* in BaPs had no effect demonstrating that C/EBPα is required for BaP specification but not for their maintenance and/or maturation. C/EBPα function is stage-specific. Exogenous expression of C/EBPα in GMPs blocked the development of both basophils and mast cells and promoted predominantly monocyte/neutrophil development. Lastly, ShRNA knockdown of C/EBPα in GMPs promoted mast cell development (Iwasaki et al. 2006).

Similar to what occurs in eosinophils, PU.1 is also required for mast cells. Mast cells are completely absent in the tissues of $PU.1^{-/-}$ mice (Walsh et al. 2002).

In addition no mature mast cell colonies were obtained when *PU.1*$^{-/-}$ fetal liver cells were cultured in methylcellulose with IL-3 and SCF. Retroviral expression of PU.1 into IL-3-dependent *PU.1*$^{-/-}$ fetal liver cells would rescue macrophage, neutrophil, and mast cell development. Interestingly though in cells expressing the conditional PUER fusion protein, macrophage but not mast cell development was detected when PU.1 was activated by OHT. In contrast to *PU.1*$^{-/-}$ progenitors, which express high levels of GATA-2, PUER cells express little *GATA2* mRNA. If PUER cells were infected with a GATA-2 retrovirus, then OHT activation of PUER resulted in mast cell differentiation (Walsh et al. 2002). These data demonstrated that early in myeloid development PU.1 antagonizes GATA-2 to direct monocyte (and neutrophil?) development. However, if GATA-2 expression is able to remain on then these two cooperate to induce mast cell development.

Similar to its inhibition of eosinophil development, FOG-1 expression blocks mast cell development (Cantor et al. 2008; Sugiyama et al. 2008). In addition mast cell progenitors in the yolk sac of *FOG1*$^{-/-}$ mice are increased approximately 47-fold compared to yolk sacs from wild-type mice (Cantor et al. 2008). Nakano and colleagues developed a tet-repressor conditional FOG-1 expression system in ES cells. Immature myeloid cells are generated after 8 days of culturing ES cells on OP9 stromal cells in the presence of IL-3. By day 14 greater than 80% of the cells become mast cells; however, if TET is removed to induce FOG-1 at day 8 then greater than 70% of the cells in culture are neutrophils (Sugiyama et al. 2008). FOG-1 disrupts the interaction between PU.1 and GATA-1, which is critical for its ability to increase neutrophil development at the expense of mast cell development. In a similar study exogenous FOG-1 expression in MCPs redirected their differentiation to the erythroid, megakaryocyte, and neutrophil lineages (Cantor et al. 2008). Mutants of FOG-1 that do not bind GATA-1 could not mediate this effect. Interestingly FOG-1 negatively regulates GATA-2 expression, and GATA-2 appears to negatively regulate FOG-1 expression (Cantor et al. 2008; Welch et al. 2004). This is another example of mutually antagonistic factors determining cell fate decisions.

Cytokines and Their Receptor in Myeloid Development

The exact role of cytokines (and growth factors) and their receptors in hematopoietic development is controversial. This debate has been termed instruction versus stochastic model of hematopoietic differentiation. The stochastic model predicts that expression of transcription factors is random yet antagonism and cooperation between factors results in big changes in factor expression patterns that result in commitment to specific hematopoietic lineages. The role of cytokines is to select for the survival and proliferation of specific lineages. The instructive model predicts that cytokines directly induce or repress the expression of specific transcription factors and thus promote the development of specific lineages.

Evidence for the stochastic model comes from numerous mouse gene targeting experiments in which cytokine genes and their receptors have been deleted (Lieschke 1997).

So far deletion of a cytokine gene, a receptor gene, or combinations of cytokines and/or receptors has not resulted in the complete absence of a hematopoietic lineage. However, these knockouts do cause perturbations in hematopoiesis with the generation of specific lineages substantially reduced. This could be due to lack of instruction but could also be due to absence of survival and proliferative signals. Indeed transgenic mice expressing the antiapoptotic protein BCL-2 in myeloid cells partially rescue the macrophage deficiency of osteopetrotic mice that have a naturally occurring mutation that creates a null allele for the cytokine M-CSF (Lagasse and Weissman 1997). In addition the multipotential cell line FDCP-Mix engineered to constitutively express BCL-2 would randomly differentiate into erythrocytes, monocytes, or granulocytes in the absence of cytokines and serum (Fairbairn et al. 1993). In total these results suggest that cytokines are required for the survival of hematopoietic cells and that lineage commitment can occur in their absence.

Evidence supporting an instructive role for cytokines comes from the laboratories of Weissman and Akashi. It was initially observed that CLPs and pro-T cells isolated from mice expressing a human IL-2Rγ chain would differentiate into myeloid cells when cultured in the presence of human IL-2 (King et al. 2002; Kondo et al. 2000). Similarly, CLPs and pro-T cells expressing hGM-CSFR would differentiate into myeloid cells in the presence of hGM-CSF (Iwasaki-Arai et al. 2003; Kondo et al. 2000). C/EBPα was upregulated by hGM-CSF in transgenic T cells, which may play a role in lineage conversion or simply be indicative of a myeloid cell fate conversion. However, the effects of GM-CSFR are cell-type specific. When hGM-CSFR was expressed in EPO-R null fetal liver cells, hGM-CSF did not increase myeloid development but instead rescued erythroid development (Hisakawa et al. 2001). This result is more consistent with the stochastic model.

Knockout studies and antiapoptotic rescue experiments clearly demonstrate that cytokine signaling is not necessary for lineage specification. However, the experiments with ectopic expression of IL-2R and GM-CSFR in lymphoid cells suggest that they can under certain cellular contexts influence cell fate. GM-CSF signaling induces the expression of PU.1 in alveolar macrophages (Shibata et al. 2001). In addition GM-CSF and other cytokines induce the expression of AP-1 transcription factors c-Jun and c-Fos (Reddy et al. 2000). As discussed previously these factors interact with C/EBPα and influence myeloid cell fate decisions. Clearly, cytokines can change transcription factor expression; however, evidence so far has come from observations of cell maturation and not from lineage specification.

Specific Cytokines and Their Receptors Involved in Monocyte and Granulocyte Development

Regardless of whether they provide instructive or permissive signals, several cytokines and their receptors are critical for myelopoiesis. The role of individual cytokines and their receptors in monocyte and granulocyte development is discussed later. The lineages that are affected by specific cytokines are shown in Fig. 6.1.

Signal transduction downstream of the receptors will not be discussed. For reviews on hematopoietic signal transduction see the following (Baker et al. 2007; Rane and Reddy 2002).

Cytokines and Receptors Involved in the Development of Myeloid Progenitors

Both GM-CSF and IL-3 act on a wide range of hematopoietic cells including stem cells, progenitor macrophages, granulocytes, erythrocytes, and megakaryocytes. IL-3, in addition, has effects on mast cells and lymphocytes. Their similarity is in part due to a shared receptor subunit, the common β chain (βc) (Gorman et al. 1992). Also, the two genes are proximal to each other on mouse chromosome 11 and human chromosome 5 suggesting that they derive from an ancient gene duplication (Lee and Young 1989). Surprisingly IL-3 deficiency does not lead to any overt phenotypes (Mach et al. 1998). Steady state hematopoiesis is normal. In addition deletion of the IL-3R subunits, IL-3Rα or β_{IL-3}, did not result in perturbations of normal hematopoiesis (Ichihara et al. 1995; Nicola et al. 1996). Similarly $GMCSF^{-/-}$ mice are viable and have no overt defects at birth, though female fertility is slightly decreased. Steady state production of monocytes and granulocytes is unperturbed. The major defect in these mice is increased respiratory system infections, pulmonary proteinosis, and disrupted lung surfactant metabolism. These defects are due to impaired function of alveolar macrophages. Similar pulmonary defects are seen in mice lacking the GM-CSF receptor subunit βc.

Since IL-3 and GM-CSF have such similar effects on hematopoietic cells it was hypothesized that functional redundancy may mask their importance in hematopoietic development. However, mice deficient for both IL-3 and GM-CSF do not have any defects in the generation of hematopoietic progenitors (Gillessen et al. 2001). The only observed hematopoietic defect is an increase in circulating eosinophils.

Macrophage Development

M-CSF (CSF-1) stimulates the development of macrophages and their precursors (Stanley 1985; Stanley et al. 1983). The importance of M-CSF was demonstrated by the phenotype of the osteopetrotic mouse mutant (op), which lacks the expression of the $MCSF$ gene (Wiktor-Jedrzejczak et al. 1990; Yoshida et al. 1990). These mice are osteopetrotic due to a severe reduction in osteoclasts, which remodel bone. The reduction in osteoclasts leads to skeletal abnormalities, hypocellularity of the bone marrow, and a failure of teeth to erupt. Additionally there is a huge decrease in peripheral macrophages. Interestingly the phenotype of op mice corrects itself with age (Begg et al. 1993). In macrophage development M-CSF effects the maturation of monocytes to macrophages. Monocyte numbers are normal in adult op/op mice (Chitu and Stanley 2006). In addition fetal liver hematopoiesis in op/op mice is comparable to that in wild-type mice. Defects in hematopoietic progenitors in op/op mice appear to be a consequence of the impaired ability to remodel the bone cavity postnatally (Nilsson and Bertoncello 1994). Deletion of the M-CSFR (c-fms) has similar effects on mice as *MCSF* deficiency (Dai et al. 2002). The mice are osteopetrotic and have reduced numbers of macrophages. The osteopetrosis is more severe in the receptor-deficient mice, and they do not survive as well as op/op mice. This suggests the existence of an alternate ligand or perhaps a ligand-independent role for the receptor.

Mice deficient for both M-CSF and GM-CSF have mainly additive defects (Lieschke et al. 1994b). The mice are osteopetrotic and produce macrophages at levels observed in M-CSF-deficient animals. Similar to GM-CSF-deficient animals the compound mutant animals have alveolar-proteinosis lung pathology. This defect is more severe in the compound mutants with increased accumulation of proteinaceous material and an increased susceptibility to infection. The double knockout mice also revealed that the macrophage recovery in older op/op mice is not due to GM-CSF as macrophage numbers increased in aged double mutants as well.

Neutrophil and Granulocyte Development

G-CSF supports the proliferation and differentiation of neutrophils. Intravenous injection of G-CSF into mice promotes granulopoiesis. In addition modified forms of recombinant G-CSF are used to treat patients who are neutropenic after chemotherapy regiments. *GCSF*-deficient mice are viable and fertile; however, they have chronic neutropenia (Lieschke et al. 1994a). Peripheral blood neutrophil numbers are decreased 20–30% compared to wildtype, and granulocyte progenitors in the bone marrow were reduced approximately 50%. Similarly mice that lack the receptor for G-CSF are neutropenic (Richards et al. 2003).

Interleukin 6 (IL-6) is another cytokine that can have effects on neutrophils. Mice injected with IL-6 have significant increases in several hematopoietic progenitors in the bone marrow and have increased numbers of neutrophils in the peripheral blood (Pojda and Tsuboi 1990). In addition IL-6 promotes the growth of granulocytic progenitors in hematopoietic colony assays with mouse bone marrow (Liu et al. 1996). Steady state granulopoiesis is normal in mice lacking IL-6 (Dalrymple et al. 1995). However, these mice cannot mount a sufficient defense against the pathogens *Listeria monocytogenes* or *Candida albicans* (Dalrymple et al. 1995; Romani et al. 1996).

Mice lacking both the *IL6* and *GCSFR* genes had a more severe neutropenia than *GCSFR*[-/-] mice, and there was a decrease in myeloid progenitors in the bone marrow (Liu et al. 1997). Mature neutrophils were still generated in these mice. GM-CSF- and G-CSF-deficient mice mostly had additive defects (Seymour et al. 1997). Although neutropenia in adult mice was similar to what is observed in

GCSF-deficient mice, compound mutant mice were more severely neutropenic postnatally than GCSF single mutant mice. Lastly, mice have been generated that lack G-CSF, M-CSF, and GM-CSF (Hibbs et al. 2007). The mice have a significantly reduced lifespan due to a compromised immunity. They have the expected osteopetrotic, alveolar proteinosis, and neutropenia seen in single mutant mice. Surprisingly the mice have more circulating monocytes than op/op mice and more circulating neutrophils than $GCSF^{-/-}$ $GMCSF^{-/-}$ mice.

Eosinophil Development

IL-5, IL-3, and GM-CSF have effects on eosinophil development (Owen et al. 1987; Tanno et al. 1987; Yamaguchi et al. 1988). All these three signal through the βc cytokine receptor subunit but have distinct α subunits. IL-5 appears to be the most critical and specific to the eosinophil lineage. Transgenic mice expressing IL-5 have a massive increase of eosinophils in peripheral blood, bone marrow, and spleen (Foster et al. 1996; Kopf et al. 1996). In addition patients with idiopathic hypereosinophilic syndrome have high concentrations of IL-5 in their serum. Deletion of IL-5 in mice does not result in a change in basal levels of eosinophils. However, an eosinophil response is not induced upon allergen or pathogenic challenge in $IL5^{-/-}$ mice (Foster et al. 1996; Kopf et al. 1996). Disruption of the IL-5Rα chain also did not effect basal production of eosinophils. IL-3 and/or GM-CSF must compensate for loss of IL-5 signaling since there is a reduced number of eosinophils and decreased eosinophil responses in mice that lack the signaling of all three cytokines (Nishinakamura et al. 1996).

Mast Cell Development

IL-3 has been frequently used to generate murine cultures of mast cells (Lantz and Huff 1995). Surprisingly deletion of IL-3 signaling in mice does not affect mast cell development. The critical cytokine requirements for mast cells were elucidated from two mutant strains of mice – white spotting (W/W^v) and steele (Sl/Sl^d) (Galli et al. 1993). Both these strains contain less than 1% of the number of mast cells found in corresponding wild-type tissues. However, both mice have progenitors that can be cultured into mast cells by IL-3. Elegant transplant experiments suggested that the W/W^v and Sl/Sl^d mutations would affect the genes for a receptor–ligand pair. This turned out to be true when the mutations were cloned and the W locus mapped to the receptor tyrosine kinase c-kit gene and the Sl locus mapped to the growth factor stem cell factor (SCF). SCF is able to stimulate the development of mast cells in several hematopoietic assays (Lantz and Huff 1995). In addition activating mutations in c-kit have been associated with human mastocytosis (Nagata et al. 1995).

Basophil Development

Unlike other hematopoietic lineages there does not seem to be a critical cytokine for basophil development (Arock et al. 2002). However, IL-3 does seem to play an important role despite the lack of observed basophil defect in IL-3-deficient mice. Injection of IL-3 into primates results in basophilia. Additionally a significant increase in basophils is seen when mice are infected with the nematode *Strongyloides venezuelensis* (Lantz et al. 1998). This basophil response is not seen in mice lacking IL-3. This is another example of a cytokine being more critical for responses to infection than in steady state hematopoiesis.

miRNAs Regulating Myelopoiesis

Micro-RNAs (miRNAs) are short 19–22 nucleotide (nt) RNA molecules that regulate gene expression by binding to the 3′UTR of messenger RNAs (mRNAs) through nonexact base pairing (Bartel 2004). In most cases this association leads to translational silencing of the mRNA. miRNAs are first transcribed as primary miRNA transcripts (pri-miRNA) and are processed into an approximately 70 nt pre-miRNA hairpin RNA structure (Lee et al. 2003). This hairpin pre-miRNA is exported to the cytoplasm and further processed to the mature 19–22 nt single stranded miRNA (Lee et al. 2004b). Most miRNAs are made from independent transcription units regulated by RNA polymerase II (Lee et al. 2004a). However, a large number of miRNAs are encoded in the introns of mRNAs (Bartel 2004). Many miRNAs are also clustered so that they are likely transcribed as a single RNA and then processed into multiple pre-miRNAs and mature miRNAs. There is much redundancy among miRNAs. Many miRNAs exist as multiple copies in the genome, and there may be significant redundancy in targeting specific mRNAs.

miRNAs in Hematopoiesis

Chen and colleagues were the first to report on the expression of miRNAs in the hematopoietic system (Chen et al. 2004). They cloned approximately 100 miRNAs from mouse bone marrow most of which had been previously identified from other mammalian tissues. Currently only five miRNAs are thought to be exclusively expressed in hematopoietic cells: mir-142, mir-144, mir-150, mir-155, and mir-223 (Landgraf et al. 2007). In Chen et al. the functions of mir-142, mir-181a, and mir-223 were examined by retroviral transduction of hematopoietic progenitors. Mir-142 and Mir-223 expression in bone marrow progenitors increased T-cell development 30–40% with no effect on B cell development in vitro. Mir-181a had the opposite effect in vitro, doubling the number of B cells with minimal effect on T-cell differentiation. When miR-181a expressing hematopoietic progenitors were transplanted into irradiated mice, the investigators observed that B-cell differentiation was significantly increased with a decrease in T-cell development. This study clearly showed that miRNAs were expressed in the hematopoietic system and that miRNA expression could affect the differentiation of hematopoietic cells.

Other studies have shown that exogenous expression of miRNAs affects the differentiation of myeloid cells. Lentiviral expression of mir-223 into the promyelocytic leukemia cell line NB4 induced granulocyte differentiation (Fazi et al. 2005). NB4 cells are normally induced to differentiate into granulocytes by treatment of all trans retinoic acid (ATRA). ATRA treatment of NB4 cells results in an increase in the expression of mir-223. The authors demonstrated that complementary oligonucleotides to mir-223 that downregulate its activity block ATRA-induced differentiation of NB4 cells.

Similarly it was shown by Fontana el al. that the three homologous miRNA clusters 17-92, mir106a-92, and 106b-25 are downregulated during normal human monocyte differentiation (Fontana et al. 2007). Overexpression of these miRNAs clusters in CD34+ progenitors blocked monocytic differentiation. Anti-miRNA oligonucleotides directed toward individual miRNAs in the cluster accelerated monocyte differentiation with the greatest effect seen when multiple miRNAs were antagonized. The authors demonstrated that miRNAs mir-17, mir-20a, and mir-106a coded for in the clusters all target the transcription factor RUNX1. Although mRNA levels of RUNX1 do not significantly change during the course of cultured differentiation of CD34+ progenitors into monocytes there is an increase in RUNX1 protein, which is barely detectable in the CD34+ progenitors. These studies demonstrate that miRNAs can influence myeloid cell fate determination.

Recently the importance of miRNAs in hematopoiesis has been demonstrated by gene targeting in mice. Deletion of the hematopoietic-specific miRNAs, mir-150 and mir-155, results in immune system defects (Rodriguez et al. 2007; Thai et al. 2007; Xiao et al. 2007). Mir-155 mutants have a significant decrease in germinal center B cells and a reduced ability to produce cytokines upon B-cell stimulation (Thai et al. 2007). Additionally T helper cells from mutant animals produced more TH2-associated cytokines than TH1 cytokines compared with wild-type animals(Rodriguez et al. 2007; Thai et al. 2007). Deletion of mir-150 resulted in a defect in B cells in which the B1 subpopulation of B cells was expanded at the expense of B2 cells. These knockout mice demonstrate that although there is much redundancy among the miRNAs, loss of an individual miRNA can have effects on the hematopoietic system.

Conclusion

Many of the basic rules of programming hematopoietic stem cells to become myeloid cells have been elucidated. Combinations of transcription factors are critical for specifying the individual myeloid lineages and generating the appropriate gene expression for these lineages. Cooperative and antagonistic interactions between these factors are clearly important, and most interestingly two factors can antago-nize each other in one context and cooperate in another. This is clearly seen with GATA-1 and PU.1, which have been shown to antagonize each other's activity for the commitment to the erythroid/megakaryocyte versus myeloid lineages. However, GATA-1 and PU.1 cooperate in the generation of eosinophils, basophils, and mast cells. Levels and timing of expression of these factors are also critical.

Ultimately the transcription factors will direct the expression of genes that will determine cell fate identity. However, the importance of cytokines and other extra-cellular signals should not be undervalued. Some clearly have a role for maintain-ing steady-state levels of hematopoietic cells by promoting survival and proliferation. In addition their ability to regulate hematopoiesis during certain infections is abso-lutely critical.

Lastly there is a lot we do not know about the role of miRNAs in the process of lineage commitment. Clearly their ability to modulate protein expression could affect cell fate acquisition through modulating the levels of key transcription fac-tors. Currently so little is known about miRNA pathways that it is hard to under-stand their absolute importance in cell fate. Additionally the redundancy of miRNAs in the genome will make elucidating their role in regulating hematopoiesis challenging. Current overexpression studies demonstrate that miRNAs can influ-ence cell fate decisions but we will have to wait for more loss-of-function studies to determine if they are essential for cell fate acquisition like transcription factors.

References

Akashi, K., Traver, D., Miyamoto, T. & Weissman, I.L. (2000). *Nature*, 404, 193–7.

Anderson, K.L., Smith, K.A., Pio, F., Torbett, B.E. & Maki, R.A. (1998). *Blood*, 92, 1576–85.

Anderson, K.L., Smith, K.A., Perkin, H., Hermanson, G., Anderson, C.G., Jolly, D.J., Maki, R.A. & Torbett, B.E. (1999). *Blood*, 94, 2310–18.

Arinobu, Y., Iwasaki, H., Gurish, M.F., Mizuno, S., Shigematsu, H., Ozawa, H., Tenen, D.G., Austen, K.F. & Akashi, K. (2005). *Proc Natl Acad Sci USA*, 102, 18105–10.

Arock, M., Schneider, E., Boissan, M., Tricottet, V. & Dy, M. (2002). *J Leukoc Biol*, 71, 557–64.

Austen, K.F. & Boyce, J.A. (2001). *Leuk Res*, 25, 511–18.

Baehner, R.L. (2000). *Hematology: Basic Principles and Practice*. Hoffman, R., Benz, E.J, Jr., Shattil, S.J., Furie, B., Cohen, H.J., Silberstein, L.E. & McGlave, P. (eds). Churchhill Livinstone: Philadelphia, PA, pp 667–686.

Baker, S.J., Rane, S.G. & Reddy, E.P. (2007). *Oncogene*, 26, 6724–37.

Bartel, D.P. (2004). *Cell*, 116, 281–97.

Begg, S.K., Radley, J.M., Pollard, J.W., Chisholm, O.T., Stanley, E.R. & Bertoncello, I. (1993). *J Exp Med*, 177, 237–42.

Brass, A.L., Zhu, A.Q. & Singh, H. (1999). *EMBO J*, 18, 977–91.

Breton-Gorius, J., Mason, D.Y., Buriot, D., Vilde, J.L. & Griscelli, C. (1980). *Am J Pathol*, 99, 413–28.

Cai, D.H., Wang, D., Keefer, J., Yeamans, C., Hensley, K. & Friedman, A.D. (2007). *Oncogene*, 27, 2772–9.

Cammenga, J., Mulloy, J.C., Berguido, F.J., MacGrogan, D., Viale, A. & Nimer, S.D. (2003). *Blood*, 101, 2206–14.

Cantor, A.B. & Orkin, S.H. (2001). *Curr Opin Genet Dev*, 11, 513–19.

Cantor, A.B., Iwaski, H., Arinobu, Y., Moran, T.B., Shigematsu, H., Sullivan, M.R., Akashi, K. & Orkin, S.H. (2008). *J Exp Med*, 205, 611–24.

Chen, C.Z., Li, L., Lodish, H.F. & Bartel, D.P. (2004). *Science*, 303, 83–6.

Chitu, V. & Stanley, E.R. (2006). *Curr Opin Immunol*, 18, 39–48.

Dahl, R., Walsh, J.C., Lancki, D., Laslo, P., Iyer, S.R., Singh, H. & Simon, M.C. (2003). *Nat Immunol*, 4, 1029–36.

Dahl, R., Iyer, S.R., Owens, K.S., Cuylear, D.D. & Simon, M.C. (2006). *J Biol Chem*, 282, 6473–83.

Dai, X.M., Ryan, G.R., Hapel, A.J., Dominguez, M.G., Russell, R.G., Kapp, S., Sylvestre, V. & Stanley, E.R. (2002). *Blood*, 99, 111–20.

Dakic, A., Metcalf, D., Di Rago, L., Mifsud, S., Wu, L. & Nutt, S.L. (2005). *J Exp Med*, 201, 1487–502.

Dalrymple, S.A., Lucian, L.A., Slattery, R., McNeil, T., Aud, D.M., Fuchino, S., Lee, F. & Murray, R. (1995). *Infect Immun*, 63, 2262–8.

de The, H., Marchio, A., Tiollais, P. & Dejean, A. (1989). *EMBO J*, 8, 429–33.

DeKoter, R.P. & Singh, H. (2000). *Science*, 288, 1439–41.

DeKoter, R.P., Walsh, J.C. & Singh, H. (1998). *EMBO J*, 17, 4456–68.

Eferl, R., Sibilia, M., Hilberg, F., Fuchsbichler, A., Kufferath, I., Guertl, B., Zenz, R., Wagner, E.F. & Zatloukal, K. (1999). *J Cell Biol*, 145, 1049–61.

Eisenbeis, C.F., Singh, H. & Storb, U. (1995). *Genes Dev*, 9, 1377–87.

Fairbairn, L.J., Cowling, G.J., Reipert, B.M. & Dexter, T.M. (1993). *Cell*, 74, 823–32.

Faller, D.V. & Mentzer, S.J. (2000). *Hematology: Basic Principles and Practice.* Hoffman, R., Benz, E.J., Jr., Shattil, S.J., Furie, B., Cohen, H.J., Silberstein, L.E. & McGlave, P. (eds). Churchill Livinstone: Philadelphia, PA, pp 686–701.

Fazi, F., Rosa, A., Fatica, A., Gelmetti, V., De Marchis, M.L., Nervi, C. & Bozzoni, I. (2005). *Cell*, 123, 819–31.

Feinberg, M.W., Wara, A.K., Cao, Z., Lebedeva, M.A., Rosenbauer, F., Iwasaki, H., Hirai, H., Katz, J.P., Haspel, R.L., Gray, S., Akashi, K., Segre, J., Kaestner, K.H., Tenen, D.G. & Jain, M.K. (2007). *EMBO J*, 26, 4138–48.

Fontana, L., Pelosi, E., Greco, P., Racanicchi, S., Testa, U., Liuzzi, F., Croce, C.M., Brunetti, E., Grignani, F. & Peschle, C. (2007). *Nat Cell Biol*, 9, 775–87.

Foster, P.S., Hogan, S.P., Ramsay, A.J., Matthaei, K.I. & Young, I.G. (1996). *J Exp Med*, 183, 195–201.

Galli, S.J., Tsai, M. & Wershil, B.K. (1993). *Am J Pathol*, 142, 965–74.

Gillessen, S., Mach, N., Small, C., Mihm, M. & Dranoff, G. (2001). *Blood*, 97, 922–8.

Gleich, G.J., Ottesen, E.A., Leiferman, K.M. & Ackerman, S.J. (1989). *Int Arch Allergy Appl Immunol*, 88, 59–62.

Gombart, A.F. & Koeffler, H.P. (2002). *Curr Opin Hematol*, 9, 36–42.

Gombart, A.F., Shiohara, M., Kwok, S.H., Agematsu, K., Komiyama, A. & Koeffler, H.P. (2001). *Blood*, 97, 2561–7.

Gorman, D.M., Itoh, N., Jenkins, N.A., Gilbert, D.J., Copeland, N.G. & Miyajima, A. (1992). *J Biol Chem*, 267, 15842–8.

Harigae, H., Takahashi, S., Suwabe, N., Ohtsu, H., Gu, L., Yang, Z., Tsai, F.Y., Kitamura, Y., Engel, J.D. & Yamamoto, M. (1998). *Genes Cells*, 3, 39–50.

Hass, R., Brach, M., Kharbanda, S., Giese, G., Traub, P. & Kufe, D. (1991). *J Cell Physiol*, 149, 125–31.

Heath, V., Suh, H.C., Holman, M., Renn, K., Gooya, J.M., Parkin, S., Klarmann, K.D., Ortiz, M., Johnson, P. & Keller, J. (2004). *Blood*, 104, 1639–47.

Hibbs, M.L., Quilici, C., Kountouri, N., Seymour, J.F., Armes, J.E., Burgess, A.W. & Dunn, A.R. (2007). *J Immunol*, 178, 6435–43.

Hirasawa, R., Shimizu, R., Takahashi, S., Osawa, M., Takayanagi, S., Kato, Y., Onodera, M., Minegishi, N., Yamamoto, M., Fukao, K., Taniguchi, H., Nakauchi, H. & Iwama, A. (2002). *J Exp Med*, 195, 1379–86.

Hisakawa, H., Sugiyama, D., Nishijima, I., Xu, M.J., Wu, H., Nakao, K., Watanabe, S., Katsuki, M., Asano, S., Arai, K., Nakahata, T. & Tsuji, K. (2001). *Blood*, 98, 3618–25.

Hock, H., Hamblen, M.J., Rooke, H.M., Traver, D., Bronson, R.T., Cameron, S. & Orkin, S.H. (2003). *Immunity*, 18, 109–20.

Hromas, R., Orazi, A., Neiman, R.S., Maki, R., Van Beveran, C., Moore, J. & Klemsz, M. (1993). *Blood*, 82, 2998–3004.

Ichihara, M., Hara, T., Takagi, M., Cho, L.C., Gorman, D.M. & Miyajima, A. (1995). *EMBO J*, 14, 939–50.

Iwama, A., Osawa, M., Hirasawa, R., Uchiyama, N., Kaneko, S., Onodera, M., Shibuya, K., Shibuya, A., Vinson, C., Tenen, D.G. & Nakauchi, H. (2002). *J Exp Med*, 195, 547–58.

Iwasaki, H., Somoza, C., Shigematsu, H., Duprez, E.A., Iwasaki-Arai, J., Mizuno, S., Arinobu, Y., Geary, K., Zhang, P., Dayaram, T., Fenyus, M.L., Elf, S., Chan, S., Kastner, P., Huettner, C.S., Murray, R., Tenen, D.G. & Akashi, K. (2005). *Blood*, 106, 1590–600.

Iwasaki, H., Mizuno, S., Arinobu, Y., Ozawa, H., Mori, Y., Shigematsu, H., Takatsu, K., Tenen, D.G. & Akashi, K. (2006). *Genes Dev*, 20, 3010–21.

Iwasaki-Arai, J., Iwasaki, H., Miyamoto, T., Watanabe, S. & Akashi, K. (2003). *J Exp Med*, 197, 1311–22.

Karsunky, H., Zeng, H., Schmidt, T., Zevnik, B., Kluge, R., Schmid, K.W., Duhrsen, U. & Moroy, T. (2002). *Nat Genet*, 30, 295–300.

Khanna-Gupta, A., Sun, H., Zibello, T., Lee, H.M., Dahl, R., Boxer, L.A. & Berliner, N. (2007). *Blood*, 109, 4181–90.

King, A.G., Kondo, M., Scherer, D.C. & Weissman, I.L. (2002). *Proc Natl Acad Sci USA*, 99, 4508–13.

Koike, M., Chumakov, A.M., Takeuchi, S., Tasaka, T., Yang, R., Nakamaki, T., Tsuruoka, N. & Koeffler, H.P. (1997). *Leuk Res*, 21, 833–9.

Kondo, M., Scherer, D.C., Miyamoto, T., King, A.G., Akashi, K., Sugamura, K. & Weissman, I.L. (2000). *Nature*, 407, 383–6.

Kopf, M., Brombacher, F., Hodgkin, P.D., Ramsay, A.J., Milbourne, E.A., Dai, W.J., Ovington, K.S., Behm, C.A., Kohler, G., Young, I.G. & Matthaei, K.I. (1996). *Immunity*, 4, 15–24.

Krishnaraju, K., Hoffman, B. & Liebermann, D.A. (2001). *Blood*, 97, 1298–305.

Krysinska, H., Hoogenkamp, M., Ingram, R., Wilson, N., Tagoh, H., Laslo, P., Singh, H. & Bonifer, C. (2006). *Mol Cell Biol*, 27, 878–87.

Kulessa, H., Frampton, J. & Graf, T. (1995). *Genes Dev*, 9, 1250–62.

Labrecque, J., Allan, D., Chambon, P., Iscove, N.N., Lohnes, D. & Hoang, T. (1998). *Blood*, 92, 607–15.

Lagasse, E. & Weissman, I.L. (1997). *Cell*, 89, 1021–31.

Laiosa, C.V., Stadtfeld, M., Xie, H., de Andres-Aguayo, L. & Graf, T. (2006). *Immunity*, 25, 731–44.

Landgraf, P., Rusu, M., Sheridan, R., Sewer, A., Iovino, N., Aravin, A., Pfeffer, S., Rice, A., Kamphorst, A.O., Landthaler, M., Lin, C., Socci, N.D., Hermida, L., Fulci, V., Chiaretti, S., Foa, R., Schliwka, J., Fuchs, U., Novosel, A., Muller, R.U., Schermer, B., Bissels, U., Inman, J., Phan, Q., Chien, M., Weir, D.B., Choksi, R., De Vita, G., Frezzetti, D., Trompeter, H.I., Hornung, V., Teng, G., Hartmann, G., Palkovits, M., Di Lauro, R., Wernet, P., Macino, G., Rogler, C.E., Nagle, J.W., Ju, J., Papavasiliou, F.N., Benzing, T., Lichter, P., Tam, W., Brownstein, M.J., Bosio, A., Borkhardt, A., Russo, J.J., Sander, C., Zavolan, M. & Tuschl, T. (2007). *Cell*, 129, 1401–14.

Lantz, C.S. & Huff, T.F. (1995). *J Immunol*, 155, 4024–9.

Lantz, C.S., Boesiger, J., Song, C.H., Mach, N., Kobayashi, T., Mulligan, R.C., Nawa, Y., Dranoff, G. & Galli, S.J. (1998). *Nature*, 392, 90–3.

Laslo, P., Spooner, C.J., Warmflash, A., Lancki, D.W., Lee, H.J., Sciammas, R., Gantner, B.N., Dinner, A.R. & Singh, H. (2006). *Cell*, 126, 755–66.

Lee, J.S. & Young, I.G. (1989). *Genomics*, 5, 359–62.

Lee, Y., Ahn, C., Han, J., Choi, H., Kim, J., Yim, J., Lee, J., Provost, P., Radmark, O., Kim, S. & Kim, V.N. (2003). *Nature*, 425, 415–19.

Lee, Y., Kim, M., Han, J., Yeom, K.H., Lee, S., Baek, S.H. & Kim, V.N. (2004a). *EMBO J*, 23, 4051–60.

Lee, Y.S., Nakahara, K., Pham, J.W., Kim, K., He, Z., Sontheimer, E.J. & Carthew, R.W. (2004b). *Cell*, 117, 69–81.

Lekstrom-Himes, J. & Xanthopoulos, K.G. (1999). *Blood*, 93, 3096–105.

Lekstrom-Himes, J.A., Dorman, S.E., Kopar, P., Holland, S.M. & Gallin, J.I. (1999). *J Exp Med*, 189, 1847–52.

Lieschke, G.J. (1997). *Ciba Found Symp*, 204, 60–74; discussion 74–7.

Lieschke, G.J., Grail, D., Hodgson, G., Metcalf, D., Stanley, E., Cheers, C., Fowler, K.J., Basu, S., Zhan, Y.F. & Dunn, A.R. (1994a). *Blood*, 84, 1737–46.

Lieschke, G.J., Stanley, E., Grail, D., Hodgson, G., Sinickas, V., Gall, J.A., Sinclair, R.A. & Dunn, A.R. (1994b). *Blood*, 84, 27–35.

Liu, F., Wu, H.Y., Wesselschmidt, R., Kornaga, T. & Link, D.C. (1996). *Immunity*, 5, 491–501.

Liu, F., Poursine-Laurent, J., Wu, H.Y. & Link, D.C. (1997). *Blood*, 90, 2583–90.

Mach, N., Lantz, C.S., Galli, S.J., Reznikoff, G., Mihm, M., Small, C., Granstein, R., Beissert, S., Sadelain, M., Mulligan, R.C. & Dranoff, G. (1998). *Blood*, 91, 778–83.

McKercher, S.R., Torbett, B.E., Anderson, K.L., Henkel, G.W., Vestal, D.J., Baribault, H., Klemsz, M., Feeney, A.J., Wu, G.E., Paige, C.J. & Maki, R.A. (1996). *EMBO J*, 15, 5647–58.

Nagata, H., Worobec, A.S., Oh, C.K., Chowdhury, B.A., Tannenbaum, S., Suzuki, Y. & Metcalfe, D.D. (1995). *Proc Natl Acad Sci USA*, 92, 10560–4.

Nerlov, C., McNagny, K.M., Doderlein, G., Kowenz-Leutz, E. & Graf, T. (1998). *Genes Dev*, 12, 2413–23.

Nicola, N.A., Robb, L., Metcalf, D., Cary, D., Drinkwater, C.C. & Begley, C.G. (1996). *Blood*, 87, 2665–74.

Nilsson, S.K. & Bertoncello, I. (1994). *Dev Biol*, 164, 456–62.

Nishinakamura, R., Miyajima, A., Mee, P.J., Tybulewicz, V.L. & Murray, R. (1996). *Blood*, 88, 2458–64.

Owen, W.F., Jr., Rothenberg, M.E., Silberstein, D.S., Gasson, J.C., Stevens, R.L., Austen, K.F. & Soberman, R.J. (1987). *J Exp Med*, 166, 129–41.

Passegue, E., Jochum, W., Schorpp-Kistner, M., Mohle-Steinlein, U. & Wagner, E.F. (2001). *Cell*, 104, 21–32.

Passegue, E., Wagner, E.F. & Weissman, I.L. (2004). *Cell*, 119, 431–43.

Person, R.E., Li, F.Q., Duan, Z., Benson, K.F., Wechsler, J., Papadaki, H.A., Eliopoulos, G., Kaufman, C., Bertolone, S.J., Nakamoto, B., Papayannopoulou, T., Grimes, H.L. & Horwitz, M. (2003). *Nat Genet*, 34, 308–12.

Pevny, L., Simon, M.C., Robertson, E., Klein, W.H., Tsai, S.F., D'Agati, V., Orkin, S.H. & Costantini, F. (1991). *Nature*, 349, 257–60.

Pojda, Z. & Tsuboi, A. (1990). *Exp Hematol*, 18, 1034–7.

Querfurth, E., Schuster, M., Kulessa, H., Crispino, J.D., Doderlein, G., Orkin, S.H., Graf, T. & Nerlov, C. (2000). *Genes Dev*, 14, 2515–25.

Radomska, H.S., Huettner, C.S., Zhang, P., Cheng, T., Scadden, D.T. & Tenen, D.G. (1998). *Mol Cell Biol*, 18, 4301–14.

Rane, S.G. & Reddy, E.P. (2002). *Oncogene*, 21, 3334–58.

Reddy, E.P., Korapati, A., Chaturvedi, P. & Rane, S. (2000). *Oncogene*, 19, 2532–47.

Reddy, V.A., Iwama, A., Iotzova, G., Schulz, M., Elsasser, A., Vangala, R.K., Tenen, D.G., Hiddemann, W. & Behre, G. (2002). *Blood*, 100, 483–90.

Rekhtman, N., Radparvar, F., Evans, T. & Skoultchi, A.I. (1999). *Genes Dev*, 13, 1398–411.

Rhodes, J., Hagen, A., Hsu, K., Deng, M., Liu, T.X., Look, A.T. & Kanki, J.P. (2005). *Dev Cell*, 8, 97–108.

Richards, M.K., Liu, F., Iwasaki, H., Akashi, K. & Link, D.C. (2003). *Blood*, 102, 3562–8.

Rodriguez, A., Vigorito, E., Clare, S., Warren, M.V., Couttet, P., Soond, D.R., van Dongen, S., Grocock, R.J., Das, P.P., Miska, E.A., Vetrie, D., Okkenhaug, K., Enright, A.J., Dougan, G., Turner, M. & Bradley, A. (2007). *Science*, 316, 608–11.

Romani, L., Mencacci, A., Cenci, E., Spaccapelo, R., Toniatti, C., Puccetti, P., Bistoni, F. & Poli, V. (1996). *J Exp Med*, 183, 1345–55.

Scott, E.W., Simon, M.C., Anastasi, J. & Singh, H. (1994). *Science*, 265, 1573–7.

Scott, L.M., Civin, C.I., Rorth, P. & Friedman, A.D. (1992). *Blood*, 80, 1725–35.

Seymour, J.F., Lieschke, G.J., Grail, D., Quilici, C., Hodgson, G. & Dunn, A.R. (1997). *Blood*, 90, 3037–49.

Sherman, M.L., Stone, R.M., Datta, R., Bernstein, S.H. & Kufe, D.W. (1990). *J Biol Chem*, 265, 3320–3.

Shibata, Y., Berclaz, P.Y., Chroneos, Z.C., Yoshida, M., Whitsett, J.A. & Trapnell, B.C. (2001). *Immunity*, 15, 557–67.

Shivdasani, R.A., Fujiwara, Y., McDevitt, M.A. & Orkin, S.H. (1997). *EMBO J*, 16, 3965–73.

Skokowa, J., Cario, G., Uenalan, M., Schambach, A., Germeshausen, M., Battmer, K., Zeidler, C., Lehmann, U., Eder, M., Baum, C., Grosschedl, R., Stanulla, M., Scherr, M. & Welte, K. (2006). *Nat Med*, 12, 1191–7.

Stanley, E.R. (1985). *Methods Enzymol*, 116, 564–87.

Stanley, E.R., Guilbert, L.J., Tushinski, R.J. & Bartelmez, S.H. (1983). *J Cell Biochem*, 21, 151–9.

Stopka, T., Amanatullah, D.F., Papetti, M. & Skoultchi, A.I. (2005). *EMBO J*, 24, 3712–23.

Sugiyama, D., Tanaka, M., Kitajima, K., Zheng, J., Yen, H., Murotani, T., Yamatodani, A. & Nakano, T. (2008). *Blood*, 111, 1924–32.

Suh, H.C., Gooya, J., Renn, K., Friedman, A.D., Johnson, P.F. & Keller, J.R. (2006). *Blood*, 107, 4308–16.

Tamura, T., Nagamura-Inoue, T., Shmeltzer, Z., Kuwata, T. & Ozato, K. (2000). *Immunity*, 13, 155–65.

Tanno, Y., Stadler, B. & Denburg, J.A. (1987). *Int Arch Allergy Appl Immunol*, 83, 1–5.

Thai, T.H., Calado, D.P., Casola, S., Ansel, K.M., Xiao, C., Xue, Y., Murphy, A., Frendewey, D., Valenzuela, D., Kutok, J.L., Schmidt-Supprian, M., Rajewsky, N., Yancopoulos, G., Rao, A. & Rajewsky, K. (2007). *Science*, 316, 604–8.

Tsai, F.Y. & Orkin, S.H. (1997). *Blood*, 89, 3636–43.

Tsai, S. & Collins, S.J. (1993). *Proc Natl Acad Sci USA*, 90, 7153–7.

Tsang, A.P., Visvader, J.E., Turner, C.A., Fujiwara, Y., Yu, C., Weiss, M.J., Crossley, M. & Orkin, S.H. (1997). *Cell*, 90, 109–19.

Walsh, J.C., DeKoter, R.P., Lee, H.J., Smith, E.D., Lancki, D.W., Gurish, M.F., Friend, D.S., Stevens, R.L., Anastasi, J. & Singh, H. (2002). *Immunity*, 17, 665–76.

Wang, D., D'Costa, J., Civin, C.I. & Friedman, A.D. (2006). *Blood*, 108, 1223–9.

Welch, J.J., Watts, J.A., Vakoc, C.R., Yao, Y., Wang, H., Hardison, R.C., Blobel, G.A., Chodosh, L.A. & Weiss, M.J. (2004). *Blood*, 104, 3136–47.

Wiktor-Jedrzejczak, W., Bartocci, A., Ferrante, A.W.Jr., Ahmed-Ansari, A., Sell, K.W., Pollard, J.W. & Stanley, E.R. (1990). *Proc Natl Acad Sci USA*, 87, 4828–32.

Xiao, C., Calado, D.P., Galler, G., Thai, T.H., Patterson, H.C., Wang, J., Rajewsky, N., Bender, T.P. & Rajewsky, K. (2007). *Cell*, 131, 146–59.

Xie, H., Ye, M., Feng, R. & Graf, T. (2004). *Cell*, 117, 663–76.

Yamaguchi, Y., Suda, T., Suda, J., Eguchi, M., Miura, Y., Harada, N., Tominaga, A. & Takatsu, K. (1988). *J Exp Med*, 167, 43–56.

Yamanaka, R., Kim, G.D., Radomska, H.S., Lekstrom-Himes, J., Smith, L.T., Antonson, P., Tenen, D.G. & Xanthopoulos, K.G. (1997). *Proc Natl Acad Sci USA*, 94, 6462–7.

Yoshida, H., Hayashi, S., Kunisada, T., Ogawa, M., Nishikawa, S., Okamura, H., Sudo, T. & Shultz, L.D. (1990). *Nature*, 345, 442–4.

Yu, C., Cantor, A.B., Yang, H., Browne, C., Wells, R.A., Fujiwara, Y. & Orkin, S.H. (2002). *J Exp Med*, 195, 1387–95.

Yuan, W., Payton, J.E., Holt, M.S., Link, D.C., Watson, M.A., DiPersio, J.F. & Ley, T.J. (2007). *Blood*, 109, 961–70.

Zhang, D.E., Zhang, P., Wang, N.D., Hetherington, C.J., Darlington, G.J. & Tenen, D.G. (1997). *Proc Natl Acad Sci USA*, 94, 569–74.

Zhang, P., Iwasaki-Arai, J., Iwasaki, H., Fenyus, M.L., Dayaram, T., Owens, B.M., Shigematsu, H., Levantini, E., Huettner, C.S., Lekstrom-Himes, J.A., Akashi, K. & Tenen, D.G. (2004). *Immunity*, 21, 853–63.

Zon, L.I., Yamaguchi, Y., Yee, K., Albee, E.A., Kimura, A., Bennett, J.C., Orkin, S.H. & Ackerman, S.J. (1993). *Blood*, 81, 3234–41.

Chapter 7
Development of T Lymphocytes

Benjamin A. Schwarz and Avinash Bhandoola

Abstract The goal of this chapter is to introduce readers to the cellular and molecular basis of T-cell differentiation. We have divided T-cell development into three major stages (early T-lineage development, β or γδ selection, positive and negative selection). These correlate with specific subsets of thymocytes (ETP-DN3, DN3-DP, DP-SP thymocytes), specific locations within the thymus (cortex, subcapsular zone, cortex and medulla), T-cell receptor rearrangement events (no functional rearrangements, β or γδ rearrangements, α rearrangements), and lineage fate decisions (T-lineage commitment, αβ vs. γδ, CD4 vs. CD8). Models for each developmental decision and the molecular players are presented.

Introduction

T lymphocytes are critical cells of the adaptive immune system. They play a central role in regulating the type and intensity of immune response as well as mediating immunity to subsequent infections. T-cell deficiencies, including HIV-induced AIDS, can lead to severe immunodeficiency and put patients at risk of opportunistic infections, whereas T-cell dysfunctions can lead to autoimmunity. Therefore, the study of T-cell development at both the cellular and molecular levels has emerged as an important field of immunology.

T lymphopoiesis differs from the development of other hematopoietic lineages in that it occurs primarily in the thymus rather than the bone marrow (BM). The thymus is a specialized hematopoietic organ whose only known function is T lymphopoiesis (Miller and Osoba 1967). The thymus is located in the anterior thorax and is derived in ontogeny from the third pharyngeal pouch (Rodewald 2008).

B.A. Schwarz and A. Bhandoola
Department of Pathology and Laboratory Medicine,
University of Pennsylvania School of Medicine,
Philadelphia, PA19104, USA
e-mail: bhandooa@mail.med.upenn.edu

A. Wickrema and B. Kee (eds.), *Molecular Basis of Hematopoiesis*,
DOI: 10.1007/978-0-387-85816-6_1, © Springer Science+Business Media, LLC 2009

Failure of normal thymic genesis, as in DiGeorge syndrome and nude mice, results in profound T-cell deficiencies (Rodewald 2008). The thymus has two symmetric lobes, each divided into a central medulla, outer cortex, and surrounding capsule (Miller and Osoba 1967; Rodewald 2008). Thymic stroma is composed predominantly of epithelial cells and dendritic cells (DCs) (Rodewald 2008). However, the vast majority of cells in the thymus are developing T-lineage progenitors of hematopoietic origin (Miller and Osoba 1967). The thymus is incapable of supporting the long-term self-renewal of these progenitors (Goldschneider et al. 1986; Scollay et al. 1986). Instead, progenitors derived from fetal liver or adult BM are imported into the thymus, where they reside transiently as developing thymocytes. The thymic stroma imposes the T-lineage fate on hematopoietic progenitors, supports T-lineage differentiation, and shapes the T-cell repertoire (Petrie and Zuniga-Pflucker 2007). Finally, the progeny of hematopoietic progenitors exits the thymus as mature T cells.

Multiple T lineages are generated within the thymus, characterized by their T-cell receptor (TCR) chain usage ($\alpha\beta$ vs. $\gamma\delta$ T cells) (Hayday and Pennington 2007; Hayes and Love 2007) and TCR coreceptor usage (CD4 vs. CD8) (Singer and Bosselut 2004; Kappes et al. 2006). In ontogeny, $\gamma\delta$ T cells are generated in waves, with each wave homing to different peripheral organs such as the epidermis or gut (Dunon et al. 1997). In adult life, $\gamma\delta$ T cells continue to be generated at a low level. Besides a role in wound healing, not much is known about the function of $\gamma\delta$ T cells (Jameson et al. 2002; Shin et al. 2005). Most peripheral T cells found in the blood or lymphatic tissues express the $\alpha\beta$ TCR. CD4+ $\alpha\beta$ T cells, referred to as helper T cells, produce cytokines that direct the type of immune response, and also interact directly with B cells. Their TCRs recognize peptide antigens in the context of MHC class II expressed by professional antigen presenting cells. A subset of CD4+ T cells, referred to as regulatory T cells, serves to dampen the immune response (Hori et al. 2003). CD8+ $\alpha\beta$ T cells, referred to as cytotoxic T lymphocytes, can directly kill host cells infected by intracellular pathogens. Their TCRs recognize peptide antigens in the context of MHC class I expressed by all nucleated host cells. Other T lineages include natural killer (NK) T cells, which recognize nonpeptide antigens presented by the nonclassical MHC molecule CD1d (Bendelac et al. 2007), and CD8$\alpha\alpha$ gut intraepithelial lymphocytes (IEL), for which a role of intrathymic development remains controversial (Lambolez et al. 2007).

The majority of thymocytes, in the postnatal thymus, are of the $\alpha\beta$ T-cell lineage. The stages of $\alpha\beta$ T-cell development are outlined in Fig. 7.1 and Table 7.1. The most mature cells express either CD4 or CD8, termed single-positive (SP) thymocytes. These are derived from cells that express both CD4 and CD8, termed double-positive (DP) thymocytes, which normally account for >80% of thymic cellularity (Egerton et al. 1990). DP thymocytes, in turn, are generated from progenitors that express neither surface CD4 nor CD8, termed double-negative (DN) thymocytes (Fowlkes et al. 1985). DN thymocytes have been traditionally divided into four subsets (DN1 through DN4) based on CD44 and CD25 expression (Table 7.1) (Pearse et al. 1989; Godfrey et al. 1993). Recently this scheme has been revised to exclude non-T-lineage progenitors. TCRβ+ DN4 thymocytes, although surface CD4– CD8–, are transcriptionally CD4+ CD8+ and can therefore be referred to as pre-DP thymocytes (Petrie et al. 1990). The earliest identified intrathymic T-lineage progenitors, on the other hand, are transcriptionally CD4– CD8–, but

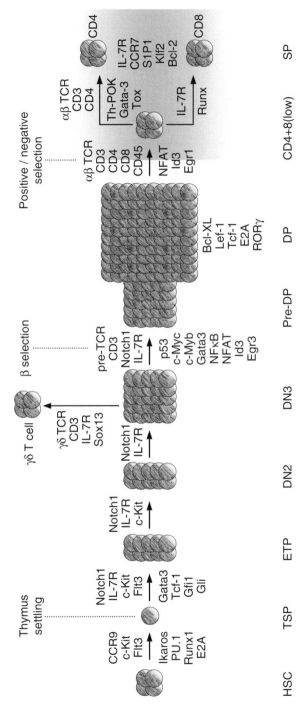

Fig. 7.1 Overview of T lymphopoiesis. Thymus settling is followed by proliferation (indicated by a relative increase in cell number) and differentiation (indicated by the labeled stages). Assembly of a functional TCR is assessed at the β and positive selection checkpoints (indicated by the *dotted lines*). Negative selection continues through the SP stage (indicated by the *shaded area*). Multiple surface receptors (listed above the *arrows*) and transcription factors (listed below the *arrows*) are implicated in each transition (signal transduction molecules and recombination machinery are not shown)

Table 7.1 Cell surface definitions of intrathymic T-lineage progenitors

Stage	CD90	TCR	CD4	CD8	CD44	CD117	CD25
ETP (DN1)	−/low	−	−/low	−	+	+	−
DN2	+	−	−	−	+	+	+
DN3	+	−	−	−	−	−	+
Pre-DP (DN4)	+	−	−	−	−	−	−
DP	+	+	+	+			
CD4 SP	+	+	+	−			
CD8 SP	+	+	−	+			

express low levels of CD4 on their surface (Wu et al. 1991b) due to passive acquisition (Michie et al. 1998). These early T-lineage progenitors (ETP) (Shortman and Wu 1996) overlap with traditional DN1 thymocytes (Bhandoola et al. 2007). However, most DN1 thymocytes are not true T-lineage progenitors. Instead, ETPs can be differentiated by surface c-Kit (CD117) expression (Ceredig and Rolink 2002; Allman et al. 2003; Porritt et al. 2004).

T-cell differentiation is coupled to intrathymic migration (Petrie and Zuniga-Pflucker 2007). Hematopoietic progenitors enter the thymus at the corticomedullary junction (CMJ) and generate the ETP pool (Kyewski 1987; Penit and Vasseur 1988; Lind et al. 2001). This is followed by DN2 migration outward through the cortex. DN3 thymocytes are located at the subcapsular zone (SCZ) of the thymus. DP thymocytes then migrate back through the cortex toward the medulla (Petrie and Zuniga-Pflucker 2007). SP thymocytes cross from the cortex into the medulla, where they reside for 4–5 days (McCaughtry et al. 2007). Finally mature thymocytes exit the thymus and contribute to the peripheral T-cell pool.

The ultimate purpose of T-cell development is to generate a diverse, functionally useful, self-tolerant T-cell repertoire. This is accomplished through organized DNA recombination to generate the diversity of TCRs (Nikolich-Zugich et al. 2004). For the most part, each T cell has only one functional TCR, thus selection for a cell's survival or proliferation can be used to select for that cell's specific TCR. Recombination is followed by selection for functionally useful TCRs (positive selection) and deletion of T cells with autoreactive TCRs (negative selection) (Starr et al. 2003; Nikolich-Zugich et al. 2004). Of the peak 50×10^6 thymocytes generated per day in mice, only about 2–3% survive these selection events and contribute to the peripheral T-cell repertoire (Egerton et al. 1990; Huesmann et al. 1991).

The goal of this chapter is to introduce readers to the cellular and molecular basis of T-cell differentiation. We have divided T-cell development into three major stages (early T-lineage development, β or γδ selection, positive and negative selection). These correlate with specific subsets of thymocytes (ETP-DN3, DN3-DP, DP-SP thymocytes), specific locations within the thymus (cortex, SCZ, cortex and medulla), T-cell receptor rearrangement events (no functional rearrangements, β or γδ rearrangements, α rearrangements), and lineage fate decisions (T-lineage commitment, αβ vs. γδ, CD4 vs. CD8) (Table 7.2). Models for each developmental decision and the molecular players are presented (Fig. 7.1).

Table 7.2 Stages of T-cell development correlate with location within the thymus, lineage commitment, TCR rearrangements, and selection for functional and self-tolerant TCRs

Stage	Location	Lineage potentials	Rearrangements	Selection
ETP	PMC	T, B, NK, DC, myeloid	$D_\beta-J_\beta$	
DN2	Cortex	T, NK, DC, myeloid	$D_\beta-J_\beta$, $V_\gamma-J_\gamma$, $V_\delta-D_\delta-J_\delta$	
DN3	SCZ	T, $\alpha\beta$ and $\gamma\delta$	$V_\beta-D_\beta J_\beta$, $V_\gamma-J_\gamma$, $V_\delta-D_\delta-J_\delta$	β selection, $\gamma\delta$ selection
DP	Cortex	$\alpha\beta$ T cell, CD4 and CD8	$V_\alpha-J_\alpha$	Positive selection, Negative selection
SP	Medulla	$\alpha\beta$ T cell, CD4 or CD8		Negative selection

Early T-Lineage Development

Prethymic Progenitors of T Cells

Like all hematopoietic lineages, T cells are ultimately derived from hematopoietic stem cells (HSCs) (Morrison et al. 1995), which in adults reside in the BM. However, T-cell development continues in the thymus (Miller and Osoba 1967). Therefore, in order to initiate T lymphopoiesis, hematopoietic progenitors must first be mobilized from the BM into the bloodstream and then they must settle within the thymus parenchyma from the bloodstream (Scollay et al. 1986; Donskoy and Goldschneider 1992; Schwarz and Bhandoola 2006). These processes are poorly understood mainly because the number of cells that undergo this migration is likely to be extremely low (Wallis et al. 1975; Donskoy and Goldschneider 1992), precluding their direct identification within the thymus. HSCs are physiologically present in adult blood (Goodman and Hodgson 1962; Wright et al. 2001; Schwarz and Bhandoola 2004). However, when placed in the bloodstream, HSCs are unable to directly settle within the thymus (Schwarz et al. 2007). This implies that prethymic differentiation steps are required to generate progenitors competent of migrating to the thymus.

As HSCs differentiate they first lose their self-renewal ability, generating multipotent progenitors (MPPs) that retain the capacity to generate all blood lineages (Morrison et al. 1995; Adolfsson et al. 2001; Christensen and Weissman 2001). MPPs were then proposed to choose between the lymphoid or myeloerythroid lineage fates (Morrison et al. 1995). Support for this model came with the identification of lymphoid-restricted common lymphoid progenitors (CLPs) (Kondo et al. 1997) and myeloid-restricted common myeloid progenitors (CMPs) (Akashi et al. 2000) in BM. However, ETPs possess myeloid and DC lineage potentials in addition to T, B, and NK lymphoid lineage potentials (Wu et al. 1991a; Ardavin et al. 1993; Shortman and Wu 1996; Allman et al. 2003). The myeloid potential of ETPs is not due to copurification of separate myeloid-committed and lymphoid-committed progenitors within the ETP population; instead clonal assays have demonstrated the existence of single cells with both T and myeloid potential within the ETP population (Bell and Bhandoola 2008; Wada et al. 2008). Furthermore, the B-lineage potential of most ETPs is lost prior to myeloid potential (Porritt et al. 2004; Benz and Bleul 2005; Sambandam et al. 2005; Ceredig et al. 2007; Bell and Bhandoola 2008; Wada et al. 2008). Therefore, the idea that all lymphocytes are generated from a common lymphoid-committed progenitor requires revision. Instead, a spectrum of progenitors may physiologically migrate to the thymus and contribute to T-cell development (Bhandoola et al. 2007), including lymphoid specified early lymphoid progenitors (ELPs), which are multipotent but express the lymphoid-specific genes *Rag1*, *Rag2*, and *TdT* (Igarashi et al. 2002) and overlap with lymphoid-primed multipotent progenitors (LMPPs), which have limited erythrocyte and megakaryocyte lineage potentials (Adolfsson et al. 2005; Forsberg et al. 2006), CLPs [which have myeloid potential in vitro (Rumfelt et al. 2006)], CLP-2 cells (Martin et al. 2003), and T-lineage committed circulating T progenitors (CTPs) (Rodewald et al. 1994; Carlyle and Zuniga-Pflucker 1998; Krueger and von Boehmer 2007). ELPs, CLPs, and CTPs all circulate (Schwarz and Bhandoola 2004; Perry et al. 2006; Krueger and von Boehmer 2007), and when placed in blood can contribute to T lymphopoiesis in the thymus (Krueger and von Boehmer 2007; Lai and Kondo 2007; Schwarz et al. 2007).

Prethymic T-Lineage Competence

Multiple transcription factors are important prethymically for the generation of T-lineage-competent ELPs or CLPs including PU.1, Runx1, and Ikaros (Laiosa et al. 2006a; Rothenberg et al. 2008). These factors are all important at the HSC stage and are required for multilineage progression. PU.1 is necessary for both myeloid and lymphoid development. Loss of PU.1 expression leads to a block in lymphopoiesis at or before the CLP stage (Dakic et al. 2005; Iwasaki et al. 2005), whereas overexpression of PU.1 at the ETP (Franco et al. 2006) or DN3 stages (Laiosa et al. 2006b) leads to lineage diversion toward myeloid or DC development. Runx1 was also found to be important for the generation of CLPs (Growney et al. 2005) and ETPs (Talebian et al. 2007). Ikaros is required for the development of both B and T lineages (Georgopoulos 2002). Ikaros null mice lack CLPs in the BM, but ETPs can be detected in the thymus of adult mice (Allman et al. 2003), suggesting that upstream progenitors of ETPs are at least partially intact. Furthermore, Ikaros is expressed at high levels in a subset of MPPs that lack megakaryocyte and erythroid potential corresponding to ELPs (Yoshida et al. 2006) and is likely to be required for efficient ELP generation. T-lineage competence is associated with the upregulation of Notch1 (Heinzel et al. 2007; Lai and Kondo 2007), critical for T-lineage progression within the thymus (Maillard et al. 2005; Tanigaki and Honjo 2007). E proteins are important for regulated expression of Notch1, suggesting a prethymic role for E proteins in ELPs (Ikawa et al. 2006).

Cytokine signaling has also been implicated in prethymic T-lineage differentiation. Prethymic HSCs, MPPs, and ELPs (Ogawa et al. 1991; Ikuta and Weissman 1992), as well as intrathymic ETPs and DN2 thymocytes, express high levels of surface c-Kit (Ceredig and Rolink 2002), the cytokine receptor for stem cell factor (SCF), whereas CLPs are c-Kit[low] (Kondo et al. 1997). Mutant c-Kit mice have an early defect in T lymphopoiesis; however, it is unclear if c-Kit is required prethymically, intrathymically, or both (Waskow et al. 2002). Flt3, the cytokine receptor for Flt3 ligand (Flt3L), is first expressed at the MPP stage of hematopoiesis (Adolfsson et al. 2001; Christensen and Weissman 2001), with both ELPs and CLPs characterized by high surface Flt3 expression dependent on Ikaros (Yoshida et al. 2006). Mice deficient in Flt3 signaling lack ELPs and CLPs in the BM (Sitnicka et al. 2002; Schwarz et al. 2007) and have a large reduction of ETPs in the thymus (Sambandam et al. 2005; Schwarz et al. 2007). However, the thymus is able to compensate for reduced ETP numbers by the DN3 stage and generate nearly normal numbers of thymocytes (Mackarehtschian et al. 1995; Sambandam et al. 2005). IL-7 receptor (IL-7R) signaling is also critical for both B and T lymphopoiesis. IL-7R is expressed first at the ELP stage (Adolfsson et al. 2005), and CLPs are defined by surface expression of this receptor (Kondo et al. 1997). It is unknown whether IL-7R is first required prethymically or intrathymically.

Mobilization and Thymus Settling

Following their generation in the BM, T-lineage-competent progenitors must next migrate to the thymus. The molecular basis for progenitor mobilization from BM to blood has been studied extensively for HSCs, but not for any relevant T-lineage-competent progenitor. Mobilization is likely to require cytokine signals, loss of responsiveness to chemokine gradients retaining progenitors in the BM, and loss of adhesion to the BM matrix (Schwarz and Bhandoola 2006). Thymus settling is likely to be analogous to the migration of mature leukocytes, involving selectin-mediated weak adhesion to blood vessel endothelial cells, followed by chemokine signaling, strong adhesion mediated by integrins, and finally transmigration (Cyster 2005).

CD44 (Lesley et al. 1985) and P-selectin glycoprotein ligand 1 (PSGL-1) (Rossi et al. 2005) expression on progenitors have both been implicated in weak adhesion to the thymic vasculature. A role for α4β1 and αLβ2 integrins has been demonstrated for CLP-2 progenitor settling (Scimone et al. 2006). Chemokine receptors implicated in thymic homing include CCR7 and CCR9. CCR7 has been shown to be important in homing to the fetal but not adult thymus, whereas CCR9 is important in both fetal and adult life (Uehara et al. 2002; Liu et al. 2005, 2006; Wurbel et al. 2006; Schwarz et al. 2007). CCR9 is expressed by a subset of ELPs, CLPs, and CLP-2 progenitors but not HSCs and therefore is one of the molecular determinants for the selectivity of thymus settling (Scimone et al. 2006; Lai and Kondo 2007; Schwarz et al. 2007). Importantly, the generation of CCR9-expressing progenitors requires Flt3 signaling (Schwarz et al. 2007).

Intrathymic Proliferation and T-Lineage Commitment

The number of progenitors that enter the thymus is likely to be extremely small, with estimates ranging from 1 to 100 cells/day (Wallis et al. 1975; Donskoy and Goldschneider 1992). Progenitors enter the thymus at the CMJ (Lind et al. 2001), where they proliferate to generate the ETP population (~20,000 cells per mouse thymus) (Allman et al. 2003). This stage is estimated to last at least 10 days (Porritt et al. 2003), with ~10 rounds of division (Shortman et al. 1990), making it the longest period of intrathymic T-cell development with the most extensive proliferation. However, the molecular requirements for this proliferation are poorly characterized. Cytokine signals, which may include SCF and IL-7, are likely important. Flt3, which is required prethymically to generate thymus-settling progenitors (TSPs) (Schwarz et al. 2007), is also important for intrathymic proliferation (Kenins et al. 2008). Flt3 is downregulated within developing ETPs and is not expressed at any subsequent stage of T lymphopoiesis (Sambandam et al. 2005). Transcription factors implicated to act at this stage include Tcf-1 downstream of the Wnt/β-catenin signaling pathway, Gfi1, and Gata3 (Hattori et al. 1996; Yucel et al. 2003), all of which play recurrent roles in T lymphopoiesis (Ho and Pai 2007; Rothenberg et al. 2008), as well as Gli factors downstream of the hedgehog signaling pathway (El Andaloussi et al. 2006).

One of the master regulators of T-cell development is Notch1 (Maillard et al. 2005; Tanigaki and Honjo 2007). Constitutively active Notch is sufficient to drive T-cell development to the DP stage even in the BM (Pui et al. 1999). Consistent with this, many T cell acute lymphocytic leukemias (T-ALL) are associated with activating Notch mutations (Aster et al. 2008). Furthermore, OP9 cell lines, which support the development of B cells in vitro, can induce T lymphopoiesis simply by transduction with the Notch ligand delta-like 1 (DL1) (Schmitt and Zuniga-Pflucker 2002). The Notch family of receptors is important in many developmental systems. Notch is expressed on the cell surface, and upon binding its ligands (DL1, DL3, DL4, Jagged1, and Jagged2) undergoes enzymatic cleavage to release the intracellular Notch domain (ICN). ICN functions as a transcription factor, binding to CSL and converting it from a transcriptional repressor to a transcriptional activator (Maillard et al. 2005; Tanigaki and Honjo 2007). Notch1 is expressed by BM progenitors and upregulated at the ELP stage (Heinzel et al. 2007), but signaling in the BM is constrained by Notch regulators including Pokemon (Maeda et al. 2007; Maillard et al. 2008). Notch is absolutely required for the development of ETPs from TSPs (Sambandam et al. 2005; Tan et al. 2005). The mechanisms by which Notch signaling plays such an important role in early T-cell development remain largely unknown. The direct effect of Notch activation is increased transcription of canonical Notch target genes including *Nrarp*, *Deltex1*, and the basic-helix-loop-helix transcription factor *Hes1*. Notch also activates genes that are cell-type and developmental stage specific, such as *ptcra* encoding the pre-TCR α chain (pTα) and *il2ra* encoding CD25 (Maillard et al. 2005; Rothenberg et al. 2008).

The most studied function of Notch at this stage of development is in the T- versus B-lineage commitment decision. *Notch1*-deficient mice have increased B lymphopoiesis in the thymus at the expense of T cells (Radtke et al. 1999), whereas ICN-transduced BM progenitors undergo T lymphopoiesis within the BM at the expense of B cells (Pui et al. 1999). One explanation for these results is that Notch acts on a multipotent progenitor with both B- and T-lineage potentials. Such cells have been identified in the thymus, within the ETP Flt3+ population, which has received the least amount of Notch signaling and may represent the most immediate progeny of TSPs (Benz and Bleul 2005; Sambandam et al. 2005). The majority of ETPs, however, have lost B-lineage potential (Benz and Bleul 2005; Sambandam et al. 2005; Ceredig et al. 2007) requiring prolonged exposure to Notch ligands (Taghon et al. 2005). Notch signaling has also been shown to be important for the loss of myeloid (Bell and Bhandoola 2008) and NK (Schmitt et al. 2004) lineage potentials during the process of T-lineage commitment. Notch signaling results in the generation of DN2 thymocytes that migrate from the perimedullary cortex (PMC) to the SCZ (Petrie and Zuniga-Pflucker 2007). The molecular basis for this migration is poorly understood. At the SCZ DN2 thymocytes progress to DN3 thymocytes, which are finally committed to the T lineage (Petrie and Zuniga-Pflucker 2007).

TCRβ and γδ Selection

TCRβ and γδ Rearrangements

The first phase of T-cell development is independent of the TCR. All subsequent development in the thymus, beyond the DN3 stage, requires the assembly of a pre-TCR for the αβ lineage or a complete γδ TCR (Dudley et al. 1994; Fehling et al. 1995). Assembly of these receptors requires DNA rearrangements of *tcrb* or *tcrg* and *tcrd*. For *tcrb*, diversity (D_β) and joining (J_β) segments are ligated first (Krangel 2007). This begins at the ETP stage and is completed by the DN3 stage. Variable (V_β)-to-$D_\beta J_\beta$ rearrangements occur within quiescent DN3 thymocytes to generate a fully rearranged *tcrb* locus (Fig. 7.2) (Krangel 2007). One-third of rearrangements are productive (in frame) and can be translated into functional TCRβ protein, which can then pair with pTα to assemble the pre-TCR (Borowski et al. 2002). Cells that succeed in producing either a pre-TCR or a complete γδ receptor (through productive rearrangements of both *tcrg* and *tcrd*) are selected to survive and differentiate further, whereas cells that fail to make either receptor die by apoptosis (Borowski et al. 2002).

TCR rearrangements are mediated by the DNA recombinases Rag-1 and Rag-2, as well as constitutively present housekeeping DNA repair enzymes (Fugmann et al. 2000). The Rag recombinases recognize conserved heptamer–nonamer DNA recombination signal sequences (RSS) with spacers of either 12 or 23 bases that flank the V(D)J cassettes (Fugmann et al. 2000). According to the 12/23 rule, a 12 RSS can only recombine with a 23 RSS. For all four TCR loci, each V region is flanked by a 23 RSS, each J region is flanked by a 12 RSS, whereas D_β and D_δ are each flanked by a 12 and 23 RSS. This creates the possibility of V-to-J recombination

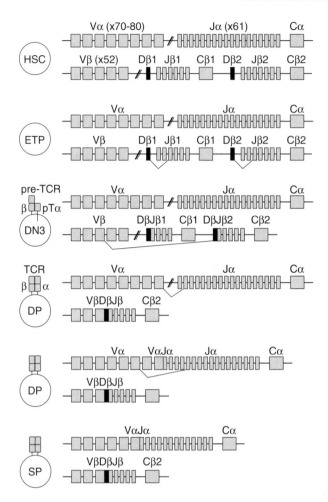

Fig. 7.2 Generation of the αβ TCR requires sequential rearrangements. The *tcra* locus is composed of 70–80 V_α segments, 61 J_α segments, and 1 C_α segment, whereas the *tcrb* locus is composed of 52 V_β segments, $D_\beta 1$ followed by 6 $J_\beta 1$ segments and $C_\beta 1$, and $D_\beta 2$ followed by 7 $J_\beta 2$ segments and $C_\beta 2$. HSCs have germline *tcra* and *tcrb* loci. ETPs initiate D_β-to-J_β rearrangements. DN3 thymocytes undergo V_β-to-$D_\beta J_\beta$ rearrangements, producing a fully rearranged *tcrb* locus. Productive rearrangements are assessed through the pairing of TCRβ with pTα to produce the pre-TCR, which permits progression to the DP stage. V_α-to-J_α rearrangements occur at the DP stage, first utilizing 3'V_α and 5'J_α regions, allowing for multiple rearrangements of each allele until a completed αβ TCR is able to interact with MHC/peptide resulting in positive selection and progression to the SP stage

without D usage for both *tcrb* and *tcrd*. However, this is avoided by temporally separating D-to-J recombination from accessibility of the V regions to Rag (Krangel 2007). For *tcrb*, an enhancer region (E_β), activated in ETPs, is important for opening chromatin spanning the D_β, J_β, and constant (C_β) gene segments, in part through activating the $D_\beta 1$ promoter, but does not allow for accessibility of the V_β regions

(Mathieu et al. 2000). At the DN3 stage, the V_β promoters become active, independent of E_β, making the V_β regions accessible to Rag and allowing V_β-to-$D_\beta J_\beta$ recombination (Krangel 2007).

Generation and Survival of DN3 Thymocytes

IL-7R, which is important for both B and T lymphopoiesis, is critical by the DN3 stage of T-cell development (Peschon et al. 1994). Interestingly, IL-7R has distinct roles for $\alpha\beta$ and $\gamma\delta$ T-cell development (Hayes and Love 2007). For the $\alpha\beta$ lineage, the primary function of IL-7R may be survival, as an IL-7R requirement can be bypassed by enforced expression of an antiapoptotic protein such as Bcl-2 (Akashi et al. 1997), although this result is controversial (Rodewald et al. 2001). Furthermore, IL-7 is likely to be one of the factors that defines the DN3 niche (Prockop and Petrie 2004). Upstream defects in T-cell development, such as deficient thymus settling, can be compensated for at the DN3 stage as progenitors proliferate to fill the DN3 compartment (Schwarz et al. 2007). IL-7R is downregulated following β-selection, depriving DP thymocytes from this survival signal until successful positive selection (Yu et al. 2006). For the $\gamma\delta$ lineage, IL-7R signaling is required for accessibility of the Tcrg locus for recombination (Schlissel et al. 2000). IL-7R may also be important for the survival and proliferation of $\gamma\delta$ progenitors (Hayes and Love 2007).

Notch also plays a recurrent role in T lymphopoiesis at the DN3 stage (Maillard et al. 2005; Ciofani and Zuniga-Pflucker 2007; Tanigaki and Honjo 2007). Notch1 has been implicated in mediating the accessibility of the *tcrb* locus (Wolfer et al. 2002). Furthermore, Notch acts as a trophic signal, important for survival of preselected DN3 thymocytes, through the activation of Akt and c-Myc (Ciofani and Zuniga-Pflucker 2005; Weng et al. 2006). Notch has also been implicated in the $\alpha\beta$ versus $\gamma\delta$ T-lineage commitment decision (Hayday and Pennington 2007). Attenuated Notch signaling appears to favor the $\gamma\delta$ lineage over the $\alpha\beta$ lineage (Wolfer et al. 2002; Tanigaki et al. 2004). However, this may be due to a continued metabolic requirement for Notch signaling after β-selection, whereas Notch signals are less critical for the survival of postselection $\gamma\delta$ T cells (Ciofani et al. 2006). Following the DN3 to DP transition in $\alpha\beta$ T-cell development, Notch receptor is downregulated and cells become dependent on TCR signaling for survival (Wolfer et al. 2002; Tanigaki et al. 2004; Ciofani and Zuniga-Pflucker 2007).

The $\alpha\beta$ Versus $\gamma\delta$ Lineage Decision

Commitment to either the $\alpha\beta$ or $\gamma\delta$ lineage correlates with production of the pre-TCR or $\gamma\delta$ TCR, respectively (Hayday and Pennington 2007). Therefore, an instructive lineage fate model has been proposed based on the identity of the receptor first expressed.

However, experimental systems have revealed that thymocytes with a γδ TCR receptor can be diverted down the αβ lineage pathway and cells with productive β rearrangement can generate functionally γδ T cells. Furthermore, the CD3 signaling complex downstream of the γδ TRC and pre-TCR is the same. How then do DN3 thymocytes know whether they have generated a productive γδ or pre-TCR receptor? A current model proposes that the strength of receptor signaling determines the αβ versus γδ lineage fate (Hayes and Love 2006; Lauritsen et al. 2006). In this model, TCR γδ mediates a stronger signal than the pre-TCR, signaling commitment to the γδ lineage. This model can account for why attenuated signaling by the γδ TCR can lead to conversion to the αβ lineage (Hayes et al. 2005), whereas strong signals mediated by αβ TCR transgenes can lead to functional γδ T-cell development (Bruno et al. 1996).

Many questions remain about the αβ versus γδ lineage decision. How is the TCR signal translated into lineage commitment? Which transcription factors are differentially required for each lineage? What are the molecular targets of these transcription factors and how do they promote one lineage while inhibiting the other? One molecule implicated in the γδ lineage choice is the HMG-box transcription factor Sox13. Overexpression of Sox13 promotes the γδ lineage at the expense of the αβ lineage, whereas Sox13 deficiency results in a block of γδ but not αβ lineage thymocytes (Melichar et al. 2007). One role of Sox13 is to inhibit the Wnt/Tcf pathway important for the development and survival of αβ T-lineage progenitors (Ioannidis et al. 2001; Melichar et al. 2007).

The Pre-TCR Checkpoint

Expression of the pre-TCR, composed of TCRβ covalently bound to pTα, is the mechanism by which thymocytes are informed that they have produced a productive TCRβ protein. The gene encoding pTα is upregulated in DN3 thymocytes, regulated by c-Myb, Notch, and E proteins (Reizis and Leder 2001, 2002; Takeuchi et al. 2001). The pTα subunit may act as a surrogate TCRα chain stabilizing the TCRβ chain and bringing it to the cell surface. However, substituting a TCRα chain for pTα at the DN3 stage results in less efficient β selection, implicating a function for the cytoplasmic tail of pTα in normal pre-TCR signaling (Borowski et al. 2004). Furthermore, the pre-TCR may be less stable and more rapidly internalized than an αβ or γδ TCR capable of binding ligand (von Boehmer 2005). Therefore the pre-TCR would produce a weak signal, consistent with the signal strength model for αβ T-lineage commitment.

The pre-TCR functions independently of ligand binding, instead requiring only an association with CD3 and the proximal TCR signaling complex localized to the cell surface in order to signal (Yamasaki and Saito 2007). Pre-TCR activation leads to proliferation, survival, *tcrb* allelic exclusion, and differentiation into DP thymocytes, whereas DN3 thymocytes that fail to produce a pre-TCR (or γδ TCR) die by neglect (Aifantis et al. 2006). Many molecules are required to successfully pass this developmental checkpoint including proteins involved in DNA rearrangement (Rag-1, Rag-2, DNA-PK), components of the pre-TCR (pTα, TCRβ), and proximal

TCR signaling molecules (CD3γ, CD3ε, Lck and Fyn, Zap70 and Syk, SLP-76, LAT). Deficiencies of any of these molecules result in a partial or complete block of T-cell development at the DN3 stage (Yamasaki and Saito 2007). Furthermore, distinct molecular pathways may account for the different functional consequences of β selection. Activation of Ras, for example, promotes thymocyte proliferation and differentiation but not allelic exclusion (Gartner et al. 1999).

Successful β selection leads to a burst of intrathymic proliferation prior to *tcra* rearrangement, insuring that each productive TCRβ rearrangement will have the chance to pair with multiple TCRα chains. DN3 cells that have passed β selection can be identified by CD27 expression (Taghon et al. 2006). Next, thymocytes upregulate expression of both CD4 and CD8, producing first pre-DP, then immature single positive (ISP), and finally DP thymocytes, which are the vast majority of cells in the thymus (Singer and Bosselut 2004). The DP population migrates back through the cortex from the SCZ toward the CMJ (Petrie and Zuniga-Pflucker 2007). DP thymocytes turn over very rapidly with ~50×10^6 DP cells generated daily and lasting on an average for 3 days (in young adult mice), which is remarkable considering the small number of progenitors that entered the thymus only a few weeks earlier, (Egerton et al. 1990).

The regulation of CD4 and the CD8αβ heterodimer expression is complex, with important differences between the DP stage, where expression is reversible, and the SP and mature T-cell stages, where expression of one coreceptor and silencing of the other is permanent and passed on epigenetically to downstream progeny (Kioussis and Ellmeier 2002). CD4 is regulated by a silencer, which is active in early T-cell development but not required after CD8 lineage commitment (Sawada et al. 1994; Zou et al. 2001). Runx1 is required for CD4 suppression in DN thymocytes, whereas Runx3 is important for establishing epigenetic silencing of CD4 in CD8 SP thymocytes (Taniuchi et al. 2002). CD8 chains, on the other hand, are regulated mainly by enhancer regions (Ellmeier et al. 1998).

The DN to DP transition is regulated by multiple transcription factors including Notch, c-Myc, c-Myb, HEB, E2A, p53, NFκB, NFAT, Tcf-1, Egr3, and RORγ (Aifantis et al. 2006). Egr3 is rapidly and transiently induced by pre-TCR signaling allowing proliferation by directly inhibiting RORγ and also inducing Id3, which inhibits the E protein E2A important for RORγ expression (Engel and Murre 2004; Xi et al. 2006). When Egr3 levels decline, RORγ and E2A are no longer suppressed, resulting in cell cycle arrest and allowing for *tcra* rearrangements (Li et al. 1996; Hernandez-Munain et al. 1999) and assembly of a complete αβ TCR.

Positive and Negative Selection

TCRα Rearrangements and Positive Selection

TCR V_α-to-J_α rearrangements occur in a progressive sequence, with the initial use of central $3'V_\alpha$ segments and $5'J_\alpha$ segments followed by more peripheral $5'V_\alpha$ and $3'J_\alpha$ segment usage (Fig. 7.2) (Petrie et al. 1993). This spares distal V_α regions and proximal J_α regions from early deletion, allowing for multiple chances to rearrange each allele. The TCRα enhancer (Eα) mediates Rag accessibility to chromatin spanning the proximal third of V_α segments and all of the J_α segments through activation of the T early-α and $J_\alpha 49$ promoters proximal to $5'J_\alpha$ segments (Krangel 2007). Once these promoters are deleted by primary recombination, introduced V_α promoters target recombination to the most 5′ remaining J_α gene segments (Krangel 2007). Interestingly, the repertoire of TCRα usage is dependent on the 3-day programmed lifespan of DP thymocytes regulated by transcription factors including RORγ, Tcf-1, and Lef-1, which control expression of the antiapoptotic factor $Bcl-x_L$ (Starr et al. 2003). If the DP lifespan is shortened there is a bias toward $5'J_\alpha$ usage, whereas if the DP lifespan is increased there is a bias toward $3'J_\alpha$ usage (Guo et al. 2002). Rearrangements of *tcra* cease following positive selection (Turka et al. 1991).

Positive selection screens for functionally useful αβ TCRs based on their ability to bind self-MHC/peptide expressed by thymic stromal cells (Starr et al. 2003). Positive selection differs from β selection in that productive *tcra* rearrangements are insufficient for selection and productive rearrangement of both *tcra* alleles is allowed (Petrie et al. 1993). Instead, the production of a TCRα that, in combination with TCRβ, can bind to self-MHC/peptide leading to a TCR signal is required for successful positive selection and progression to the SP stage. Cells that fail to produce a TCR that can bind ligand die by neglect. Thymocyte development is blocked at the DP stage in the absence of TCRα, both MHC-I and MHC-II, or signaling molecules downstream of the TCR that do not block T-cell development at the upstream DN3 stage (Starr et al. 2003). Positive selection greatly influences the T-cell repertoire, imposing self-MHC restriction (Lo and Sprent 1986) and creating a repertoire that will recognize foreign MHC as *altered self*, accounting for the robust alloreactivity of T cells.

One important question is how the highly variable binding region of the TCR is specified to interact only with MHC. Is the ability to bind MHC programmed within the TCR gene loci (Huseby et al. 2005), based on selection (Van Laethem et al. 2007), or both? An insight to this question was provided by the role of the CD4 and CD8 coreceptors (Van Laethem et al. 2007). CD4 binds directly to MHC-II whereas CD8 binds to MHC-I, stabilizing the TCR-MHC interaction (Doyle and Strominger 1987; Norment et al. 1988). Another function of these coreceptors is to bind the tyrosine kinase Lck through their intracellular domains, which is important for TCR signaling (Turner et al. 1990). Therefore, in order for the TCR to effectively signal, the TCR and either CD4 or CD8 must be brought in close proximity. This is accomplished through both the TCR and a coreceptor binding simultaneously to MHC, thus imposing MHC specificity on the positive selection signal (Van Laethem et al. 2007).

The CD4 Versus CD8 Lineage Decision

The CD4 versus CD8 lineage decision is also based on the specificity of the TCR for either MHC-II or MHC-I, respectively. This was shown using mice transgenic for a TCR restricted to either MHC-II, in which mainly CD4+ T cells were generated, or MHC-I, in which mainly CD8+ T cells were produced (Teh et al. 1988; Kaye et al. 1989). This has led to both instructive (CD4 coreceptor binding to MHC-II or CD8 binding to MHC-I instructs lineage commitment) and stochastic (progenitors randomly chosen to become CD4 or CD8 SPs and if they lose the ability to bind MHC, due to downregulating the wrong coreceptor, they are deleted) models of lineage commitment. A further distinction exists between qualitative and quantitative models of lineage commitment (Yasutomo et al. 2000). A current kinetic signaling model incorporates components of earlier schemes (Singer and Bosselut 2004). This model postulates that following positive selection, DP thymocytes downregulate CD8, generating a CD4+8^{low} population (Brugnera et al. 2000). Cells that continue to receive TCR signaling despite the loss of CD8 differentiate into CD4 SPs, whereas cells that lose TCR signaling due to a CD8 coreceptor requirement undergo coreceptor reversal and differentiate into CD8 SPs (Brugnera et al. 2000). Therefore, in this model, TCR signaling alone is sufficient to specify the CD4 versus CD8 fate and this decision is temporally separated from prior positive selection (Fig. 7.1).

TCR signaling at this stage is again translated into a lineage commitment decision. A recent insight into this process came with the discovery of the transcription factor Th-POK, which is both necessary and sufficient for CD4 lineage commitment (He et al. 2005; Sun et al. 2005). Runx transcription factor complexes, on the other hand, are required for CD8 lineage progression (Woolf et al. 2003; Setoguchi et al. 2008). Th-POK is first expressed at the CD4+8^{low} stage and may be regulated by a distal response element containing a Runx-dependent silencer (He et al. 2008; Setoguchi et al. 2008). Furthermore, Th-POK promotes CD4 expression by antagonizing Runx-dependent silencing of CD4 (Wildt et al. 2007). Enforced expression of Th-POK in mature CD8+ T cells results in downregulation of CD8 and other genes important for cytotoxic T-cell function, whereas helper T-cell-specific genes were upregulated (Jenkinson et al. 2007). These studies suggest a binary molecular commitment switch regulated by antagonism of Th-POK and Runx. Gata3 is also required for the generation of CD4 but not CD8 SPs from DPs; however, Gata3 overexpression does not lead to lineage diversion. Instead, Gata3 induced by TCR signaling may act as a CD4-specific survival or maturation factor (Hernandez-Hoyos et al. 2003; Pai et al. 2003). The HMG-box transcription factor Tox is similarly required for CD4 but not CD8 SP generation (Aliahmad and Kaye 2008). Tox deficiency leads to a block upstream of the CD4+8^{low} stage necessary for CD4 but not CD8 lineage commitment. For the generation of CD8 SPs, IL-7R serves as a prosurvival receptor via upregulation of the antiapoptotic factor Bcl-2, and may also contribute to efficient CD4 repression (Yu et al. 2003).

Negative Selection

Random TCR rearrangements can also produce self-reactive TCRs, which can strongly recognize self-peptides presented by self-MHC, potentially leading to autoimmunity. Therefore, another important function of the thymus is imposing central tolerance through the negative selection of autoreactive T cells (Starr et al. 2003; Mathis and Benoist 2007). Notably, the same signaling complex (TCR and CD3) mediates both the positive and negative selection of DP thymocytes. This has led to the model that the strength of TCR signaling, based on avidity for self-MHC/peptide, determines whether progenitors die by neglect (too little), die by negative selection (too much), or are positively selected (just right) (White et al. 1989; Ashton-Rickardt et al. 1994; Sebzda et al. 1994; Goldrath and Bevan 1999). Strong TCR signaling that does not lead to negative selection may result in the development of regulatory T cells (Jordan et al. 2001) through induction of the Foxp3 transcription factor (Fontenot et al. 2003).

The molecular mechanism by which the strength of TCR signals mediates these very different outcomes is unclear (Starr et al. 2003). Disruption of signaling molecules downstream of the TCR usually leads to blocks at the β selection checkpoint or failed positive selection, making this step very difficult to study. One proposed model is that weak positive selection signals lead to partial phosphorylation of the adaptor protein LAT, whereas strong negatively selecting signals lead to full LAT phosphorylation, recruitment of Grb2, and activation of the Erk, p38, and Jnk pathways (Starr et al. 2003). Another molecule implicated specifically in negative selection is the proapoptotic factor Bim (Bouillet et al. 2002); however, how strong TCR signals activate Bim is unknown.

An alternative model proposes that negative selection signals are differentiated from positive selection signals based on thymic location (Laufer et al. 1996; Hogquist et al. 2005). Mature SP thymocytes are generated in the thymic cortex and then migrate into the medulla (Petrie and Zuniga-Pflucker 2007) where they reside for 4–5 days (McCaughtry et al. 2007). In this model, TCR signaling to cortical DPs leads to positive selection, whereas signaling in the medulla to SP thymocytes results in negative selection (Laufer et al. 1996). Thymocytes may be able to differentiate their location based on the expression of costimulatory molecules, including CD40, B7-1, and B7-2, by medullary stroma including DCs and medullary thymic epithelial cells (mTECs) (Foy et al. 1995; Gao et al. 2002). Furthermore, mTECs express AIRE, a transcription factor essential for promiscuous expression of organ-specific self-antigens, important for central tolerance through negative selection and/or the generation of regulatory T cells (Anderson et al. 2002; Mathis and Benoist 2007). AIRE deficiency is associated with autoimmune polyendocrinopathy-candidiasis-ectodermal dystrophy (APECED) (Mathis and Benoist 2007). Migration into the medulla is dependent on the CCR7 chemokine receptor expressed by SP thymocytes (Ueno et al. 2004). Interestingly, CCR7 deficiency also leads to autoimmunity (Kurobe et al. 2006). Of note, cortical and medullary negative selection are not mutually exclusive and both can occur under experimental conditions (Hogquist et al. 2005).

Thymic Egress

The final step of T lymphopoiesis is the emigration of mature SP lymphocytes from the thymus. Emigration is regulated by the Klf2 transcription factor, which activates a number of trafficking molecules on SP thymocytes, including CCR7, CD62L, β_7 integrin, and sphingosine-1-phosphate receptor 1 ($S1P_1$), important for the migration of naïve lymphocytes to secondary lymphatic organs (Carlson et al. 2006). Of these, $S1P_1$ was shown to be required for thymocyte egress (Matloubian et al. 2004).

Conclusion

Mature T cells leave the thymus very different from the relatively undifferentiated hematopoietic progenitors that entered only a few weeks before. T lymphopoiesis is shaped by a complicated relationship between signals intrinsic to the developing hematopoietic cells and environmental cues. Early thymocyte development is influenced primarily by cytokine and Notch signals, which direct the survival, proliferation, and T-lineage commitment of BM-derived TSPs. At the DN3 stage, the TCR usurps this role. The TCR's ability to interact with the thymic environment then determines the ultimate fate of the thymocyte. Most cells produced in the thymus will never leave, but will die within the thymus, whereas the few that complete this journey are carefully selected to be part of the T-cell repertoire. An even smaller subset of T cells will at some point encounter antigen, and with that a new chapter of T-cell development begins.

References

Adolfsson, J., Borge, O.J., Bryder, D., Theilgaard-Monch, K., Astrand-Grundstrom, I., Sitnicka, E., Sasaki, Y., and Jacobsen, S. E. 2001. Immunity 15:659–669.

Adolfsson, J., Mansson, R., Buza-Vidas, N., Hultquist, A., Liuba, K., Jensen, C.T., Bryder, D., Yang, L., Borge, O.J., Thoren, L.A., Anderson, K., Sitnicka, E., Sasaki, Y., Sigvardsson, M., and Jacobsen, S. E. 2005. Cell 121:295–306.

Aifantis, I., Mandal, M., Sawai, K., Ferrando, A., and Vilimas, T. 2006. Immunol Rev 209:159–169.

Akashi, K., Kondo, M., von Freeden-Jeffry, U., Murray, R., and Weissman, I.L. 1997. Cell 89:1033–1041.

Akashi, K., Traver, D., Miyamoto, T., and Weissman, I.L. 2000. Nature 404:193–197.

Aliahmad, P. and Kaye, J. 2008. J Exp Med 205:245–256.

Allman, D., Sambandam, A., Kim, S., Miller, J. P., Pagan, A., Well, D., Meraz, A., and Bhandoola, A. 2003. Nat Immunol 4:168–174.

Anderson, M.S., Venanzi, E.S., Klein, L., Chen, Z., Berzins, S.P., Turley, S.J., von Boehmer, H., Bronson, R., Dierich, A., Benoist, C., and Mathis, D. 2002. Science 298:1395–1401.

Ardavin, C., Wu, L., Li, C.L., and Shortman, K. 1993. Nature 362:761–763.

Ashton-Rickardt, P.G., Bandeira, A., Delaney, J.R., Van Kaer, L., Pircher, H.P., Zinkernagel, R.M., and Tonegawa, S. 1994. Cell 76:651–663.

Aster, J.C., Pear, W.S., and Blacklow, S.C. 2008. Annu Rev Pathol 3:587–613.

Bell, J.J. and Bhandoola, A. 2008. Nature 452:764–767.

Bendelac, A., Savage, P.B., and Teyton, L. 2007. Annu Rev Immunol 25:297–336.

Benz, C. and Bleul, C.C. 2005. J Exp Med 202:21–31.

Bhandoola, A., von Boehmer, H., Petrie, H.T., and Zuniga-Pflucker, J.C. 2007. Immunity 26:678–689.

Borowski, C., Martin, C., Gounari, F., Haughn, L., Aifantis, I., Grassi, F., and von Boehmer, H. 2002. Curr Opin Immunol 14:200–206.

Borowski, C., Li, X., Aifantis, I., Gounari, F., and von Boehmer, H. 2004. J Exp Med 199:607–615.

Bouillet, P., Purton, J.F., Godfrey, D.I., Zhang, L.C., Coultas, L., Puthalakath, H., Pellegrini, M., Cory, S., Adams, J. M., and Strasser, A. 2002. Nature 415:922–926.

Brugnera, E., Bhandoola, A., Cibotti, R., Yu, Q., Guinter, T.I., Yamashita, Y., Sharrow, S.O., and Singer, A. 2000. Immunity 13:59–71.

Bruno, L., Fehling, H.J., and von Boehmer, H. 1996. Immunity 5:343–352.

Carlson, C.M., Endrizzi, B.T., Wu, J., Ding, X., Weinreich, M.A., Walsh, E. R., Wani, M.A., Lingrel, J.B., Hogquist, K. A., and Jameson, S.C. 2006. Nature 442:299–302.

Carlyle, J.R. and Zuniga-Pflucker, J.C. 1998. Immunity 9:187–197.

Ceredig, R. and Rolink, T. 2002. Nat Rev Immunol 2:888–897.

Ceredig, R., Bosco, N., and Rolink, A. G. 2007. Eur J Immunol 37:830–837.

Christensen, J. L. and Weissman, I.L. 2001. Proc Natl Acad Sci USA 98:14541–14546.

Ciofani, M. and Zuniga-Pflucker, J.C. 2005. Nat Immunol 6:881–888.

Ciofani, M. and Zuniga-Pflucker, J.C. 2007. Annu Rev Cell Dev Biol 23:463–493.

Ciofani, M., Knowles, G.C., Wiest, D.L., von Boehmer, H., and Zuniga-Pflucker, J.C. 2006. Immunity 25:105–116.

Cyster, J.G. 2005. Annu Rev Immunol 23:127–159.

Dakic, A., Metcalf, D., Di Rago, L., Mifsud, S., Wu, L., and Nutt, S.L. 2005. J Exp Med 201:1487–1502.

Donskoy, E. and Goldschneider, I. 1992. J Immunol 148:1604–1612.

Doyle, C. and Strominger, J.L. 1987. Nature 330:256–259.

Dudley, E.C., Petrie, H.T., Shah, L.M., Owen, M.J., and Hayday, A.C. 1994. Immunity 1:83–93.

Dunon, D., Courtois, D., Vainio, O., Six, A., Chen, C. H., Cooper, M.D., Dangy, J.P., and Imhof, B.A. 1997. J Exp Med 186:977–988.

Egerton, M., Scollay, R., and Shortman, K. 1990. Proc Natl Acad Sci USA 87:2579–2582.

El Andaloussi, A., Graves, S., Meng, F., Mandal, M., Mashayekhi, M., and Aifantis, I. 2006. Nat Immunol 7:418–426.

Ellmeier, W., Sunshine, M.J., Losos, K., and Littman, D.R. 1998. Immunity 9:485–496.

Engel, I. and Murre, C. 2004. EMBO J 23:202–211.

Fehling, H.J., Krotkova, A., Saint-Ruf, C., and von Boehmer, H. 1995. Nature 375:795–798.

Fontenot, J.D., Gavin, M.A., and Rudensky, A.Y. 2003. Nat Immunol 4:330–336.

Forsberg, E. C., Serwold, T., Kogan, S., Weissman, I.L., and Passegue, E. 2006. Cell 126:415–426.

Fowlkes, B.J., Edison, L., Mathieson, B.J., and Chused, T.M. 1985. J Exp Med 162:802–822.

Foy, T. M., Page, D.M., Waldschmidt, T.J., Schoneveld, A., Laman, J.D., Masters, S. R., Tygrett, L., Ledbetter, J. A., Aruffo, A., Claassen, E., Xu, J.C., Flavell, R.A., Oehen, S., Hedrick, S.M., and Noelle, R. J. 1995. J Exp Med 182:1377–1388.

Franco, C.B., Scripture-Adams, D.D., Proekt, I., Taghon, T., Weiss, A.H., Yui, M.A., Adams, S.L., Diamond, R.A., and Rothenberg, E.V. 2006. Proc Natl Acad Sci USA 103:11993–11998.

Fugmann, S.D., Lee, A.I., Shockett, P.E., Villey, I.J., and Schatz, D.G. 2000. Annu Rev Immunol 18:495–527.

Gao, J.X., Zhang, H., Bai, X.F., Wen, J., Zheng, X., Liu, J., Zheng, P., and Liu, Y. 2002. J Exp Med 195:959–971.

Gartner, F., Alt, F.W., Monroe, R., Chu, M., Sleckman, B.P., Davidson, L., and Swat, W. 1999. Immunity 10:537–546.

Georgopoulos, K. 2002. Nat Rev Immunol 2:162–174.

Godfrey, D.I., Kennedy, J., Suda, T., and Zlotnik, A. 1993. J Immunol 150:4244–4252.

Goldrath, A.W. and Bevan, M.J. 1999. Nature 402:255–262.

Goldschneider, I., Komschlies, K.L., and Greiner, D.L. 1986. J Exp Med 163:1–17.

Goodman, J.W., and Hodgson, G. S. 1962. Blood 19:702–714.

Growney, J.D., Shigematsu, H., Li, Z., Lee, B.H., Adelsperger, J., Rowan, R., Curley, D.P., Kutok, J.L., Akashi, K., Williams, I.R., Speck, N.A., and Gilliland, D.G. 2005. Blood 106:494–504.

Guo, J., Hawwari, A., Li, H., Sun, Z., Mahanta, S.K., Littman, D.R., Krangel, M.S., and He, Y.W. 2002. Nat Immunol 3:469–476.

Hattori, N., Kawamoto, H., Fujimoto, S., Kuno, K., and Katsura, Y. 1996. J Exp Med 184:1137–1147.

Hayday, A.C. and Pennington, D. J. 2007. Nat Immunol 8:137–144.

Hayes, S.M. and Love, P.E. 2006. Immunol Rev 209:170–175.

Hayes, S.M. and Love, P.E. 2007. Immunol Rev 215:8–14.

Hayes, S.M., Li, L., and Love, P.E. 2005. Immunity 22:583–593.

He, X., He, X., Dave, V.P., Zhang, Y., Hua, X., Nicolas, E., Xu, W., Roe, B. A., and Kappes, D.J. 2005. Nature 433:826–833.

He, X., Park, K., Wang, H., He, X., Zhang, Y., Hua, X., Li, Y., and Kappes, D.J. 2008. Immunity 28:346–358.

Heinzel, K., Benz, C., Martins, V.C., Haidl, I.D., and Bleul, C.C. 2007. J Immunol 178:858–868.

Hernandez-Hoyos, G., Anderson, M.K., Wang, C., Rothenberg, E.V., and Alberola-Ila, J. 2003. Immunity 19:83–94.

Hernandez-Munain, C., Sleckman, B.P., and Krangel, M.S. 1999. Immunity 10:723–733.

Ho, I.C. and Pai, S.Y. 2007. Cell Mol Immunol 4:15–29.

Hogquist, K.A., Baldwin, T.A., and Jameson, S.C. 2005. Nat Rev Immunol 5:772–782.

Hori, S., Nomura, T., and Sakaguchi, S. 2003. Science 299:1057–1061.

Huesmann, M., Scott, B., Kisielow, P., and von Boehmer, H. 1991. Cell 66:533–540.

Huseby, E.S., White, J., Crawford, F., Vass, T., Becker, D., Pinilla, C., Marrack, P., and Kappler, J.W. 2005. Cell 122:247–260.

Igarashi, H., Gregory, S., Yokota, T., Sakaguchi, N., and Kincade, P. 2002. Immunity 17:117–130.

Ikawa, T., Kawamoto, H., Goldrath, A.W., and Murre, C. 2006. J Exp Med 203:1329–1342.

Ikuta, K. and Weissman, I.L. 1992. Proc Natl Acad Sci USA 89:1502–1506.

Ioannidis, V., Beermann, F., Clevers, H., and Held, W. 2001. Nat Immunol 2:691–697.

Iwasaki, H., Somoza, C., Shigematsu, H., Duprez, E.A., Iwasaki-Arai, J., Mizuno, S., Arinobu, Y., Geary, K., Zhang, P., Dayaram, T., Fenyus, M.L., Elf, S., Chan, S., Kastner, P., Huettner, C.S., Murray, R., Tenen, D.G., and Akashi, K. 2005. Blood 106:1590–1600.

Jameson, J., Ugarte, K., Chen, N., Yachi, P., Fuchs, E., Boismenu, R., and Havran, W. 2002. Science 296:747–749.

Jenkinson, S.R., Intlekofer, A.M., Sun, G., Feigenbaum, L., Reiner, S.L., and Bosselut, R. 2007. J Exp Med 204:267–272.

Jordan, M.S., Boesteanu, A., Reed, A.J., Petrone, A.L., Holenbeck, A.E., Lerman, M.A., Naji, A., and Caton, A.J. 2001. Nat Immunol 2:301–306.

Kappes, D.J., He, X., and He, X. 2006. Immunol Rev 209:237–252.

Kaye, J., Hsu, M.L., Sauron, M.E., Jameson, S.C., Gascoigne, N.R., and Hedrick, S.M. 1989. Nature 341:746–749.

Kenins, L., Gill, J.W., Boyd, R.L., Hollander, G.A., and Wodnar-Filipowicz, A. 2008. J Exp Med 205:523–531.

Kioussis, D. and Ellmeier, W. 2002. Nat Rev Immunol 2:909–919.

Kondo, M., Weissman, I.L., and Akashi, K. 1997. Cell 91:661–672.

Krangel, M.S. 2007. Nat Immunol 8:687–694.

Krueger, A. and von Boehmer, H. 2007. Immunity 26:105–116.

Kurobe, H., Liu, C., Ueno, T., Saito, F., Ohigashi, I., Seach, N., Arakaki, R., Hayashi, Y., Kitagawa, T., Lipp, M., Boyd, R.L., and Takahama, Y. 2006. Immunity 24:165–177.

Kyewski, B.A. 1987. J Exp Med 166:520–538.

Lai, A.Y., and Kondo, M. 2007. Proc Natl Acad Sci USA 104:6311–6316.

Laiosa, C.V., Stadtfeld, M., and Graf, T. 2006a. Annu Rev Immunol 24:705–738.

Laiosa, C.V., Stadtfeld, M., Xie, H., de Andres-Aguayo, L., and Graf, T. 2006b. Immunity 25:731–744.

Lambolez, F., Kronenberg, M., and Cheroutre, H. 2007. Immunol Rev 215:178–188.

Laufer, T.M., DeKoning, J., Markowitz, J. S., Lo, D., and Glimcher, L. H. 1996. Nature 383:81–85.

Lauritsen, J.P., Haks, M.C., Lefebvre, J.M., Kappes, D.J., and Wiest, D.L. 2006. Immunol Rev 209:176–190.

Lesley, J., Hyman, R., and Schulte, R. 1985. Cell Immunol 91:397–403.

Li, Z., Dordai, D. I., Lee, J., and Desiderio, S. 1996. Immunity 5:575–589.

Lind, E. F., Prockop, S.E., Porritt, H.E., and Petrie, H.T. 2001. J Exp Med 194:127–134.

Liu, C., Ueno, T., Kuse, S., Saito, F., Nitta, T., Piali, L., Nakano, H., Kakiuchi, T., Lipp, M., Hollander, G.A., and Takahama, Y. 2005. Blood 105:31–39.

Liu, C., Saito, F., Liu, Z., Lei, Y., Uehara, S., Love, P., Lipp, M., Kondo, S., Manley, N., and Takahama, Y. 2006. Blood 108:2531–2539.

Lo, D. and Sprent, J. 1986. Nature 319:672–675.

Mackarehtschian, K., Hardin, J.D., Moore, K.A., Boast, S., Goff, S.P., and Lemischka, I.R. 1995. Immunity 3:147–161.

Maeda, T., Merghoub, T., Hobbs, R.M., Dong, L., Maeda, M., Zakrzewski, J., van den Brink, M.R., Zelent, A., Shigematsu, H., Akashi, K., Teruya-Feldstein, J., Cattoretti, G., and Pandolfi, P.P. 2007. Science 316:860–866.

Maillard, I., Fang, T., and Pear, W.S. 2005. Annu Rev Immunol 23:945–974.

Maillard, I., Koch, U., Dumortier, A., Shestova, O., Xu, L., Sai, H., Pross, S.E., Aster, J.C., Bhandoola, A., Radtke, F., and Pear, W.S. 2008. Cell Stem Cell 2:356–366.

Martin, C.H., Aifantis, I., Scimone, M.L., Von Andrian, U.H., Reizis, B., Von Boehmer, H., and Gounari, F. 2003. Nat Immunol 4:866–873.

Mathieu, N., Hempel, W.M., Spicuglia, S., Verthuy, C., and Ferrier, P. 2000. J Exp Med 192:625–636.

Mathis, D. and Benoist, C. 2007. Nat Rev Immunol 7:645–650.

Matloubian, M., Lo, C.G., Cinamon, G., Lesneski, M.J., Xu, Y., Brinkmann, V., Allende, M.L., Proia, R.L., and Cyster, J. G. 2004. Nature 427:355–360.

McCaughtry, T.M., Wilken, M.S., and Hogquist, K.A. 2007. J Exp Med 204:2513–2520.

Melichar, H.J., Narayan, K., Der, S. D., Hiraoka, Y., Gardiol, N., Jeannet, G., Held, W., Chambers, C.A., and Kang, J. 2007. Science 315:230–233.

Michie, A.M., Carlyle, J.R., and Zuniga-Pflucker, J.C. 1998. J Immunol 160:1735–1741.

Miller, J. F. and Osoba, D. 1967. Physiol Rev 47:437–520.

Morrison, S.J., Uchida, N., and Weissman, I.L. 1995. Annu Rev Cell Dev Biol 11:35–71.

Nikolich-Zugich, J., Slifka, M.K., and Messaoudi, I. 2004. Nat Rev Immunol 4:123–132.

Norment, A.M., Salter, R.D., Parham, P., Engelhard, V.H., and Littman, D.R. 1988. Nature 336:79–81.

Ogawa, M., Matsuzaki, Y., Nishikawa, S., Hayashi, S., Kunisada, T., Sudo, T., Kina, T., and Nakauchi, H. 1991. J Exp Med 174:63–71.

Pai, S. Y., Truitt, M.L., Ting, C.N., Leiden, J.M., Glimcher, L.H., and Ho, I. C. 2003. Immunity 19:863–875.

Pearse, M., Wu, L., Egerton, M., Wilson, A., Shortman, K., and Scollay, R. 1989. Proc Natl Acad Sci USA 86:1614–1618.

Penit, C. and Vasseur, F. 1988. J Immunol 140:3315–3323.

Perry, S.S., Welner, R.S., Kouro, T., Kincade, P.W., and Sun, X.H. 2006. J Immunol 177:2880–2887.

Peschon, J.J., Morrissey, P.J., Grabstein, K.H., Ramsdell, F.J., Maraskovsky, E., Gliniak, B.C., Park, L.S., Ziegler, S.F., Williams, D.E., Ware, C.B., et al. 1994. J Immunol 180:1955–1960.

Petrie, H.T. and Zuniga-Pflucker, J. C. 2007. Annu Rev Immunol 25:649–679.

Petrie, H.T., Pearse, M., Scollay, R., and Shortman, K. 1990. Eur J Immunol 20:2813–2815.

Petrie, H.T., Livak, F., Schatz, D.G., Strasser, A., Crispe, I.N., and Shortman, K. 1993. J Exp Med 178:615–622.

Porritt, H.E., Gordon, K., and Petrie, H.T. 2003. J Exp Med 198:957–962.

Porritt, H.E., Rumfelt, L.L., Tabrizifard, S., Schmitt, T. M., Zuniga-Pflucker, J.C., and Petrie, H.T. 2004. Immunity 20:735–745.

Prockop, S.E. and Petrie, H.T. 2004. J Immunol 173:1604–1611.

Pui, J.C., Allman, D., Xu, L., DeRocco, S., Karnell, F. G., Bakkour, S., Lee, J.Y., Kadesch, T., Hardy, R.R., Aster, J.C., and Pear, W. S. 1999. Immunity 11:299–308.

Radtke, F., Wilson, A., Stark, G., Bauer, M., van Meerwijk, J., MacDonald, H.R., and Aguet, M. 1999. Immunity 10:547–558.

Reizis, B. and Leder, P. 2001. J Exp Med 194:979–990.

Reizis, B. and Leder, P. 2002. Genes Dev 16:295–300.

Rodewald, H.R. 2008. Annu Rev Immunol 26:355–388.

Rodewald, H.R., Kretzschmar, K., Takeda, S., Hohl, C., and Dessing, M. 1994. EMBO J 13:4229–4240.

Rodewald, H.R., Waskow, C., and Haller, C. 2001. J Exp Med 193:1431–1437.

Rossi, F.M., Corbel, S.Y., Merzaban, J.S., Carlow, D.A., Gossens, K., Duenas, J., So, L., Yi, L., and Ziltener, H.J. 2005. Nat Immunol 6:626–634.

Rothenberg, E.V., Moore, J.E., and Yui, M.A. 2008. Nat Rev Immunol 8:9–21.

Rumfelt, L. L., Zhou, Y., Rowley, B.M., Shinton, S.A., and Hardy, R.R. 2006. J Exp Med 203:675–687.

Sambandam, A., Maillard, I., Zediak, V. P., Xu, L., Gerstein, R.M., Aster, J.C., Pear, W.S., and Bhandoola, A. 2005. Nat Immunol 6:663–670.

Sawada, S., Scarborough, J.D., Killeen, N., and Littman, D.R. 1994. Cell 77:917–929.

Schlissel, M.S., Durum, S.D., and Muegge, K. 2000. J Exp Med 191:1045–1050.

Schmitt, T.M. and Zuniga-Pflucker, J. C. 2002. Immunity 17:749–756.

Schmitt, T.M., Ciofani, M., Petrie, H. T., and Zuniga-Pflucker, J.C. 2004. J Exp Med 200:469–479.

Schwarz, B.A. and Bhandoola, A. 2004. Nat Immunol 5:953–960.

Schwarz, B.A. and Bhandoola, A. 2006. Immunol Rev 209:47–57.

Schwarz, B.A., Sambandam, A., Maillard, I., Harman, B.C., Love, P.E., and Bhandoola, A. 2007. J Immunol 178:2008–2017.

Scimone, M.L., Aifantis, I., Apostolou, I., von Boehmer, H., and von Andrian, U.H. 2006. Proc Natl Acad Sci USA 103:7006–7011.

Scollay, R., Smith, J., and Stauffer, V. 1986. Immunol Rev 91:129–157.

Sebzda, E., Wallace, V.A., Mayer, J., Yeung, R.S., Mak, T.W., and Ohashi, P.S. 1994. Science 263:1615–1618.

Setoguchi, R., Tachibana, M., Naoe, Y., Muroi, S., Akiyama, K., Tezuka, C., Okuda, T., and Taniuchi, I. 2008. Science 319:822–825.

Shin, S., El-Diwany, R., Schaffert, S., Adams, E.J., Garcia, K.C., Pereira, P., and Chien, Y.H. 2005. Science 308:252–255.

Shortman, K. and Wu, L. 1996. Annu Rev Immunol 14:29–47.

Shortman, K., Egerton, M., Spangrude, G.J., and Scollay, R. 1990. Semin Immunol 2:3–12.

Singer, A. and Bosselut, R. 2004. Adv Immunol 83:91–131.

Sitnicka, E., Bryder, D., Theilgaard-Monch, K., Buza-Vidas, N., Adolfsson, J., and Jacobsen, S.E. 2002. Immunity 17:463–472.

Starr, T.K., Jameson, S.C., and Hogquist, K.A. 2003. Annu Rev Immunol 21:139–176.

Sun, G., Liu, X., Mercado, P., Jenkinson, S.R., Kypriotou, M., Feigenbaum, L., Galera, P., and Bosselut, R. 2005. Nat Immunol 6:373–381.

Taghon, T., Yui, M.A., Pant, R., Diamond, R.A., and Rothenberg, E.V. 2006. Immunity 24:53–64.

Taghon, T.N., David, E.S., Zuniga-Pflucker, J.C., and Rothenberg, E.V. 2005. Genes Dev 19:965–978.

Takeuchi, A., Yamasaki, S., Takase, K., Nakatsu, F., Arase, H., Onodera, M., and Saito, T. 2001. J Immunol 167:2157–2163.

Talebian, L., Li, Z., Guo, Y., Gaudet, J., Speck, M. E., Sugiyama, D., Kaur, P., Pear, W.S., Maillard, I., and Speck, N.A. 2007. Blood 109:11–21.

Tan, J.B., Visan, I., Yuan, J.S., and Guidos, C.J. 2005. Nat Immunol 6:671–679.

Tanigaki, K. and Honjo, T. 2007. Nat Immunol 8:451–456.

Tanigaki, K., Tsuji, M., Yamamoto, N., Han, H., Tsukada, J., Inoue, H., Kubo, M., and Honjo, T. 2004. Immunity 20:611–622.

Taniuchi, I., Osato, M., Egawa, T., Sunshine, M.J., Bae, S.C., Komori, T., Ito, Y., and Littman, D.R. 2002. Cell 111:621–633.

Teh, H.S., Kisielow, P., Scott, B., Kishi, H., Uematsu, Y., Bluthmann, H., and von Boehmer, H. 1988. Nature 335:229–233.

Turka, L.A., Schatz, D.G., Oettinger, M. A., Chun, J.J., Gorka, C., Lee, K., McCormack, W.T., and Thompson, C.B. 1991. Science 253:778–781.

Turner, J.M., Brodsky, M.H., Irving, B.A., Levin, S.D., Perlmutter, R.M., and Littman, D.R. 1990. Cell 60:755–765.

Uehara, S., Grinberg, A., Farber, J.M., and Love, P. E. 2002. J Immunol 168:2811–2819.

Ueno, T., Saito, F., Gray, D.H., Kuse, S., Hieshima, K., Nakano, H., Kakiuchi, T., Lipp, M., Boyd, R.L., and Takahama, Y. 2004. J Exp Med 200:493–505.

Van Laethem, F., Sarafova, S.D., Park, J.H., Tai, X., Pobezinsky, L., Guinter, T.I., Adoro, S., Adams, A., Sharrow, S. O., Feigenbaum, L., and Singer, A. 2007. Immunity 27:735–750.

von Boehmer, H. 2005. Nat Rev Immunol 5:571–577.

Wada, H., Masuda, K., Satoh, R., Kakugawa, K., Ikawa, T., Katsura, Y., and Kawamoto, H. 2008. Nature 452:768–772.

Wallis, V.J., Leuchars, E., Chwalinski, S., and Davies, A.J. 1975. Transplantation 19:2–11.

Waskow, C., Paul, S., Haller, C., Gassmann, M., and Rodewald, H. 2002. Immunity 17:277–288.

Weng, A.P., Millholland, J.M., Yashiro-Ohtani, Y., Arcangeli, M.L., Lau, A., Wai, C., Del Bianco, C., Rodriguez, C.G., Sai, H., Tobias, J., Li, Y., Wolfe, M.S., Shachaf, C., Felsher, D., Blacklow, S.C., Pear, W.S., and Aster, J.C. 2006. Genes Dev 20:2096–2109.

White, J., Herman, A., Pullen, A.M., Kubo, R., Kappler, J.W., and Marrack, P. 1989. Cell 56:27–35.

Wildt, K. F., Sun, G., Grueter, B., Fischer, M., Zamisch, M., Ehlers, M., and Bosselut, R. 2007. J Immunol 179:4405–4414.

Wolfer, A., Wilson, A., Nemir, M., MacDonald, H.R., and Radtke, F. 2002. Immunity 16:869–879.

Woolf, E., Xiao, C., Fainaru, O., Lotem, J., Rosen, D., Negreanu, V., Bernstein, Y., Goldenberg, D., Brenner, O., Berke, G., Levanon, D., and Groner, Y. 2003. Proc Natl Acad Sci USA 100:7731–7736.

Wright, D.E., Wagers, A.J., Gulati, A.P., Johnson, F.L., and Weissman, I.L. 2001. Science 294:1933–1936.

Wu, L., Antica, M., Johnson, G.R., Scollay, R., and Shortman, K. 1991a. J Exp Med 174:1617–1627.

Wu, L., Scollay, R., Egerton, M., Pearse, M., Spangrude, G.J., and Shortman, K. 1991b. Nature 349:71–74.

Wurbel, M.A., Malissen, B., and Campbell, J. J. 2006. Eur J Immunol 36:73–81.

Xi, H., Schwartz, R., Engel, I., Murre, C., and Kersh, G. J. 2006. Immunity 24:813–826.

Yamasaki, S. and Saito, T. 2007. Trends Immunol 28:39–43.

Yasutomo, K., Doyle, C., Miele, L., Fuchs, C., and Germain, R. N. 2000. Nature 404:506–510.

Yoshida, T., Ng, S.Y., Zuniga-Pflucker, J.C., and Georgopoulos, K. 2006. Nat Immunol 7:382–391.

Yu, Q., Erman, B., Bhandoola, A., Sharrow, S.O., and Singer, A. 2003. J Exp Med 197:475–487.

Yu, Q., Park, J. H., Doan, L.L., Erman, B., Feigenbaum, L., and Singer, A. 2006. J Exp Med 203:165–175.

Yucel, R., Karsunky, H., Klein-Hitpass, L., and Moroy, T. 2003. J Exp Med 197:831–844.

Zou, Y.R., Sunshine, M.J., Taniuchi, I., Hatam, F., Killeen, N., and Littman, D.R. 2001. Nat Genet 29:332–336.

Chapter 8
Development of B Lymphocytes

Steven A. Corfe and Christopher J. Paige

Abstract B cells are a central component of the humoral arm of the immune system and are essential in protecting against infection and disease. They mediate their effector functions through the production of neutralizing antibody, induction of antibody-dependent cell-mediated cytotoxicity, and activation of the complement system. B cells are produced throughout the duration of our life, and the efficiency with which they are generated is affected by extrinsic and intrinsic factors including chemokines, cytokines, and transcription factors. Severe immunodeficiency is observed in cases where B cell development is reduced or absent and results in persistent infection. This chapter will focus on the molecular mechanisms regulating the survival, commitment, differentiation, and selection of B cells as they develop from hematopoietic progenitors to fully functional mature effector B cells.

Introduction

B lymphopoiesis occurs in a series of steps whereby progenitor cells undergo rearrangement of their immunoglobulin (Ig) loci, and this leads to the eventual expression of a functional B cell receptor (BCR) that iscapable of responding to antigen (Fig. 8.1). Expression of the pre-B cell receptor (pre-BCR) and the mature BCR are critical events during maturation and mediate transition through checkpoints that select for functional cells that are not self-reactive. Ig recombination is initiated at the heavy chain locus in committed B cell progenitors (pro-B cells) and results in the generation of the μ heavy chain proteins (μHC). μHCs associate with surrogate light chain (SLC) proteins V_{preB} and λ5 to form the pre-BCR, which is expressed on

S.A. Corfe(✉) and C.J. Paige
Department of Immunology,
University of Toronto, Ontario Cancer Institute,
8-105 Princess Margaret Hospital, University Health Network,
Toronto, ON, Canada M5G 2M9
e-mail: steve.corfe@utoronto.ca

A. Wickrema and B. Kee (eds.), *Molecular Basis of Hematopoiesis*,
DOI: 10.1007/978-0-387-85816-6_1, © Springer Science+Business Media, LLC 2009

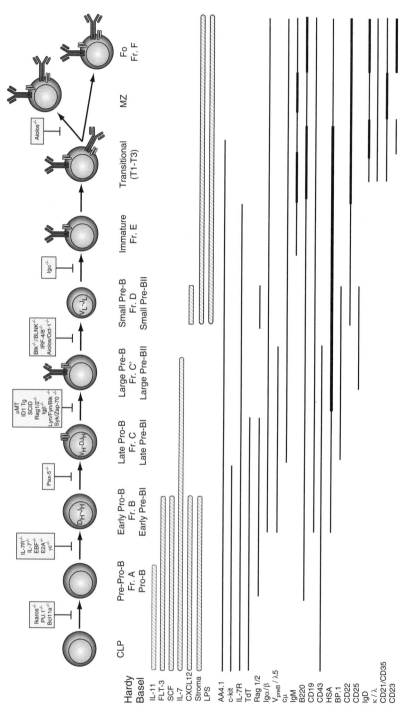

Fig. 8.1 B cell developmental scheme: Developmental progression of B lineage cells from CLPs to mature B cells in mice. *Hatched bars* denote growth factor and stroma-dependent stages. *Solid lines* represent expression of cell phenotype markers, with line thickness indicating relative expression levels. Developmental blocks arising from mutant or transgenic gene expression are denoted at appropriate stages (*see Color Plates*)

the surface of newly formed pre-B cells. Pre-B cells express µHCs on their cell surface with only one specificity, and further rearrangement at the HC locus is prevented by a process referred to as allelic exclusion. Successful pre-BCR expression and signalling activate survival, proliferation, and differentiation pathways that lead to the selection and expansion of large pre-B cells. Proliferating large pre-B cells exit the cell cycle and become resting small pre-B cells that begin rearrangement of the light chain (LC) proteins. LC proteins associate with µHCs to form the BCR, which is first expressed on immature cells in the bone marrow (BM). Immature B cells are positively and negatively selected based on their antigen specificity and migrate from the BM to the spleen. Immature B cells arriving in the spleen pass through a series of transitional stages prior to developing into mature follicular (Fo), marginal zone (MZ), and B-1 B cells (Murphy 2007).

Mature B cells circulate throughout the periphery and localize in the BM, spleen, lymph nodes (LN), and peritoneal cavities. Antigen-activated mature B cells proliferate and differentiate into plasmablasts, which secrete high levels of antibodies. Antibodies can coat pathogens and cause them to agglutinate, which limits their mobility and replication, and prevents their entry into host cells. Antibody-bound pathogens can also be recognized by complement, which initiates the complement cascade. This cascade can result in both opsonization, whereby a pathogen is ingested and destroyed by phagocytes, and direct killing by means of the complement membrane attack complex. Antibody binding can also trigger antibody-dependent cell-mediated cytotoxicity (ADCC), a mechanism by which natural killer cells, monocytes, and/or eosinophils release compounds such as granzymes, IFN-γ, major basic protein, and perforin, all of which can lyse infected cells and kill invading pathogens. In addition to fighting off immediate infection, B cells also develop into long-lived memory cells that remain in the body to protect against future infections.

B cell Ontogeny

As with all hematopoietic lineages, B cells are derived from self-renewing hematopoietic stem cells (HSC). B lymphopoiesis can first be detected in the fetal liver of mice and humans on day 14 and week 8 of gestation, respectively. Development continues in the fetal BM, which is seeded with progenitor cells from the fetal liver. BM is also the predominant site of post-natal hematopoietic B lymphopoiesis, where developing cells are closely associated with stromal cells, a class of large adherent BM cells that include fibroblasts, reticular cells, preadipocytes, endothelial cells, and macrophages. This microenvironment provides essential support for hematopoiesis, and development is severely inhibited in cases where bone structure is abnormal.

In the BM, progression from HSC to committed B cell follows a path in which a series of stochastic decisions result in cells that progressively develop B cell traits while repressing traits of other lineages. Cells are biased towards a certain lineage fate by the expression and interaction of transcription factors (TF) in a process referred to as specification. However, final commitment occurs only after all other lineage potentials are fully arrested. During murine development, HSCs are characterized by their lineage (lin⁻), Sca-1⁺, c-Kit^{hi} (LSK) phenotype as well as self-renewing capacity and transition into non-self renewing multipotent progenitors (MPP). A subset of these LSK cells express high levels of Flt-3 and are designated lymphoid primed MPPs (LMPP). Expression of recombination of activation genes (*Rag*) as well as terminal deoxynucleotidyltransferase (TdT) denotes the gradual transition to early lymphoid progenitors (ELP), which subsequently gives rise to common lymphoid progenitors (CLPs) (Welner et al. 2008). CLPs are the first cells to express the inter-leukin-7 receptor (IL-7R) and possess B, T, natural killer (NK), and dendritic (DC) cell potential in vivo. B220⁺ (CD45) pre-pro-B cells develop from CLPs and are the first cells specified to the B cell fate. Pro-B cells, identified by CD19 expression and increased levels of the TF Pax-5, are fully committed to the B lineage and are dependent on interleukin-7 (IL-7) for their proliferation and survival.

Markers of B cell Differentiation

B cell development has been described by various conventions that identify similar B cell populations but differ in nomenclature (Fig. 8.1) (Hardy et al. 1991; Melchers 1995). Described by the Basel nomenclature, Ig recombination is initiated by Rag proteins in c-Kit⁺ pro-B cells that subsequently develop into pre-BI cells, which are rearranging their HC genes and express SLC proteins. Large cycling Pre-BII cells express the pre-BCR, downregulate c-Kit, and begin to express the IL-2α receptor (CD25). Pre-BCR expression is difficult to detect on the cell surface and thus it is convenient to use CD25 in conjunction with other B cell markers to identify pre-B cells. Small pre-B cells re-express Rag proteins that allow LC rearrangement to begin. Detailed examination of the surface of small pre-B cells revealed that the pre-BCR is downregulated and internalized during the large pre-B to small pre-B transition (Wang et al. 2002). The mature BCR is first expressed on the surface of CD25 negative immature B cells.

Hardy et al. constructed a developmental scheme that distinguishes B cell frac-tions based on their expression of surface markers CD43, B220, BP.1, HSA (CD24), and IgM (Fig. 8.1). Early B cell populations (Fr. A-C′) are identified as B220⁺ CD43⁺ cells, and are further subdivided by increased expression of HSA and BP.1 as they develop from the pre-pro-B stage (Fr. A) to the large pre-B stage (Fr. C′). Maturation to the small pre-B (Fr. D) stage coincides with loss of CD43, and immature B cells (Fr. E) are distinguished by their expression of surface IgM. Mature re-circulating B cells (Fr. F) in the BM can be identified by their co-expression of IgD and IgM. AA4.1 expression on early subsets has led to further distinction between AA4.1⁻ (A1) and AA4.1⁺ (A2) populations, which differ in their response to IL-7. Only AA4.1⁺ cells express Rag proteins and are IL-7-responsive, and it is this population that is believed to represent true B cell precursors (Miller et al. 2002). It should be noted that the aforementioned classification systems define stages of development as snapshots of a continual process. In fact, populations are much more dynamic and not phenotypically absolute.

Human hematopoietic precursors can be identified by CD34 expression and, similar to mouse development, human HSCs pass through lineage restriction stages that bias them towards either myeloid or lymphoid fate. CD45RA$^+$ CD10$^+$ CD19$^-$ IL-7R$^+$ cells are specified B cell progenitors that possess DJ$_H$ rearrangements and express components of the pre-BCR (Igα/β, VpreB), events similarly observed in specified murine B cell progenitors (reviewed in LeBien 2000). CD34$^+$ CD10$^+$ human pro-B cells expressing CD19 are committed to the B lineage. Subsequent stages of human B cell development closely follow those characterized in mice, with pro-B cells developing into large cycling pre-B cells (CD34$^-$ CD10$^+$ CD19$^+$), small pre-B cells (also CD34$^-$ CD10$^+$ CD19$^+$), and immature B cells (CD34$^-$ CD10$^+$ CD19$^+$ CD40$^+$ sIgM$^+$).

In mice and humans, immature B cells exiting the BM go through a series of transitional stages before becoming mature functional B cells. These cells all express AA4.1 and low/intermediate levels of CD21/CD35. Transitional cells can be further subdivided into recent splenic emigrants, termed T1 cells (IgM^{hi}IgD^{lo}CD23$^-$), which give rise to T2 cells (IgM^{hi} IgD^{hi} CD23$^+$) that subsequently develop either directly into Fo and MZ cells or go through a T3 (IgM^{lo} IgD^{hi} CD23$^+$) intermediate stage. Mature B cells lose expression of AA4.1 and can be distinguished by their IgM^{lo} IgD^{hi} CD21^{int} CD23^{hi} (Fo) or IgM^{hi} IgD^{lo} CD21^{hi} CD23$^{lo/-}$ (MZ) phenotypes (reviewed in Allman and Pillai 2008).

Mechanisms of Antibody Diversity

The BCR is composed of two identical disulphide-linked HCs, each of which is disulphide linked to identical kappa (κ) or lambda (λ) LCs. Both heavy and light chains consist of C-terminal constant regions and N-terminal variable regions. The constant region determines the class (IgM, IgD, IgG, IgA, and IgE for HC, and Igκ or Igλ for LC) and effector function of the receptor. The variable region, which is composed of three complimentary determining regions (CDRs), determines the receptor specificity. The variable regions of the receptor heavy chains are formed by the joining of variable (V), diversity (D), and joining (J) gene segments in a process known as V(D)J recombination (reviewed in Bassing et al. 2002). LC rearrangement occurs at either the κ or λ locus in a similar fashion, except that there are no diversity gene segments. κ rearrangement normally precedes that of λ, and individual cells exhibit isotypic exclusion, meaning that they express only κ or λ LCs. Expression of κ/λ is not represented equally on B cells or in serum Ig, and in the majority of cases κ is favoured (humans 65%, mice 95%).

Ig gene segments are spaced out along the same chromosome (12 for mice and 14 for human HCs; 6 and 16 for mice κ and λLCs; 2 and 22 for human κ and λ LCs) in such a manner as to allow recombination machinery to bind two coding segments and join them together. This process occurs in a reproducible manner with D$_H$ and J$_H$ joining preceding V$_H$ to DJ$_H$ joining. Rag proteins initiate recombination by catalyzing double-strand DNA breaks, which lead to the generation of a hairpin coding end and a blunt signal end. The nuclease Artemis opens the hairpin coding end allowing for the binding of the non-homologous end joining (NHEJ) machinery (DNA-PK, Ku-70, Ku86, ligase IV, and XRCC4), which ligates the gene segments to yield a new coding sequence (reviewed in Weterings and Chen 2008). Mutations in components of the recombination machinery result in severe combined immunodeficiency disease (SCID) in mice and humans, whereby most or all lymphoid cells are absent. In fact, many of the molecules involved in recombination were first identified in patients displaying B cell defects, which highlights the importance of investigating genetic diseases to link mutations to their molecular origins.

The diversity of the antibody repertoire observed in the periphery is the result of several layers of complexity. Germline-encoded diversity is achieved by the random selection of V gene germline sequences, which directly encode the first two CDRs. The third CDR is encoded by the random assortment of the aforementioned V (~150 options), D (12–13 options), and J (4 options) gene segments and results in combinatorial somatic diversity. Finally, the joining of these gene segments is imprecise by nature, such that addition or deletion of nucleotides occurs and changes the specificity of the receptors encoded by identical V, D, and J segments. The DNA polymerase TdT is important in contributing to this final level of antigen receptor diversity as it functions by adding add non-templated (N) nucleotides to the coding junctions. TdT is only expressed in adult life and can increase the diversity of the genome almost infinitely, but is not essential for receptor formation.

Characterization of B Lymphopoiesis

Much of our understanding regarding the functional mechanisms of B cell development has come from in vitro studies. Experimentation with different cytokine cocktails led to the discovery that IL-11 and stem cell factor (SCF) could support the generation of bi-potent macrophage-B cell progenitors from fetal liver cells independent of stromal cells, and that the efficiency of this maturation is improved by the addition of Flt-3L (Fig. 8.1) (Kee et al. 1994; Ray et al. 1996). Stromal cells are important for later stages of B cell development, and no combination of cytokines that mediate this transition have been discovered. However, maturation to the antibody secreting stage can occur independent of stromal cell support when IL-7-expanded B cell precursors are cultured with mitogenic stimulus under conditions that facilitate homotypic interactions between progenitors (Ray et al. 1998). Addition of blocking antibodies to these cultures led to the discovery that components of the pre-BCR were critical for this development, either by preventing ligand binding or receptor oligomerization (Stoddart et al. 2001). The identification of heparan sulphate as a pre-BCR binding partner, as well as the observation that its addition to cultures increases the amount of antibody secreted, suggest that it plays a role during this stage of maturation (Bradl et al. 2003; Milne et al. 2008).

Knockout and transgenic mice are also key tools in understanding the mechanisms that regulate B lymphopoiesis. By analyzing the development and functionality of B cell populations in these mice it is possible to identify molecules that are required during specific stages of maturation. Additionally, in vitro experimentation with cells isolated from these mice can lead to the discovery of novel protein binding partners and redundant family members, as well as a better understanding of the signal transduction pathways that regulate development.

Factors Regulating B cell Development

The development of non-committed progenitor cells is influenced by both environmental conditions and cell-intrinsic factors. TFs are DNA-binding proteins, which, by induction or repression of target genes, control many of the events leading to lineage specification and commitment. These factors can be shared or lineage-specific and, typically, no single factor leads to lineage commitment. Instead it is the quantity, combination, and cross competition between them that regulate the gene expression patterns that activate lineage-directed programs. TFs in turn are regulated by growth factors, cytokines, and chemokines, which are produced by supportive cells in developmental niches such as the fetal liver, BM, spleen, and thymus. It is the combination of these factors in the appropriate environmental context that provides the necessary cues for B cell development to occur.

TFs and B Cell Specification

MPPs express a variety of TFs that prime cells for lineage commitment, including the zinc-fingered domain proteins Ikaros and Bcl11a, as well as the ETS family member PU.1. Mice deficient for any of these factors display a defect at the CLP stage, likely due to failed expression of Flt3 and/or IL-7Rα. However, ectopic expression of IL-7Rα in PU.1$^{-/-}$ progenitors does not fully rescue B lymphopoiesis as PU.1 has alternative roles during B cell development, including the repression of T-cell and NK-cell development (DeKoter et al. 2002; Kamath et al. 2008).

In addition to directing the early stages of lymphoid specification, Ikaros and PU.1 also influence later B lineage development. In contrast to fully deficient animals, mice exhibiting reduced levels of Ikaros develop normally past the CLP stage but are impaired at the pro-B to pre-B transition (Kirstetter et al. 2002). Ikaros also acts in conjunction with related family member Aiolos to suppress *Igll1* (λ5) expression. Studies in which levels of PU.1 were manipulated showed that high-level PU.1 expression correlates with the development of myeloid cells at the expense of B cells, while low levels failed to support myeloid development and led to increased generation of B cells (DeKoter and Singh 2000). However, lymphoid versus myeloid cell fate does not appear to be this straightforward. Subsequent studies have shown that PU.1 levels are similar in precursor populations and that high levels of PU.1 required for both myeloid and B cell development (Houston et al. 2007). Further commitment or differentiation of B cells is not dependent on PU.1, as conditional knockout of PU.1 at the pro-B stage does not alter subsequent B cell development (Iwasaki et al. 2005).

TFs and B cell Specification and Commitment

Activation of the helix-loop-helix domain family members E2A and early B cell factor-1 (EBF) is a critical step in B cell specification. Defects in either E2A or EBF block development prior to B cell commitment, with cells having yet to begin Ig recombination (Lin and Grosschedl 1995; Bain et al. 1997). E2A is composed of E-protein splice variants E12 and E47 that homo- or heterodimerize prior to binding of the consensus DNA sequence CANNTG, denoted the E-box motif. E proteins regulate the expression of several important factors during B cell development including the IgH enhancer and EBF (reviewed in Kee et al. 2000). The other E-box proteins (E2-2 and HEB) have both redundant and independent functions during B cell development. Deletion of either decreases pro-B cells by about half, and mice heterozygous for any two E-protein family members also display B cell defects (Zhuang et al. 1996). E-proteins can also bind inhibitors of DNA-binding proteins (ID 1–4), which are structurally similar to E-proteins but lack the DNA-binding domain. Dimerization of ID and E-proteins abolishes E-protein DNA-binding potential and inactivates its function. Experimentally, over-expression of ID1 leads to a block at the pro-B stage of development. This suggests that ID proteins work to regulate the levels of E-proteins during development (Sun 1994).

The ability of EBF to control B cell specification has been demonstrated in studies where ectopic expression skewed the differentiation of HSCs towards the B lineage (Zhang et al. 2003). EBF expression can also partially or fully rescue B cell development in mice deficient for PU.1, E2A, IL-7, or the IL-7Rα (reviewed in Nutt and Kee 2007). EBF partners with E2A to specify cells to the B cell lineage through the induction of Ig rearrangement and the regulation of Igα, Igβ, V_{preB}, and λ5 (O'Riordan and Grosschedl 1999). Another critical target of EBF is Pax-5, a paired homeodomain TF necessary for B-lineage commitment. Numerous factors regulate EBF expression, including PU.1, E2A, IL-7, Pax-5 and even EBF itself (reviewed in Nutt and Kee 2007). This multileveled regulation allows for the control of EBF expression and provides positive feedback loops to maintain its expression.

Pax-5 functions not only to induce gene expression patterns leading to B cell commitment, but also to repress other lineage options. The N-terminal paired domain motif of Pax-5 binds DNA and positively regulates gene transcription. Pax-5 targets include *mb-1* (Igα), BLNK, CD19, *Iglll* (λ5), as well as numerous TFs such as interferon-regulatory factors 4 and 8 (IRF-4/8) and Aiolos (Pridans et al. 2008). Pax-5 also represses non-B lineage genes myeloperoxidase (*MPO*), *Notch-1*, *M-CSFR*, and *Flt3*. This gene repression is essential in maintaining B cell commitment as Pax-5$^{-/-}$ cells retain the ability to differentiate into other hematopoietic lineages when cultured under permissive conditions (Nutt et al. 1999). Additionally, injection of Pax-5$^{-/-}$ pro-B cells into Rag-2$^{-/-}$ mice leads to reconstitution of the thymus and generation of T cells (Rolink et al. 1999). This T-cell development is likely due to the failure to repress Notch-1, a critical T-cell transcription factor. Continual expression of Pax-5 is necessary throughout B cell development, as conditional deletion of Pax-5 in pro-B or later stage mature cells leads to the reactivation of many repressed genes and reversion to other lineage types (Horcher et al. 2001; Mikkola et al. 2002). However, recent work has identified novel roles for EBF in B cell commitment independent of Pax-5. Sustained expression of EBF in Pax-5$^{-/-}$ hematopoietic progenitor cells restricted their ability to differentiate into myeloid or T cells *in vivo* and, *in vitro,* EBF repressed myeloid and T-cell genes in Pax-5$^{-/-}$ pro-B cells (Pongubala et al. 2008). The authors proposed that the lack of lineage commitment observed in Pax-5$^{-/-}$ cells is actually the result of a failure of these cells to maintain EBF expression.

A variety of abnormalities in the Pax-5 locus have been identified in numerous cases of B cell progenitor acute lymphoblastic leukemia (BCP-ALL). In one such case, the t(9;12)(q11;p13) translocation generated a fusion protein composed of Pax-5 and the Ets transcription factor TEL (Fazio et al. 2008). This hybrid molecule possessed the DNA-binding potential of Pax-5 and the transcriptional repressor function of TEL. When tested *in vitro,* the Pax-5/TEL fusion protein downregulated Pax-5 targets CD19, BLNK, *mb-1,* and Flt3, led to increased CXCL12-induced migration, and improved survival after IL-7 withdrawal or TGF-β treatment. The apparent contribution of genomic alterations of B lineage transcription factors in BCP-ALL is becoming increasingly recognized. Characterization of single nucleotide polymorphism (SNP) arrays from over 200 BCP-ALL samples showed that 40% of them contained such defects (Mullighan et al. 2007). While Pax-5 cases were the most prevalent (>30%), aberrations were also observed for other B cell transcription factors, including E2A, EBF, Ikaros, and Aiolos.

TFs and B cell Development

While E2A and EBF work together to positively regulate *Igll1* (λ5) expression, Aiolos and Ikaros function to downregulate *Igll1* expression. Aiolos is a zinc-finger transcription factor of the Ikaros family and its protein levels are significantly increased at the pre-B cell stage. This protein upregulation is induced by pre-BCR signalling via IRF4/8 and the adaptor protein BLNK (Thompson et al. 2007; Ma et al. 2008). In the absence of Aiolos, suppression of *Igll1* (λ5) is initiated but full repression does not occur. Aiolos also works with Oct binding factor-1 (OBF-1) to regulate BM and peripheral B cell development. Aiolos$^{-/-}$/OBF-1$^{-/-}$ mice display a block at the pre-B stage that results in an almost complete absence of immature cells (Sun et al. 2003). In the periphery, Aiolos$^{-/-}$ mice exhibit reduced thresholds of receptor activation. This leads to spontaneous formation of germinal centres and failure to generate MZ B cells (Wang et al. 1998). This defect is lost in Aiolos$^{-/-}$/OBF-1$^{-/-}$ mice suggesting that, in the periphery, these factors may play opposing roles in regulating the threshold of BCR activation.

IRF-4 and IRF-8 are structurally related and partially redundant proteins that are expressed during both lymphoid and myeloid development. IRF-4 and IRF-8 exhibit little direct DNA-binding activity but rather works in a complex with other factors such as PU.1 and E2A to regulate the expression of B cell genes. IRF-4$^{-/-}$/8$^{-/-}$ mice exhibit a block in development at the large pre-B stage, and these cells continue to cycle, exhibit increased expression of the pre-BCR, and fail to initiate LC recombination (Lu et al. 2003). These defects are likely due to sustained production of the SLC proteins V$_{preB}$ and (lambda symbol)5, which are normally downregulated by IRF-4/8 induction of Ikaros and Aiolos. IRF-4 and IRF-8 regulate LC rearrangement by targeting the κ and λ enhancers and have also been implicated in the indirect promotion of Rag expression and E2A activity through the induction of CXCR4 and the subsequent migration of cells away from IL-7-producing stromal cells (Johnson et al. 2008). IRF-4$^{-/-}$ mice also exhibit defects in LC receptor editing, class switch recombination, and plasma cell differentiation due to sub-optimal expression of activation-induced cytidine deaminase (AID) (Sciammas et al. 2006).

Chemokines and B cell Niches

The distribution of B cells within the marrow is largely dependent on their interaction with different types of stroma. Stromal cells attract hematopoietic cells by means of adhesion molecules and chemokines such as CXCL12 (SDF-1) and its ligand CXCR4. Deletion of either prevents B cell development at the earliest precursor stages (Egawa et al. 2001). Experimentally, pre-pro-B cells have been found associated mainly with CXCL12^{hi} reticular cells (Tokoyoda et al. 2004). Maturation into pro-B cells coincided with their migration away from these cells and localization with high IL-7 producing stromal cells Pre-B cells increase expression of CXCR4, and this has been suggested to be important in moving cells away from IL-7-producing stromal cells (Milne et al. 2005; Johnson et al. 2008). Immature B cells migrate away from stromal cells, are released from the BM, and enter the periphery.

Antigen-induced activation of mature peripheral B cells in secondary lymphoid tissues (spleen, LNs, and Peyer's patches) results in their differentiation into plasma and memory cells that re-express CXCR4. This expression results in cell homing to and retention in the BM, where mature cells can be found associated with CXCL12^{hi} reticular cells (Hargreaves et al. 2001). The localization of B cell precursors to specific niches and cell types provides them with the necessary growth factors and cell–cell interactions and shields them from factors that would support their development towards other lineage fates.

Cytokines and Early B cell Development

Stem cell factor (SCF, c-Kit-Ligand), which exists in either soluble or membrane-bound form, binds the surface-expressed tyrosine kinase receptor KIT (c-Kit). SCF is required in the earliest stages of hematopoietic development and SCF[-/-] or c-Kit[-/-] mice die within a week of birth due to anaemia. SCF also has specific influences on developing B cells. In Vicked mice (viable c-kit deficient) or Wepo mice (rescued by erythropoietin over-expression), B cell development is normal during fetal life but numbers of pro-B and pre-B cells greatly diminish as mice age (Waskow et al. 2002). *In vitro,* SCF can act synergistically with IL-7 to increase the numbers of pro-B cells in cultures. This explains the decreased number of B cells observed in knockout mice. However, the increase in B cell populations *in vitro* is more likely due to the fact that SCF promotes the survival and proliferation of B cell progenitors in culture and thus increases the input of IL-7-responsive pro-B cells. Flt-3 and its receptor Flt-3L also play critical roles in enhancing the survival and proliferation of early progenitors. In Flt-3[-/-] mice, pre-pro-B and pro-B cell numbers are significantly reduced while subsequent pre-B and immature populations are relatively normal (Mackarehtschian et al. 1995). When Flt-3 was added to *in vitro* cultures, it synergized with IL-7 and led to an increase in the survival and proliferation of pre-pro-B cells (Hunte et al. 1996).

IL-7 and B cell Development

The growth factor IL-7 is produced by cells in the BM, spleen, thymus, and fetal liver and provides signals required for the survival, proliferation, and differentiation of developing B cells. Mice with targeted deletions of IL-7 or the IL-7R display a severe block at the early pro-B cell stage of development (reviewed in Milne and Paige 2006). Transgenic expression of IL-7 in mice results in increased immature and mature B cell populations in the BM as well as the appearance of pro-B and pre-B populations in the spleen, blood, and lymph nodes that ultimately lead to lymphoproliferative disorders in these mice (Rich et al. 1993). Regulated IL-7 signalling is also important during B cell commitment as high-level expression of the IL-7R on multipotent stem cells leads to a block in B cell development prior to expression of CD19, and blocked cells express decreased levels of EBF and Pax-5 (Purohit et al. 2003).

IL-7 Signalling

The IL-7R is a heterodimer composed of the α chain and the common (γ) chain (γc) utilized by the IL-7 and thymic stromal lymphopoietin (TSLP) receptors, and the γc chain is a shared component of the receptors for IL-2, IL-4, IL-7, the IL-9, IL-15, and IL-21. IL-7 receptor binding leads to the heterodimerization of the α and γc chains, which in turn allows for the trans-phosphorylation of constitutively associated JAK1 and JAK3 proteins and subsequent IL-7α chain phosphorylation (Fig. 8.2). Receptor phosphorylation leads to activation of Src kinases (LYN/FYN/Blk) and also creates docking sites for SH2 containing proteins PI3K and STAT (Jiang et al. 2005). PI3K

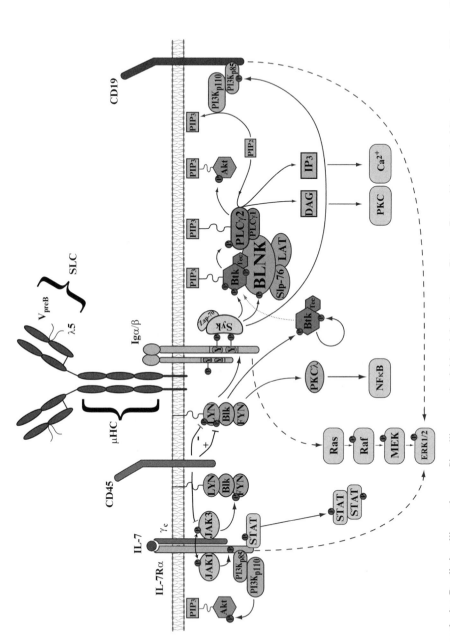

Fig. 8.2 Developing B cell signalling pathways: Signalling cascades initiated during the pro-B to pre-B transition. *Dashed lines* denote a multistep pathway. See text for detailed description.

is recruited to the IL-7Rα chain and activates the anti-apoptotic molecule AKT. Inhibition of PI3K activation led to the observation that it is important for cell proliferation and survival but dispensable for IL-7-mediated differentiation events (Corcoran et al. 1996). Recruitment of STATs to the IL-7Rα chain results in their hetero- and homodimerization prior to nuclear translocation. Deletion of STAT5 leads to a block in development at the pro-B stage, similar to that observed in IL-7R$^{-/-}$ mice, and is a result of failed expression of EBF and Pax-5 (Dai et al. 2007). Conversely, STAT5 over-expression can overcome IL-7R deficiency and demonstrates STAT5s essential role in mediating IL-7-induced events (Goetz et al. 2004). STAT3 is also critical during B cell development: the impaired differentiation and survival displayed by STAT3-/- B cells leads to a decrease in pro-, pre-, and immature B cell populations (Chou et al. 2006). Finally, it has been shown that stimulation with IL-7 leads to the activation of the MAP kinase ERK, and inhibition of this pathway prevents proliferation and survival of pro- and pre-B cells (Fleming and Paige 2001).

IL-7 and B cell Commitment

IL-7 signalling is both trophic and instructive in nature. Over-expression of Bcl-2 in IL-7$^{-/-}$ mice does not correct the B cell defect; this indicates that IL-7 is required for more than just survival of developing B cells. Experimentally, CLPs isolated from IL-7$^{-/-}$ animals have a greatly reduced ability to generate B cells compared with wild-type cells (Dias et al. 2005). This CLP defect is likely the result of ineffective activation of critical IL-7-induced TFs as enforced expression of EBF restores CLP differentiation. In addition to activating EBF, IL-7 signalling also regulates E2A and Pax-5 expression (Hirokawa et al. 2003). IL-7 also promotes V(D)J recombination, a hallmark trait of lymphoid commitment. Signals downstream of the IL-7R via STAT5 lead to hyperacetylation at the distal V_H segments and thus allow for heavy chain loci accessibility in B cells (Bertolino et al. 2005).

IL-7 and B cell Survival, Proliferation, and Maturation

Cell death and survival is regulated by the localization and interaction of pro- and anti-apoptotic factors. IL-7 signalling in pro-B cells upregulates the pro-survival factor Bcl-$_{XL}$ and increases the ratio of anti-apoptotic factor Bcl-2 to pro-apoptotic factor Bax (Banerjee and Rohtman 1998; Lu et al. 1999). Proliferative responses to IL-7 are mediated, at least in part, through myc family members, and IL-7 stimulation of pre-B cells results in c-myc and N-myc activation (Morrow et al. 1992). Myc-mediated IL-7 proliferation is partially dependent on expression of E2A as B-lymphocyte progenitors from E2A$^{-/-}$ mice possess reduced proliferative capacity due to sub-optimal levels of the IL-7 target N-myc (Seet et al. 2004). Myc activation leads to increased intracellular Ca^{2+} and is required for B cell proliferation and differentiation.

As pro-B cells proliferate in response to IL-7, they increase in number and dilute out the local IL-7 concentrations. Eventually this dilution may bring about their death. However, pre-BCR$^+$ cells are able to proliferate in reduced concentrations of IL-7, and this unique ability could provide a mechanism to select for the survival and expansion of cells that have functionally expressed the pre-BCR (Fleming and Paige 2001). It has been proposed that, after pre-B cell expansion, attenuation of IL-7 signalling is necessary to allow for LC rearrangement and the further maturation of B cells. This theory comes from the observation that withdrawal of IL-7 from *in vitro* cultures results in increased Rag expression, LC rearrangement, and percentage of immature IgM$^+$ cells (Rolink et al. 2000). However, it has since been demonstrated that these increases are relative and not absolute (Milne et al. 2004). Instead of inducing maturation, removal of IL-7 from cultures merely prevents the expansion of pro/large pre-B populations. This lack of IL-7-dependent proliferation results in an apparent increase in the number of small pre-B and immature cells, as determined by their overall culture percentage.

IL-7-Independent B cell Development

The strict requirement for IL-7 observed during adult B cell development in mice is not found during murine fetal or human B cell development. Mice with targeted deletions of IL-7 or the γc chain have a complete block in adult B cell development but peripheral B cells can still be detected and display characteristics of B cells generated during fetal life. It is believed that TSLP, or other cytokines, can substitute for IL-7 during fetal development and this will be discussed in more detail later. In humans, it has been shown that *in vitro* B cell cultures can be generated from CD34$^+$ BM in the absence of IL-7 (Pribyl and LeBien 1996). Genetic evidence has confirmed these observations: in mice, (gamma symbol)c chain deficiency results in a complete block at the pro-B stage, while in humans, X-linked severe combined immunodeficiency (XSCID), the corresponding deficiency, leads only to a reduction in T and NK cell numbers. However, several studies have shown that human B cells and their precursors not only express the IL-7R but also respond to IL-7 with increased survival and proliferation (reviewed in LeBien 2000). Hence, while not essential, IL-7 may still have important roles during human B cell development. It is unknown what other factors support human B cell development in the absence of IL-7 but, in contrast to mice, TSLP is not a candidate as patients with genetic mutations in components of the TSLP receptor exhibit normal B cell development (Giliani et al. 2005).

TSLP

TSLP, a cytokine that is produced by epithelial cells, shares shares a number of similarities to IL-7. The TSLP receptor is a heterodimer composed of the IL-7Rα chain and the specific TSLP receptor chain. Receptor ligation initiates signal transduction independently of JAK proteins but still leads to STAT activation and nuclear translocation (Levin et al. 1999). In B cells, TSLP is believed to work in combination with, or substitute for, IL-7 signals during fetal and adult life. Over-expression of TSLP in mice results in the early exit of precursor B cells from the BM as well as increased B cell numbers in the periphery, which eventually leads to autoimmunity (Astrakhan et al. 2007). A more recent study found that transgenic expression of TSLP in IL-7$^{-/-}$ mice restored B and T-cell development during both fetal and adult life (Chappaz et al. 2007). Knockout studies have shown that B lymphopoiesis in IL-7R$\alpha^{-/-}$ mice is reduced by more than tenfold as compared to IL-7Rc$\gamma^{-/-}$ mice. These results imply that deletion of the TSLP/IL-7 receptor is more detrimental to B cell development than disruption of IL-7 signalling alone (Vosshenrich et al. 2003). Additionally, while both adult and fetal pre-B cells proliferate in response to TSLP, only fetal pro-B cells possess this ability. From these observations the authors concluded that TSLP is sufficient to mediate the development of fetal B cells in the absence of IL-7. In contrast to these results, TSLP$^{-/-}$ mice fail to show any B cell abnormalities (Carpino et al. 2004). Additionally, IL-7$^{-/-}$/TSLP$^{-/-}$ mice show no defect in numbers of fetal pro- or pre-B cells and only exhibit a slight reduction in immature B cell numbers (Jensen et al. 2007). Instead the authors show that fetal and adult B cell development was completely blocked in Flt-3L$^{-/-}$/IL-7$^{-/-}$ mice and TSLP is unable to compensate in this situation. This led to the conclusion that Flt-3, rather than TSLP, is the key factor responsible for B cell development in IL-7$^{-/-}$ mice.

Regulation of B cell Development by the Pre-BCR and BCR

Receptor Composition

Expression of the pre-BCR and mature BCR are critical events during B lymphopoiesis. In mice where the pre-BCR is either unable to form or insert into a lipid bilayer (Rag 1/2$^{-/-}$, μMT, and SCID), B cell development is fully arrested at the pro-B stage. Naturally occurring human mutations of various receptor components (Igα, Igβ, μHC, or λ5) also block development at the pro-B stage and result in hypo- or agammaglobulinemia. The pre-BCR is similar to the mature BCR and uses HC proteins with the same rearrangement. pre-BCR components V_{preB} and λ5 possess structural similarities to the variable and constant domains of IgLC, respectively; however, they contain non-Ig chain tails instead of a CDR3 region (Fig. 8.2). $V_{preB}^{-/-}$ and λ5$^{-/-}$ single or double deficient mice display an incomplete block in development at the pro-B to pre-B transition (reviewed in Martensson et al. 2002). Early LC rearrangement in SLC$^{-/-}$ pro-B cells is possible, and LC pairing with μHCs may occur to provide a mechanism for cells to bypass the pre-BCR checkpoint. Alternatively, it has been suggested that escape of μHC to the cell surface, either alone or in conjunction with an unknown binding partner, could also initiate signals that allow for development (Martensson et al. 2007). The pre-BCR complex also contains the transmembrane proteins Igα and Igβ, which are non-covalently associated with μHCs. The cytoplasmic tails of Igα/β contain immunoreceptor tyrosine activation motifs (ITAMs), which are the critical regions mediating signal transduction and act as docking sites for SH2-containing proteins. Prior to expression with the pre-BCR, Igα/β can be detected on the surface of pro-B cells in a structure referred to as the pro-BCR. This complex contains Igα/β associated either with the ER chaperone calnexin and SLC, or cadherin-17 (Karasuyama et al. 1993; Ohnishi et al. 2005). The functional role of the pro-BCR during development remains unclear, as disruption of the genes encoding Igα/β (*mb1* and *B29*, respectively) does not alter pro-B cell development. Instead, deletion of *B29* results in a block at the pro-B to pre-B transition with cells containing DJ_H rearrangements but few VDJ_H rearrangements (Gong and Nussenzweig 1996). The observation that *mb1* deficiency results in a block at the later immature stage demonstrates that both Igα and Igβ are important during B cell development, but that they mediate different effects (Torres et al. 1996).

Receptor Activation

It is currently unknown whether ligand binding is required to initiate pre-BCR signal transduction. Studies supporting the ligand-independent model have shown that cells expressing a truncated extracellular domain μHC still carry out allelic exclusion, surface marker change, and LC gene transcription (Shaffer and Schlissel 1997). Surface pre-BCR expression can also be bypassed by transgenic expression of LMP2A, an EBV virally encoded protein that contains an ITAM, or by direct targeting of the cytoplasmic domain of Igα/β to the cell surface (Teh and Neuberger 1997; Caldwell et al. 2000). These observations led to the hypothesis that signalling is initiated either through lipid raft localization or through ligand-independent receptor aggregation. Lipid rafts are glycosphingolipid- and cholesterol-enriched plasma membrane microdomains that either include or exclude signalling proteins and that are believed to act as signalling platforms that stabilize protein complexes. For the BCR, it has been observed that the Src protein tyrosine LYN is constitutively associated in the raft domain while CD45 is excluded (reviewed in Pierce 2002). Src is a positive regulator of pre-BCR signalling and CD45 can negatively regulate Src's activation. Hence, it is predicted that the inclusion and exclusion of these proteins in pre-BCR lipid rafts would allow for ligand-independent receptor phosphorylation that leads to signal transduction. Additionally, in humans it has been observed that the pre-BCR is constitutively associated within the lipid raft fraction (Guo et al. 2000). In contrast to the lipid raft model, targeting of Igα/β to either raft or non-raft domains initiated pre-BCR signalling and development equally well (Fuentes-Panana et al. 2005).

Evidence demonstrating that truncated or modified receptors can mediate the pro-B to pre-B cell transition does not exclude the possibility that ligand engagement of the pre-BCR is important. Levels of surface expression of the aforementioned mutated/modified receptors often far exceed those of physiological conditions and may result in unnatural aggregation and activation. Additionally, many of these modifications bypass pre-BCR and BCR checkpoints, which is suggestive of constitutive signalling. Support for the ligand-dependent model comes from studies that have isolated molecules capable of binding to the pre-BCR. Experiments that used a soluble pre-BCR-like molecule to screen stromal cell lysates lead to the discovery that heparan sulphate binds to the unique tail of λ5 (Bradl et al. 2003). Another pre-BCR ligand, human stromal cell molecule galectin-1, binds through its interaction with the unique tail of λ5 on the human pre-BCR (Gauthier et al. 2002). The pre-BCR may also be its own ligand. Mutations of the unique tail of λ5 on the B cell surface reduced aggregation of the pre-BCR and impaired pre-BCR internalization (Ohnishi and Melchers 2003).

Receptor Signalling

Regardless of the method of activation, signal transduction is dependent on receptor aggregation and leads to the formation of the surface signalling complex. The complex includes the pre-BCR and co-activators, such as CD19, but excludes CD45 (Fig. 8.2). Signalling intermediates are recruited to the complex through their interaction with ITAMs and immunoreceptor tyrosine inhibition motifs (ITIMs), which are contained within the cytoplasmic tails of surface molecules. Signal transduction is mediated by kinases and phosphatases, which function to phosphorylate and dephosphorylate proteins, and by adaptors that lack intrinsic kinase activity but instead operate to bring molecules together and allow for their interaction.

Pre-BCR signals are initiated by Src kinase phosphorylation of the ITAM regions of Igα/β (Fig. 8.2). Src kinases also induce NF-κB signalling via PKCλ, and phosphorylate the tyrosine kinase Btk, which leads to its subsequent autophosphorylation and activation. The Src family of kinases contain members including Src, Lyn, Blk, and Fyn with Blk being the only B cell exclusive member. Due to the redundant nature of these molecules, mice deficient for any one of the members show no defect in signalling. However, Lyn/Fyn/Blk triple-deficient mice display a dramatic decrease in pre-B cell numbers but normal numbers of pro-B cells. This observation demonstrates the important role this family plays in transducing signals downstream of the pre-BCR. Phosphorylated Igα/β ITAMs serve as docking sites for the SH2-containing Syk kinase, which is subsequently able to activate several downstream pathways through signalling mediators PI3K, Btk, and BLNK. Deletion of Syk revealed that it is not the only family member that mediates mediates transduction in developing B cells and that Zap-70 can compensate for its loss. A full block was observed when both of these family members were deleted. This was a surprising discovery since it had previously been reported that Zap-70 can only expressed in T and NK cells (Schweighoffer et al. 2003). Similar redundancy is observed for BLNK, Btk, and PLCγ2 as LAT and SLP-76 can compensate for BLNK, Tec can partially compensate for Btk, and PLCγ1 performs similar function as PLCγ2 (Fig. 8.2).

Signalling intermediates are recruited to the receptor complex by way of receptor motifs and adaptors. They can also be targeted to the complex by interaction of their pleckstrin homology (PH) domains with phosphatidylinositol lipids present in biological membranes. Phosphatidylinositol 3, 4,5-triphosphate (PIP3) is a key mediator of membrane localization and is generated by PI3K, which converts phosphatidylinositol 4,5-bisphosphate (PIP2) to PIP3. PI3K is a dimer composed of a p110 subunit that possesses catalytic activity, and a SH2-containing p85 subunit that is important in targeting the kinase to the signalling complex. Mice deficient for either subunit display an incomplete block in development at the pre-B stage and a significant reduction in mature splenocytes (Donahue and Fruman 2004). This B cell defect is due to the ineffective targeting of signalling molecules Btk, PLCγ2, and protein kinase B (PKB, also referred to as AKT) to the receptor complex. PIP2 also functions as a substrate for PLCγ2, which converts PIP2 into diacyglycerol (DAG) and inositoltriphosphate (IP3). DAG and IP3 are secondary messengers that lead to protein kinase C (PKC) activation and calcium (Ca^{2+}) mobilization, respectively. PLCγ2 and Btk are both recruited to the receptor complex by the adaptor protein BLNK. This protein assembly allows Btk to phosphorylate PLCγ2, and leads to its full activation. BLNK has also been shown to be a binding partner for Syk in mature B cells. This association results in a positive feedback loop that is necessary for ERK, NFκB, and Ca^{2+} responses but not for Akt activation (Kulathu et al. 2008).

Additional Surface Proteins Regulating Receptor Signalling

B220 is a 220-kDa isoform of CD45 that is present on all mouse B cells except terminally differentiated plasma blasts. CD45 is expressed on the earliest B cell precursors, as well as in varying isoforms on DC, T cells, macrophages, and NK cells. CD45 is a phosphatase that can activate the Src family kinase LYN by dephosphorylating inhibitory tyrosine residues. CD45$^{-/-}$ mice exhibit reduced proliferation of mature B cells in response to foreign antigens but not to mitogens such as lipopolysaccharide (LPS) (Benatar et al. 1996). Conversely, other studies have shown that CD45 can dephosphorylate both positive and negative tyrosine residues and that, in its absence, LYN is constitutively phosphorylated. Thus, CD45 appears to play both positive and negative roles during BCR signalling and may function as a modulator of signalling thresholds (reviewed in Huntington and Tarlinton 2004) (Fig. 8.2). During B development, CD45 also regulates the IL-7 signalling threshold for B cells. CD45$^{-/-}$ mice exhibit increased numbers of pro-B cells (Fleming et al. 2004). This B cell defect is due to CD45 dephosphorylation of Jak proteins and, in its absence, Jak and STAT phosphorylation is prolonged after IL-7 stimulation.

CD19 is a B cell specific marker that is expressed on pro-B cells and all subsequent B cell populations except plasma cells. It is an accessory molecule that reduces the signalling threshold for the BCR by recruiting PI3K to the signalling complex, which in turn allows for the activation of Btk (Buhl and Cambier 1999) (Fig. 8.2). During B cell development, CD19 functions to regulate pre-BCR signalling: CD19$^{-/-}$ mice display defects in the transition from the pre-B to immature stage of development that are partially due to deficient phosphorylation of Btk and the MAP kinase ERK (Otero and Rickert 2003).

Receptor Function

Pre-BCR regulation of apoptotic factors is mediated through downstream pathways including, the PI3K/AKT pathway and the Ras/Raf/MEK/ERK pathway. Mice deficient for or over-expressing the anti-apoptotic protein Bcl-$_{XL}$ display reduced or elevated numbers of pre-B cells, respectively (reviewed in Lu and Osmond 2000). While Bcl-$_{XL}$ provides survival signals to pre-B cells, Bcl-2, another anti-apoptotic molecule, exerts its effect on immature cells, and Bcl-2 transgenic mice have increased IgM$^+$ cell numbers.

Cell proliferation, another outcome of pre-BCR signalling, is mediated in large part through the activation of the Ras-ERK pathway. The importance of this pathway in B cell development has been highlighted in ERK1/2$^{-/-}$ mice, which exhibit diminished pre-BCR-mediated expansion and a block at the pro-B to pre-B stage of development (Yasuda et al. 2008). Engagement of the pre-BCR or IL-7R results in downstream phosphorylation of ERK, and these pathways synergistically activate ERK to allow for the proliferation of large pre-BCR$^+$ cells in picogram concentrations of IL-7 (Fleming and Paige 2001).

Pre-BCR activation of BLNK and Btk limits the proliferation of large pre-B cells and allows for further maturation. BLNK$^{-/-}$ or Btk$^{-/-}$ mice show a partial block in development at the large pre-B stage with increased expression of SLC components, increased surface pre-BCR expression, and enhanced proliferative capacity (Middendorp et al. 2002; Flemming et al. 2003). Transcriptional downregulation of SLC expression via BLNK, Btk, and IRF-4/8 is believed to diminish individual cell surface pre-BCR expression and thus limit large pre-B cell signalling and proliferation. However, over-expression studies have demonstrated that SLC silencing is not absolutely required to limit the expansion of pre-B cells but is necessary to prevent constitutive B cell activation (van Loo et al. 2007).

BLNK$^{-/-}$ or Btk$^{-/-}$ pre-B cells also exhibit normal allelic exclusion and reduction of Rag and TdT expression, which suggests that these effects are mediated by Btk- and BLNK-independent mechanisms and possibly are not even pre-BCR-dependent events. This theory is strengthened by the observation that allelic exclusion is observed in SLC$^{-/-}$ mice (Galler et al. 2004). However, studies using μMT mice have shown that insertion of the μHC into the lipid bilayer is essential. The signalling mechanisms that result in allelic exclusion have not been fully elucidated, but it has been demonstrated that Syk$^{-/-}$/Zap-70$^{-/-}$ and PLCγ1$^{+/-}$/PLCγ2$^{-/-}$ mice do not display allelic exclusion while BLNK$^{-/-}$ mice retain this ability (Xu et al. 2000; Schweighoffer et al. 2003; Wen et al. 2004). Consequently, allelic exclusion at the HC locus is initiated only after membrane expression of a productively rearranged μHC protein initiates downstream signals through Syk and PLCγ family members.

Signalling Defects and Clinical Disease

Bruton's tyrosine kinase (Btk) deficiency in humans results in X-linked agamma-globulinemia (XLA), the most common genetic defect observed in patients with early B cell defects. Patients with XLA or defects in BLNK exhibit low levels of serum Ig of all classes and are typically diagnosed within the first few years of life after exhibiting susceptibility to recurrent bacterial infections. Btk and BLNK deficiencies have also been observed in approximately 50% of childhood pre-B acute lymphoblastic leukemias (ALL) (Goodman et al. 2003; Jumaa et al. 2003). Unlike mice deficient in Btk or BLNK, which exhibit an expansion in pre-B cells, humans deficient in these molecules exhibit a reduction in pre-B cells. (Nomura et al. 2000). Defective pro-B to pre-B cell development corresponds with the agam-maglobulinemia phenotype but it does not fit with the leukemic nature of these cells. In such cases it is believed that oncogenesis is the result of BCR-ABL trunca-tion of BLNK and/or Btk, which results in constitutively activated signalling path-ways. BCR-ABL is a common translocation observed in ALL patients. It results from the fusion of the break cluster region (BCR), not to be confused with the B cell receptor, and the c-abl proto-oncogene (reviewed in Wong and Witte 2004). Cells exhibiting this translocation display increased survival and proliferation, independent of pre-B cell receptor expression, and are immortalized with either an unresponsive or a non-productive/non-functional pre-BCR (Klein et al. 2004).

Selection of B Cells

Although It has been estimated that 10^8 B cells are produced in the murine BM every day, only a small percentage of these precursors ever develop into functional mature B cells. Approximately 75% of cells fail to transit past the pre-B checkpoint and only 10% of immature cells make it into the periphery (Osmond 1991). Cell apoptosis is usually the result of unsuccessful HC or LC rearrangement, or of ineffi-cient SLC-HC or LC–HC binding. It has also been estimated that half of the μHCs generated cannot associate with SLC. This selective binding acts as a mechanism to test the capacity of μHCs to bind with potential LCs and thus maximize the forma-tion of mature receptors (ten Boekel et al. 1998). Failure of μHCs to bind SLC results in further rearrangement at the HC locus until a functional μHC, capable of combining with SLC to produce a functional pre-BCR is generated.

While all pre-BCR+ cells enter the cell cycle, individual cell expansion may not be uniform. It has been proposed that the relative binding strength between SLC and μHCs may alter signal intensity and result in greater proliferation and selection of cells possessing ideal receptor pairing (Melchers et al. 2000). The clonal expan-sion of pre-B cells with the same productive HC allows independent daughter cells to generate different LCs and thus increase receptor diversity in the periphery.

Approximately half of all immature cells posses self-reactive BCRs and must be prevented from exiting into the periphery. Ligation of the BCR by self-antigen on immature cells can result in deletion, anergy, or receptor editing, the latter of which can occur at both the HC and LC locus and which allows for the possibility of generating alternative receptors that are not self-reactive. Several studies have shown that high-affinity receptor interactions on immature cells lead to deletion, while lower affinity reactions result in anergy. Alternatively, some studies suggest that low-affinity binding of the BCR may direct cells towards the B1 B cell fate. BCR engagement on transitional cells no longer leads to deletion or anergy; instead, BCR signal intensity directs cells towards Fo, MZ, and B-1 B cell fates as described in the next section.

Mature B Cells

B cells are classically defined as adaptive immune cells that mediate their effects through the recognition of specific antigenic epitopes. Fo B cells are the main effector cells of adaptive B cell immunity. Fo B cells are relatively long lived (~ 5 months), make up the majority of the peripheral B cell pool, and express highly diverse receptors that typically respond to T-cell-dependent (TD) protein antigens. The engagement of the BCR on these cells, in conjunction with signals from helper T cells, leads to the generation of germinal centres (GC) in the spleen. GCs are temporary structures that provide the microenvironment for B cells to develop through their interaction with antigen-presenting follicular dendritic cells. During this maturation, Fo B cells begin a process of somatic hypermutation, which causes mutations in the heavy and light chains and results in the formation of higher- and lower-affinity BCRs. Newly specified receptors then go through a process of affinity maturation, which allows for the selection of cells and leads to increased BCR specificity and a more robust response to foreign antigen. These mature B cells can also switch their effector function by converting to other antibody isotypes (IgA, IgE, and IgG). Class-switched Fo B cells express the same BCR specificity as their IgM+ parent cells and migrate throughout the periphery where they become plasmablasts and long-lived memory cells. Plasmablasts produce large amounts of antibody, which is a secreted version of the BCR (reviewed in Murphy 2007).

The majority of antibodies produced in the body are the result of the affinity maturation process; however, low-affinity antibodies are also generated both naturally and in response to antigen. These antibodies are produced by MZ and B1 B cells, tend to be less diverse than those produced by their Fo cell counterparts, and typically recognize carbohydrate- or glycolipid-containing T-cell-independent (TI) antigens associated with bacterial membranes, such as LPS. There is also evidence that, to some extent, MZ and B-1B cells are selected based on positive interaction with self-antigen (Hardy et al. 2004; Wen et al. 2005). MZ B cells that reside in the marginal sinus region of the spleen are very long lived (>1 year) and possess low-affinity/high-avidity BCRs. Activation of these cells occurs independently of T cells and leads to their proliferation and development into plasmablasts. Immediate and innate responses from MZ and B1 B cells provide a first line of defence that limits pathogen spread and provides the necessary time (~1 week) for more specific adaptive Fo B cell responses to develop (reviewed in Murphy 2007).

B-1 B cells, which typically reside in the peritoneal and pleural cavities, produce low-affinity IgM antibodies that can be detected in individuals independent of infection or immunization. B-1 B cells account for approximately 5% of the total B cell pool and differ from conventional B-2 B cells (Fo and MZ) in their development and activation. Although these cells were originally defined by their $CD11b^+$ $CD5^+$ IgM^{hi} IgD^{lo} phenotype, the discovery of $CD5^-$ counterparts led to the further classification of B-1a ($CD11b^+$ $CD5^+$ IgM^{hi} IgD^{lo}) and B-1b ($CD11b^+$ $CD5^-$ IgM^{hi} IgD^{lo}) subsets (reviewed in Dorshkind and Montecino-Rodriguez 2007). The main functional difference between these two subsets is that B-1a B cells spontaneously produce antibody and are important in the early stages of a response, while B-1b B cells are induced to secrete antibody and are more prevalent in the later stages of pathogen clearance. There has been much debate as to the origin of B-1 B cells and two leading hypotheses currently exist. One model suggests that both B-1 and B-2 B cells are generated in a similar fashion in the BM but environmental signals and thresholds of antigen receptor binding in the periphery bias B-1 versus B-2 fate. This hypothesis is supported by the observation that stimulation of B-2 B cells with certain antigens leads to their expression of CD5 and gives them the ability to respond to phorbol esters, both of which are traditionally B-1 B cell responses (Berland and Wortis 2002). The alternate model of B-1 B cell origin is one of early lineage commitment: it proposes that B-1 B cell specification occurs early in the developmental process and is mainly a function of fetal development. This model is supported by the fact that when irradiated or immunodeficient adult mice are reconstituted with cells from fetal tissues, predominantly B-1 B cell development is observed (Kantor et al. 1992). However, when reconstitution is performed using adult BM, B-2 B cells are the predominant product.

The restricted Ab specificity of B-1 B cells could be, in large part, a consequence of cell development within the fetal environment. Since TdT is not expressed during fetal development, B cells generated during this time display decreased diversity in the CDR3 region (Li et al. 1993). Additionally, Ig recombination in the absence of TdT results in joining of specific V, D, and J gene segments that contain complimentary ends (Feeney 1992). Both of these mechanisms result in a restricted receptor repertoire that is more representative of germline sequences and leads to the possibility of evolutionary conserving sequences that possess specific antigen binding domains targeted against common pathogen epitopes. It is tempting to speculate that the production of B-1 B cells during fetal development would provide a pool of natural neutralizing antibody capable of providing a quick response to common bacterial infections prior to the development of the adaptive immune system. In line with this, forced expression of TdT during fetal life can prevent the generation of B cells possessing anti-phosphorylcholine specificity, which is important for protection against *streptococcus pneumoniae* infection (Benedict and Kearney 1999).

The generation of B-1a/b, MZ, and Fo B cells is dependent on BCR signal strength as well as on the environmental context in which signals are received. In a model proposed by Hardy et al., strong BCR signals during fetal development would direct cells towards the B-1a cell fate, intermediate signals would lead to B-1b or MZ B cells, and weak BCR interactions would result in receptor editing or death. By contrast, during adult development, strong BCR signals would elicit cell death or receptor editing, intermediate signals would lead to MZ and B-1b cell fate, and weak signals would generate Fo (B-2) B cells (Hardy 2006).

Conclusion

The immune system is composed of different types of cells that mediate the development and function of immune effectors and together protect the body against infection. B cells are an integral part of this system and are the sole producers of antibody. During their development, B cells exhibit complex gene networks that are regulated by intricate signalling pathways influenced by surface receptors and intrinsic transcription factors. Decisions regarding cell survival, proliferation, and differentiation are both instructive and stochastic in nature. Selection of cells occurs at developmental checkpoints and ensures the generation of functional mature B cells, which display remarkable receptor diversity and virtually no self-reactivity. Diverse B cell effectors can elicit immediate innate-like responses, delayed highly specific antigen targeting and clearance, and long-term memory protection against foreign pathogens. Severe immunodeficiency is observed in individuals who possess reduced or deficient B cells, and therefore built-in redundancy and compensatory pathways within B cells exist to minimize the deleterious effect that genetic defects have on cell development and function.

References

Allman, D. and Pillai, S. (2008). *Curr Opin Immunol.* Apr;20(2):149–57.
Astrakhan, A., Omori, M., Nguyen, T., Becker-Herman, S., Iseki, M., Aye, T., Hudkins, K., Dooley, J., Farr, A., Alpers, C. E., Ziegler, S. F. and Rawlings, D. J. (2007). *Nature Immunology* **8**(5): 522–531.
Bain, G., Maandag, E. C. R., Riele, H., Feeney, A. J., Sheehy, A., Schlissel, M., Shinton, S. A., Hardy, R. R. and Murre, C. (1997). *Immunity* **6**(2): 145–154.
Banerjee, A. and Rohtman, P. (1998). *Journal of Immunology* **161**(9): 4611–4617.
Bassing, C. H., Swat, W. and Alt, F. W. (2002). *Cell* **109**: S45–S55.
Benatar, T., Carsetti, R., Furlonger, C., Kamalia, N., Mak, T. and Paige, C. J. (1996). *Journal of Experimental Medicine* **183**(1): 329–334.
Benedict, C. L. and Kearney, J. F. (1999). *Immunity* **10**(5): 607–617.
Berland, R. and Wortis, H. H. (2002). *Annual Review of Immunology* **20**: 253–300.
Bertolino, E., Reddy, K., Medina, K. L., Parganas, E., Ihle, J. and Singh, H. (2005). *Nature Immunology* **6**(8): 836–843.
Bradl, H., Wittmann, J., Milius, D., Vettermann, C. and Jack, H. M. (2003). *Journal of Immunology* **171**(5): 2338–2348.
Buhl, A. M. and Cambier, J. C. (1999). *Journal of Immunology* **162**(8): 4438–4446.

Caldwell, R. G., Brown, R. G. and Longnecker, R. (2000). *Journal of Virology* **74**(3): 1101–1113.

Carpino, N., Thierfelder, W. E., Chang, M. S., Saris, C., Turner, S. J., Ziegler, S. F. and Ihle, J. N. (2004). *Molecular and Cellular Biology* **24**(6): 2584–2592.

Chappaz, S., Flueck, L., Farr, A. G., Rolink, A. G. and Finke, D. (2007). *Blood* **110**(12): 3862–3870.

Chou, W. C., Levy, D. E. and Lee, C. K. (2006). *Blood* **108**(9): 3005–3011.

Corcoran, A. E., Smart, F. M., Cowling, R. J., Crompton, T., Owen, M. J. and Venkitaraman, A. R. (1996). *EMBO Journal* **15**(8): 1924–1932.

Dai, X. Z., Chen, Y. H., Di, L., Podd, A., Li, G. Q., Bunting, K. D., Hennighausen, L., Wen, R. R. and Wang, D. M. (2007). *Journal of Immunology* **179**(2): 1068–1079.

DeKoter, R. P. and Singh, H. (2000). *Science* **288**(5470): 1439–1441.

DeKoter, R. P., Lee, H. J. and Singh, H. (2002). *Immunity* **16**(2): 297–309.

Dias, S., Silva, H., Cumano, A. and Vieira, P. (2005). *Journal of Experimental Medicine* **201**(6): 971–979.

Donahue, A. C. and Fruman, D. A. (2004). *Seminars in Cell and Developmental Biology* **15**(2): 183–197.

Dorshkind, K.and Montecino-Rodriguez, E. (2007). *Nature Reviews Immunology* **7**(3): 213–219.

Egawa, T., Kawabata, K., Kawamoto, H., Amada, K., Okamoto, R., Fujii, N., Kishimoto, T., Katsura, Y. and Nagasawa, T. (2001). *Immunity* **15**(2): 323–334.

Fazio, G., Palmi, C., Rolink, A., Biondi, A. and Cazzaniga, G. (2008). *Cancer Research* **68**(1): 181–189.

Feeney, A. J. (1992). *Journal of Immunology* **149**(1): 222–229.

Fleming, H. E. and Paige, C. J. (2001). *Immunity* **15**(4): 521–531.

Fleming, H. E., Milne, C. D. and Paige, C. J. (2004). *Journal of Immunology* **173**(4): 2542–2551.

Flemming, A., Brummer, T., Reth, M. and Jumaa, H. (2003). *Nature Immunology* **4**(1): 38–43.

Fuentes-Panana, E. M., Bannish, G., van der Voort, D., King, L. B. and Monroe, J. G. (2005). *Journal of Immunology* **174**(3): 1245–1252.

Galler, G. R., Mundt, C., Parker, M., Pelanda, R., Martensson, I. L. and Winkler, T. H. (2004). *Journal of Experimental Medicine* **199**(11): 1523–1532.

Gauthier, L., Rossi, B., Roux, F., Termine, E. and Schiff, C. (2002). *Proceedings of the National Academy of Sciences of the United States of America* **99**(20): 13014–13019.

Giliani, S., Mori, L., Basile, G. D., Le Deist, F., Rodriguez-Perez, C., Forino, C., Mazzolari, E., Dupuis, S., Elhasid, R., Kessel, A., Galambrun, C., Gil, J., Fischer, A., Etzioni, A. and Notarangelo, L. D. (2005). *Immunological Reviews* **203**: 110–126.

Goetz, C. A., Harmon, I. R., O'Neil, J. J., Burchill, M. A. and Farrar, M. A. (2004). *Journal of Immunology* **172**(8): 4770–4778.

Gong, S. C. and Nussenzweig, M. C. (1996). *Science* **272**(5260): 411–414.

Goodman, P. A., Wood, C. M., Vassilev, A. O., Mao, C. and Uckun, F. M. (2003). *Leukemia and Lymphoma* **44**(6): 1011–1018.

Guo, B. C., Kato, R. M., Garcia-Lloret, M., Wahl, M. I. and Rawlings, D. J. (2000). *Immunity* **13**(2): 243–253.

Hardy, R. R. (2006). *Journal of Immunology* **177**(5): 2749–2754.

Hardy, R. R., Carmack, C. E., Shinton S. A., Kemp, J. D. and Hayakawa, K. (1991). *Journal of Experimental Medicine* **173**(5): 1213–1225.

Hardy, R. R., Wei, C. J. and Hayakawa, K. (2004). *Immunological Reviews* **197**: 60–74.

Hargreaves, D. C., Hyman, P. L., Lu, T. T., Ngo, V. N., Bidgol, A., Suzuki, G., Zou, Y. R., Littman, D. R. and Cyster, J. G. (2001). *Journal of Experimental Medicine* **194**(1): 45–56.

Hirokawa, S., Sato, H., Kato, B. and Kudo, A. (2003). *European Journal of Immunology* **33**(7): 1824–1829.

Horcher, M., Souabni, A. and Busslinger, M. (2001). *Immunity* **14**(6): 779–790.

Houston, I. B., Kamath, M. B., Schweitzer, B. L., Chlon, T. M. and DeKoter, R. P. (2007). *Experimental Hematology* **35**(7): 1056–1068.

Hunte, B. E., Hudak, S., Campbell, D., Xu, Y. M. and Rennick, D. (1996). *Journal of Immunology* **156**(2): 489–496.

Huntington, N. D. and Tarlinton, D. M. (2004). *Immunology Letters* **94**(3): 167–174.

Iwasaki, H., Somoza, C., Shigematsu, H., Duprez, E. A., Iwasaki-Arai, J., Mizuno, S., Arinobu, Y., Geary, K., Zhang, P., Dayaram, T., Fenyus, M. L., Elf, S., Chan, S., Kastner, P., Huettner, C. S., Murray, R., Tenen, D. G. and Akashi, K. (2005). *Blood* **106**(5): 1590–1600.

Jensen, C. T., Kharazi, S., Boiers, C., Liuba, K. and Jacobsen, S. E. W. (2007). *Nature Immunology* **8**(9): 897.

Jiang, Q., Li, W. Q., Aiello, F. B., Mazzucchelli, R., Asefa, B., Khaled, A. R. and Durum, S. K. (2005). *Cytokine and Growth Factor Reviews* **16**(4–5): 513–533.

Johnson, K., Hashimshony, T., Sawai, C. M., Pongubala, J. M., Skok, J. A., Aifantis, I. and Singh, H. (2008). *Immunity*. Mar;28(3):335–45.

Jumaa, H., Bossaller, L., Portugal, K., Storch, B., Lotz, M., Flemming, A., Schrappe, M., Postila, V., Riikonen, P., Pelkonen, J., Niemeyer, C. M. and Reth, M. (2003). *Nature* **423**(6938): 452–456.

Kamath, M. B., Houston, I. B., Janovski, A. J., Zhu, X., Gowrisankar, S., Jegga, A. G. and Dekoter, R. P. (2008). *Leukemia*. Jun;22(6):1214–25.

Kantor, A. B., Stall, A. M., Adams, S., Herzenberg, L. A. and Herzenberg, L. A. (1992). *Annals of the New York Academy of Sciences* **651**: 168–169.

Karasuyama, H., Rolink, A. and Melchers, F. (1993). *Journal of Experimental Medicine* **178**(2): 469–478.

Kee, B. L., Cumano, A., Iscove, N. N. and Paige, C. J. (1994). *International Immunology***6**(3): 401–407.

Kee, B. L., Quong, M. W. and Murre, C. (2000). *Immunological Reviews* **175**: 138–149.

Kirstetter, P., Thomas, M., Dierich, A., Kastner, P. and Chan, S.(2002). *European Journal of Immunology* **32**(3): 720–730.

Klein, F., Feldhahn, N., Harder, L., Wang, H., Wartenberg, M., Hofmann, W. K., Wernet, P., Siebert, R. and Muschen, M. (2004). *Journal of Experimental Medicine* **199**(5): 673–685.

Kulathu, Y., Hobeika, E., Turchinovich, G. and Reth, M. (2008). *EMBO Journal* **27**(9): 1333–1344.

LeBien, T. W. (2000). *Blood* **96**(1): 9–23.

Levin, S. D., Koelling, R. M., Friend, S. L., Isaksen, D. E., Ziegler, S. F., Perlmutter, R. M. and Farr, A. G. (1999). *Journal of Immunology* **162**(2): 677–683.

Li, Y. S., Hayakawa, K. and Hardy, R. R. (1993). *Journal of Experimental Medicine* **178**(3): 951–960.

Lin, H. H. and Grosschedl, R. (1995). *Nature* **376**(6537): 263–267.

Lu, L. W. and Osmond, D. G. (2000). *Immunological Reviews* **175**: 158–174.

Lu, L. W., Chaudhury, P. and Osmond, D. G. (1999). *Journal of Immunology* **162**(4): 1931–1940.

Lu, R. Q., Medina, K. L., Lancki, D. W.and Singh, H. (2003). *Genes and Development* **17**(14): 1703–1708.

Ma, S. B., Pathak, S., Trinh, L. and Lu, R. Q. (2008). *Blood* **111**(3): 1396–1403.

Mackarehtschian, K., Hardin, J. D., Moore, K. A., Boast, S., Goff, S. P. and Lemischka, I. R. (1995). *Immunity* **3**(1): 147–161.

Martensson, I. L., Rolink, A., Melchers, F., Mundt, C., Licence, S. and Shimizu, T. (2002). *Seminars in Immunology* **14**(5): 335–342.

Martensson, I. L., Keenan, R. A. and Licence, S. (2007). *Current Opinion in Immunology* **19**(2): 137–142.

Melchers, F. (1995). *Clinical Immunology and Immunopathology* **76**(3): S188–S191.

Melchers, F., ten Boekel, E., Seidl, T., Kong, X. C., Yamagami, T., Onishi, K., Shimizu, T., Rolink, A. G. and Andersson, J. (2000). *Immunological Reviews* **175**: 33–46.

Middendorp, S., Dingjan, G. M. and Hendriks, R. W. (2002). *Journal of Immunology* **168**(6): 2695–2703.

Mikkola, I., Heavey, B., Horcher, M. and Busslinger, M.(2002). *Science* **297**(5578): 110–113.

Miller, J. P., Izon, D., DeMuth, W., Gerstein, R., Bhandoola, A. and Allman, D. (2002). *Journal of Experimental Medicine* **196**(5): 705–711.

Milne, C. D. and Paige, C. J. (2006). *Seminars in Immunology* **18**(1): 20–30.

Milne, C. D., Fleming, H. E. and Paige, C. J. (2004). *European Journal of Immunology* **34**(10): 2647–2655.

Milne, C. D., Zhang, Y. and Paige, C. J. (2005). *Scandinavian Journal of Immunology* **62**: 67–72.

Milne, C. D., Corfe, S. A. and Paige, C. J. (2008). *J Immunol* **180**(5): 2839–2847.

Morrow, M. A., Lee, G., Gillis, S., Yancopoulos, G. D. and Alt, F. W. (1992). *Genes and Development* **6**(1): 61–70.

Mullighan, C. G., Goorha, S., Radtke, I., Miller, C. B., Coustan-Smith, E., Dalton, J. D., Girtman, K., Mathew, S., Ma, J., Pounds, S. B., Su, X. P., Pui, C. H., Relling, M. V., Evans, W. E., Shurtleff, S. A. and Downing, J. R. (2007). *Nature* **446**(7137): 758–764.

Murphy, K. M., Paul, T., Walport, M. (2007). Janeway's Immunobiology. New York: Garland Science.

Nomura, K., Kanegane, H., Karasuyama, H., Tsukada, S., Agematsu, K., Murakami, G., Sakazume, S., Sako, M., Tanaka, R., Kuniya, Y., Komeno, T., Ishihara, S., Hayashi, K., Kishimoto, T. and Miyawaki, T. (2000). *Blood* **96**(2): 610–617.

Nutt, S. L. and Kee, B. L. (2007). *Immunity* **26**(6): 715–725.

Nutt, S. L., Heavey, B., Rolink, A. G. and Busslinger, M. (1999). *Nature* **401**(6753): 556–562.

Ohnishi, K. and Melchers, F. (2003). *Nature Immunology* **4**(9): 849–856.

Ohnishi, K., Melchers, F. and Shimizu, T. (2005). *European Journal of Immunology* **35**(3): 957–963.

O'Riordan, M. and Grosschedl, R. (1999). *Immunity* **11**(1): 21–31.

Osmond, D. G. (1991). *Current Opinion in Immunology* **3**(2): 179–185.

Otero, D. C. and Rickert, R. C. (2003). *J Immunol* **171**(11): 5921–5930.

Pierce, S. K. (2002). *Nature Reviews Immunology* **2**(2): 96–105.

Pongubala, J. M. R., Northrup, D. L., Lancki, D. W., Medina, K. L., Treiber, T., Bertolino, E., Thomas, M., Grosschedl, R., Allman, D. and Singh, H. (2008). *Nature Immunology* **9**(2): 203–215.

Pribyl, J. A. R. and LeBien, T. W. (1996). *Proceedings of the National Academy of Sciences of the United States of America* **93**(19): 10348–10353.

Pridans, C., Holmes, M. L., Polli, M., Wettenhall, J. M., Dakic, A., Corcoran, L. M., Smyth, G. K. and Nutt, S. L. (2008). *Journal of Immunology* **180**(3): 1719–1728.

Purohit, S. J., Stephan, R. P., Kim, H. G., Herrin, B. R., Gartland, L. and Klug, C. A. (2003). *EMBO Journal* **22**(20): 5511–5521.

Ray, R. J., Paige, C. J., Furlonger, C., Lyman, S. D. and Rottapel, R. (1996). *European Journal of Immunology* **26**(7): 1504–1510.

Ray, R. J., Stoddart, A., Pennycook, J. L., Huner, H. O., Furlonger, C., Wu, G. E. and Paige, C. J. (1998). *Journal of Immunology* **160**(12): 5886–5897.

Rich, B. E., Campostorres, J., Tepper, R. I., Moreadith, R. W. and Leder, P. (1993). *Journal of Experimental Medicine* **177**(2): 305–316.

Rolink, A. G., Nutt, S. L., Melchers, F. and Busslinger, M. (1999). *Nature* **401**(6753): 603–606.

Rolink, A. G., Winkler, T., Melchers, F. and Andersson, J. (2000). *Journal of Experimental Medicine* **191**(1): 23–31.

Schweighoffer, E., Vanes, L., Mathiot, A., Nakamura, T. and Tybulewicz, V. L. J. (2003). *Immunity* **18**(4): 523–533.

Sciammas, R., Shaffer, A. L., Schatz, J. H., Zhao, H., Staudt, L. M. and Singh, H. (2006). *Immunity* **25**(2): 225–236.

Seet, C. S., Brumbaugh, R. L. and Kee, B. L. (2004). *Journal of Experimental Medicine* **199**(12): 1689–1700.

Shaffer, A. L. and Schlissel, M. S. (1997). *Journal of Immunology* **159**(3): 1265–1275.

Stoddart, A., Fleming, H. E. and Paige, C. J.(2001). *European Journal of Immunology* **31**(4): 1160–1172.

Sun, J., Matthias, G., Mihatsch, M. J., Georgopoulos, K. and Matthias, P.(2003). *Journal of Immunology* **170**(4): 1699–1706.

Sun, X. H. (1994). *Cell* **79**(5): 893–900.

Teh, Y. M. and Neuberger, M. S. (1997). *Journal of Experimental Medicine* **185**(10): 1753–1758.

ten Boekel, E., Melchers, F. and Rolink, A. G. (1998). *Immunity* **8**(2): 199–207.

Thompson, E. C., Cobb, B. S., Sabbattini, P., Meixlsperger, S., Parelho, V., Liberg, D., Taylor, B., Dillon, N., Georgopoulos, K., Jumaa,H., Smale, S. T., Fisher, A. G. and Merkenschlager,M. (2007). *Immunity* **26**(3): 335–344.

Tokoyoda, K., Egawa, T., Sugiyama, T., Choi, B. I. and Nagasawa, T. (2004). *Immunity* **20**(6): 707–718.

Torres, R. M., Flaswinkel, H. Reth, M. and Rajewsky, K. (1996). *Science* **272**(5269): 1804–1808.

van Loo, P. F. Dingjan, G. M., Maas, A. and Hendriks, R. W. (2007). *Immunity* 27(3): 468–480.

Vosshenrich, C. A. J., Cumano, A., Muller, W., Di Santo, J. P. and Vieira, P.(2003). *Nature Immunology* **4**(8): 773–779.

Wang, J. H., Avitahl, N., Cariappa, A., Friedrich, C., Ikeda, T., Renold, A., Andrikopoulos, K., Liang, L. B., Pillai, S., Morgan, B. A. and Georgopoulos, K. (1998). *Immunity* **9**(4): 543–553.

Wang, Y. H., Stephan, R. P., Scheffold, A., Kunkel, N., Karasuyama, H., Radbruch, A. and Cooper, M. D. (2002). *Blood* **99**(7): 2459–2467.

Waskow, C., Paul, S., Haller, C., Gassmann, M. and Rodewald, H. R. (2002). *Immunity* **17**(3): 277–288.

Welner, R. S., Pelayo, R. and Kincade, P. W. (2008). *Nature Reviews Immunology* **8**(2): 95–106.

Wen, L. J., Brill-Dashoff, J., Shinton, S. A., Asano, M., Hardy, R. R. and Hayakawa, K. (2005). *Immunity* **23**(3): 297–308.

Weterings, E. and Chen, D. J. (2008). *Cell Research* **18**(1): 114–124.

Wong, S. and Witte, O. N. (2004). *Annual Review of Immunology* **22**: 247–306.

Xu, S. L., Wong, S. C. and Lam, K. P. (2000). *Journal of Immunology* **165**(8): 4153–4157.

Yasuda, T., Sanjo, H., Pages, G., Kawano, Y., Karasuyama, H., Pouyssegur, J., Ogata, M. and Kurosaki, T. (2008). *Immunity* **28**(4): 499–508.

Zhang, Z., Cotta, C. V., Stephan, R. P., deGuzman, C. G. and Klug, C. A. (2003). *EMBO Journal* **22**(18): 4759–4769.

Zhuang, Y., Cheng, P. F. and Weintraub, H. (1996). *Molecular and Cellular Biology* **16**(6): 2898–2905.

Chapter 9
Development of Natural Killer cells

Francesco Colucci

Abstract Natural killer cells are found in blood, lymphoid organs, liver, lungs and uterus, where they participate in several aspects of health and disease. During development, NK cells express a set of genes that encode for cell surface receptors, which interact with other cell surface molecules within the individual, between individuals and across genomes. Examples of the elements recognized by NK cells are self-MHC antigens during NK cell maturation, stress-inducible ligands during infections or tumour transformation, donor antigens on tissue grafts, paternal antigens at the feto–maternal interface and viral products. The nature of these interactions sets the threshold for NK cell activation, which in turn has downstream consequences on innate immunity and adaptive responses. Being endowed with these important recognition systems and instant effector function potential, NK cells have taken centre stage in modern medicine as they participate in infection, reproduction, transplantation, autoimmunity and cancer. This chapter reviews the basics of NK cell development, with an emphasis on murine cells.

Introduction

NK cells were discovered in the seventies and named after their propensity to spontaneously recognize and kill tumor cells in test tubes. Strong evidence has since been gathered in support of the natural anti-tumor potential of NK cells in experimental animal models (Karre et al. 1986) and in some clinical settings (Ruggeri et al. 2002). For example, mice instantly reject syngeneic transplanted lymphoma cells that lack adequate MHC-I expression, and rejection is dependent on the presence of NK cells. However, NK cells are not just killers. They produce

F. Colucci
Lymphocyte Signalling and Development Laboratory,
The Babraham Institute, CB22 3AT
Cambridge, UK
e-mail: francesco.colucci@bbsrc.ac.uk

A. Wickrema and B. Kee (eds.), *Molecular Basis of Hematopoiesis,*
DOI: 10.1007/978-0-387-85816-6_1, © Springer Science+Business Media, LLC 2009

cytokines and chemokines and thus help modulating immune responses. Moreover, NK cells can produce either pro- or anti-inflammatory cytokines, thus participating not only in antimicrobial immunity but also in modulating autoimmunity and immunopathology. Furthermore, cytokines produced by specialized NK cells at the feto–maternal interface contribute to vascular modifications that lead to successful pregnancy (Croy et al. 2006).

The developmental program of NK cells can be studied ex vivo using human tissues and in vivo in animal models. The mouse offers a good and well-established model that is amenable to targeted genetic manipulation. Whilst human and mouse NK cells share striking similarities, they also have important differences. For example, the best-studied cell surface receptors that regulate NK cell biology are structurally different in the two species: killer cell immunoglobulin-like receptors (KIR) in humans and lectin-like Ly49 in mice. In spite of the fact that the genes encoding for these two sets of receptors are unrelated, the receptors have remarkably similar functions in the two species: they bind MHC class I and regulate NK cell activation by sending activating or inhibitory signals within the NK cells in both species, thus representing a remarkable example of convergent evolution (Colucci et al. 2002).

Understanding NK cell development is important as it can help refine the modulation of this cell lineage that participates in immunity to infectious disease, cancer, transplantation, autoimmunity and reproduction. Having a short life span, NK cells have to be constantly replenished throughout life. Cell culture experiments with human cells and genetic modifications in mice have helped to identify key factors for NK cell differentiation, which include environmental components and cell-intrinsic cues.

NK cells are derived from hematopoietic stem cells, and it is generally accepted that NK cell development unfolds in the bone marrow (BM) (Di Santo 2006). Experimental myeloablation suppresses NK cell development in mice (Haller et al. 1977; Seaman et al. 1979). Moreover, the mouse BM contains the earliest NK-cell committed precursors, which are called NKPs, as well as all the known developmental intermediates, including mature NK cells (Rosmaraki et al. 2001). NKPs are defined by the absence of any lineage-specific marker and the expression of CD122, also known as IL2Rβ, which is one of the two subunits shared by the receptors for IL-2 and IL-15, the other being the common gamma chain (γc, or CD132). However, CD122 is not easily detectable in human hematopoietic precursors, therefore a combination of alternative markers is used to distinguish human NK cell precursors.

NK Cell Subsets

Do all NK cells have the potential to deliver different functions or are there subsets of specialized NK cells? And if so, do functional subsets represent separate lineages or do they stem from common progenitors during differentiation? Is the acquisition of a specific functional competence inbuilt in the developmental program and does it occur at defined stages of development? The best example to explain the background to these questions is perhaps represented by the two major subsets of human NK cells (Lanier et al. 1986). All human NK cells express CD56, also known as N-CAM, or neuronal cell-adhesion molecule, which is however not expressed homogeneously. In the blood two subsets of NK cells can be distinguished by the expression intensity of CD56 (Cooper et al. 2001). The vast majority, around 90% or more, are CD56^{dim} NK cells, which are also marked by the expression of the low-affinity CD16 receptor for immunoglobulins, and by high amounts of KIRs. CD56^{dim}CD16+ NK cells develop strong cytolytic activity in vitro. On the contrary, CD56^{bright} NK cells, which are CD16$^-$ and have, on an aggregate, fewer KIRs on the surface, represent about 10% or less of the circulating NK cells and are relatively less potent killers, but relatively more potent cytokines producers (Cooper et al. 2001). Whilst the identity of the two subsets is quite straightforward in terms of cell surface markers and functions, their developmental relationship is less clear. Some evidence suggests that CD56^{bright} and CD56^{dim} NK cells correspond to two consecutive stages along the same developmental pathway, with the latter representing the more mature one. For example, CD56^{bright} NK cells, more rapidly than CD56^{dim} NK cells, reconstitute the host, upon bone marrow transplantation. Moreover, certain in vitro culture conditions promote the generation of CD56^{dim} NK cells starting from CD56^{bright} NK cells. Despite being suggestive of the developmental relationship between these two subsets, the available evidence does not formally establish this relationship and therefore does not rule out the possibilities that CD56^{bright} and CD56^{dim} NK cells represent either two separate lineages, or are generated in response to different tissue-specific clues. Contrary to the blood, lymph nodes are enriched in CD56^{bright} NK cells, which, by virtue of their potential to robustly produce cytokines, can directly influence T-cell responses and conversely can respond to IL-2, since they express CD25, the high-affinity alpha chain of the IL-2 receptor. Recent work in the Caligiuri lab has generated evidence that human lymph nodes and tonsils are reservoirs of de novo NK cell development, where immunoregulatory CD56^{bright} NK cells could be produced (Freud et al. 2005).

A special class of NK cells are found in the uterus, and therefore called uNK cells. In humans, uterine NK cells are present in the non-pregnant endometrium, where they expand at each menstrual cycle. Subsequently, depending on whether fertilization occurs, they die and are shed with the menses, or expand throughout the first trimester of pregnancy and represent up to 70% of the leukocytes in the decidua, where they secrete a wide range of cytokines, including angiogenic factors (Moffett and Loke 2006). The origin of uNK cells is not clear. Like the CD56^{bright} CD16$^-$ NK cells in the blood, uNK cells are only weakly cytolytic and, although for some aspects they do resemble the CD56^{bright} CD16$^-$ NK cells in the blood, their relationship with this peripheral subset is unclear. The gene expression profile of the CD56^{bright} CD16$^-$ uNK cells is different from that of peripheral CD56^{bright} CD16$^-$ NK cells (Koopman et al. 2003). Moreover, unlike the CD56^{bright} CD16$^-$ NK cells in the blood, uNK cells express KIRs, although their repertoire during pregnancy is unique and does not match that of peripheral NK cells, suggesting that they have unique MHC-binding properties and they may recognize ligands on trophoblast cells and possibly maternal cells too.

Mouse NK cells do not express CD56; however, they can also be divided into subsets, based on the expression of other cell surface markers such as the TNF-family CD27 receptor and the $\alpha_M\beta2$ integrin CD11b, also known as Mac-1 (Hayakawa and Smyth 2006). Mouse NK cell subsets can be purified and transplanted into secondary hosts to directly verify their developmental liaisons. Thus, it appears that CD11b+ mature NK cells in the mouse progress from CD27^{lo} to CD27^{hi}, whereas CD11b– cells are immature (see later). Like their human counterparts, mouse CD27^{lo} and CD27^{hi} subsets have functional differences, although the tendency towards cytolytic activity or cytokine secretion is not polarized. Thus, the CD27^{hi} subset appears to be endowed with the most robust killing activity as well as cytokine production, whereas the CD27^{lo} subset might be a less functional, post-activation stage of terminal differentiation.

Uterine NK cells are also a constant feature of mouse gestation, and, like in humans, their origin is unknown. Nevertheless, mouse uNK cells present some unique characteristics, perhaps the most prominent of which is specific binding to a *Dolichos biflorus* agglutinin (DBA), which is indeed used as a marker of mouse uNK cells (Croy et al. 2006; Paffaro et al. 2003). During pregnancy, cells localize at the site of implantation in the decidua basalis and further proliferate until the second third of gestation, which lasts for 19–21 days in mice. By day 14 many of the NK cells are dying and by parturition at day 19, few are left at the site of implantation. As for human uNK cells, the MHC receptor repertoire and the cell surface phenotype of mouse uNK cells are unique (Yadi et al. 2008).

Developmental Stages

What is the developmental relationship among these functionally distinct NK cell subsets in the two species? Although the definitive answer to this question awaits further research, the work conducted in several laboratories over the past few years has generated evidence that, on an aggregate, helps to draft a working model of the major steps of human and NK cell development, as well as the factors necessary for their differentiation. Briefly, human CD34+ hematopoietic stem and precursor cells derived from fetal (thymus and cord blood) or adult (BM, blood and secondary lymphoid organs) tissues can give rise to functional NK cells if cultured with IL-2 or IL-15, which both signal through the βγ complex of the IL2/15 receptor (Freud and Caligiuri 2006). Although the BM is considered to be the primary site of NK cell development, secondary lymphoid organs as well as the thymus may contribute to conventional NK cells or to specialized subsets. Figure 9.1 shows the current models of NK cell development in the mouse.

Various models have been proposed for mouse NK cell development (Colucci et al. 2003; Di Santo 2006; Di Santo and Vosshenrich 2006; Yokoyama et al. 2004) Common to these schemes are three milestones: (1) emergence of NKPs, (2) generation of intermediate/immature NK cells, which start to acquire functional competence, and (3) acquisition of MHC-binding Ly49 receptors (Colucci et al. 2003). The most comprehensive model of mouse NK cell development identifies six discrete stages based on

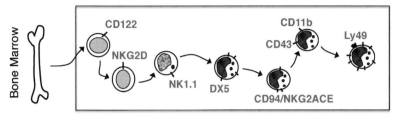

Fig. 9.1 Diagram of NK cell development in the mouse. This oversimplified figure represents the key events in murine NK cell development. NK committed precursors are generated in the bone marrow and complete development towards the NK cell lineage either in the bone marrow or in secondary lymphoid organs. The expression of CD122 marks the commitment to the NK cell lineage. NKG2D is thought to be one of the earliest receptors expressed in precursor NK cells. The subsequent and cumulative expression of the indicated cell surface markers is not linear, yet developmental intermediates can be identified by a combination of markers. Immature NK cells express NK1.1 in the C57BL/6 strain. The expression of DX5 coincides with the acquisition of cytotoxic potential (indicated by the accumulation of multiple cytotoxic granules). CD94/NKG2ACE receptors are expressed before Ly49 receptors. CD11b and CD43 mark mature NK cells. For more details and definition of NK cell subsets see the text

flow cytometry analysis of BM cells (Di Santo and Vosshenrich 2006). The early NK-committed precursor is negative for lineage-specific markers (Gr-1, CD3, CD4, CD19, TERR-119), positive for CD122, but negative for NK1.1 and DX5. CD122 expression is maintained across all stages of differentiation, whereas other receptors, as well as integrins and MHC-binding receptors are sequentially acquired along the differentiation pathway. For example, the expression of DX5 ($\alpha 2\beta 1$) coincides with the acquisition of killing activity. The expression of Ly49 receptors occurs late in the pathway and requires the contribution of stromal cells. The Ly49 family of receptors, like the KIR family, is composed of both inhibitory and activating members, which regulate NK cell functions. Inhibitory receptors are important to restrain the destructive and inflammatory potential of NK cells.

Regulation of NK Cell Activation During Development

The absence of MHC-binding inhibitory receptors during the early stages of NK cells is puzzling, if one considers that this absence persists beyond the acquisition of functional competence for killing. In other words, human KIR-negative cells and mouse Ly49-negative cells can potentially kill self-targets, yet they are effectively restrained to do so. One explanation is that another class of the lectin-like receptors belonging to the CD94/NKG2 family can mediate functional inhibition by binding non-classical MHC-I HLA-E in humans and Qa1 in mice. In line with this, NK cells in newborn mice do not express Ly49 receptors. The Ly49 receptor repertoire is fully mature only after several weeks of life. Instead, all NK cells in newborn mice express CD94/NKG2 heterodimers (Salcedo et al. 2000). Another explanation is that not all NK cell regulation revolves around MHC and its receptors. Instances of MHC-independent recognition of self have been described in experimental murine systems, as well as human in vitro cultures. A few examples may illustrate this concept. For example, the Ig-like 2B4 receptor (CD244) can trigger both activation or inhibition, depending on which signalling molecules it pairs up with. The default association is with the SHP phosphatase, thus generating inhibition. The SAP adaptor outcompetes SHP-1, thus converting 2B4 into an activating receptor. Moreover, early in ontogeny NK cells do not express inhibitory KIRs and also fail to express SAP, and thus 2B4 mediates inhibition, emerging as a key fail-safe regulatory receptor early in ontogeny (Sivori et al. 2002). In mice too 2B4 can function as an MHC-independent inhibitory receptor that recognizes self and regulates NK cell activation (Kumar and McNerney 2005). Moreover, the NKR-P1 family of cell surface receptors contains inhibitory and activating members that bind to Crl and regulate MHC-independent self-tolerance (Carlyle et al. 2004).

Having established that MHC-independent self-recognition is part of NK cell biology, we will nevertheless see later that developing NK cells engage in interactions with self-MHC that set the conditions for acquisition of functional competence.

Sites of NK Cell Development

The BM is the primary site for NK cell development. Evidence supporting this notion was generated in early studies conducted in myelo-ablated mice (Haller et al. 1977; Seaman et al. 1979). However, extramedullary NK cell lymphopoiesis can occur, for example in the spleen of myelo-ablated mice, although the NK cells generated under these conditions are not fully mature. Since the discovery of committed NKPs in the BM (Rosmaraki et al. 2001), cells with a similar phenotype have been described in other tissues, suggesting the non-mutually exclusive possibilities that NKPs are generated in situ in other tissues other than the BM, or that they are exported from the BM to complete differentiation in such tissues.

The Caligiuri lab has recently provided a clear example of NK cell development in secondary lymphoid tissues. They could identify distinct developmental stages, on the basis of flow cytometry analysis of secondary lymphoid tissues, including tonsils and lymph nodes, which appear to be naturally enriched in NK cell precursors. Using CD34, CD3, CD117 and CD94, four distinct stages could be defined, which are all CD3–. The earlier two stages express CD34 but lose it along the differentiation path; the second and third stages are CD117+, but decrease expression in the later stage, which is the only one to express CD94 (Freud and Caligiuri 2006).

Given some of the similarities between NK and T-cell biology, such as shared progenitors, responses to IL-2 and potential for cellular cytotoxicity, it is reasonable to postulate that the thymus could be a reservoir of NK cell precursors. However, the notion that NK cells are normal in human and mice that have no thymus (DiGeorge syndrome patients or *nude* mice) has taken the attention away from this organ. Nevertheless, Vosshenrich et al. have recently provided an example of thymus-derived NK cells in the mouse (Vosshenrich et al. 2006). Athymic nude mice lack a small subset of CD127 (IL7Rα)+ NK cells, which in normal mice is more abundant in lymph nodes and shares some features with the CD56[bright] subset of human NK cells. Thus, CD127+ NK cells have reduced expression of Ly49 receptors and weaker cytolytic activity as compared with the CD127– counterpart. Moreover, human CD56[bright] cells express the transcription factor GATA-3, and mouse CD127+ cells depend on this transcription factor to develop. The functions of these thymic NK cells in vivo have not been established.

Cytokines

External cytokines are necessary to promote NK cell differentiation in vitro. In a minimalistic view, these cytokines can be grouped, on one hand, into early acting cytokines, which include FLT3L, SCF and IL-7, and, on the other hand, cytokines acting in later steps of development, which include chiefly IL-15, but may be extended to IL-2, -12, -18 and -21 (Colucci et al. 2003). The early acting cytokines promote the emergence of committed NK cell progenitors, although the exact role of each of these three factors during commitment to NK cell lineage is still unknown and a certain degree of redundancy should be inbuilt in the system. Nevertheless, the combined action of these three factors induces the emergence of the earliest NK-cell committed precursor, which is the NKP. Once generated, NKPs are responsive to IL-15 and strictly depend on this cytokine for survival and further differentiation. Molecules that participate in the transduction of intracellular signals generated by the IL-15 receptor are absolutely required for NK cell development. Thus, the lack of IL-15 itself, or the beta (IL2Rβ, CD122) or gamma (IL2Rγ, CD135) chain of the receptor causes a drastic reduction of NK cell numbers, which are barely detectable in mutant mice (Huntington et al. 2007). Vitamin D3-upregulated protein-1 (VDUP-1) controls expression of CD122 and ultimately influences IL-15 responsiveness. Mice lacking VDUP-1 have reduced numbers of peripheral NK cells and a more severe reduction in NKPs, providing another example of the importance of IL-15 in NK cell development (Lee et al. 2005). Nevertheless, NKPs are generated in IL-15 mutant mice and some even make it to maturity and display reduced, yet detectable effector functions (Vosshenrich et al. 2005a). Thus, although the all-important role of IL-15 in NK cells is indisputable, there may be a yet undefined IL-15-independent developmental pathway.

Cell–Cell Interactions

Addition of external cytokines to NK cell cultures is not sufficient to generate fully mature NK cells. Contact with stromal cells is necessary to provide signals that promote the expression of KIRs in human cells and Ly49s in mouse cells. The nature of these signals is still not fully understood, but at least a set of these signals has been identified. Mice lacking the Tyro3/Axl/Mer (TAM) receptor family fail to develop fully differentiated NK cells. TAM-deficient NK cells displayed an abnormal receptor repertoire and showed functional defects in both cytolytic activity and cytokine production (Caraux et al. 2006). TAM receptors belong to the family of receptor tyrosine kinases. Upon binding to their endogenous ligands Gas6 and protein-S, which are found on BM stromal cells, it is likely that TAM receptors generate intracellular signals in developing NK cells that either directly or indirectly promote their functional maturation.

Stromal cells express the lymphotoxin (LT)-β receptor, whereas its cognate ligand LTα1β2 is expressed by hematopoietic cells. Thus, ligand and receptor engage in interactions that represent another set of cell–cell dependent communications required for normal NK cell development. These interactions are also required for the normal development of secondary lymphoid organs, which is orchestrated by lymphoid tissue inducer cells, a special subset of lymphoid cells that emerge early in ontogeny. The available evidence is consistent with a model in which, through LT-LTβR cross-talks, NK cell precursors stimulate stromal cells to produce IL-15, which is presented to the NK cell precursors in a membrane-bound form on the surface of the stromal cells (Di Santo 2006). Later in the life of mature NK cells a similar cell–cell interaction is established with dendritic cells (DC), which regulate NK cell effector functions. Indeed, following infectious challenges or stimulation of Toll-like receptors, DCs produce IL-15 and present it in *trans* to NK cells, which become in this way primed and fully functional (Lucas et al. 2007). Thus, IL-15 controls not only the early developmental stages of NK cells, but it is also required by mature cells in order to deliver effector functions.

Transcription Factors

Specific transcription factors (TF) have been associated with lymphocyte lineages. For example T-bet, Gata-3, Foxp3, RORγt and Pax5 are markers of Th1, Th2, T$_{regs}$, Th17 and B cells, respectively. Although a specific TF that marks the NK cell lineage has not been identified, it is clear that sets of TFs regulate different phases of NK cell development. Thus, a dynamic view of transcriptional regulation of NK cell development is emerging from studies using gene KO mice. NKPs are reduced in mice that lack PU.1, Ikaros, and Ets-1, thus suggesting that transcriptional regulation of genes targeted by these factors is necessary for the generation of committed NKPs (Vosshenrich et al. 2005b).

Once NKPs are generated, another set of TFs appears to be important for subsequent steps leading to further maturation. These include Id2, E-proteins, IRF-2, GATA-3 and T-bet. Some TFs work in opposition during lymphoid development, and NK cells are no exception. Id2, on one hand, and E-box proteins, on the other, provide an example of this. Id2 is a key transcription factor for the generation of mature NK cells. Mice lacking Id2 have very few peripheral NK cells, however, the number of NKPs in the BM is nearly normal, suggesting that Id2 does not dictate NK cell lineage specification. As NK cells mature, Id2 is required in mature NK cells to counteract the activity of E proteins, which suppresses the generation of mature NK cells. Thus, normal NK cell numbers are restored in the BM of mice lacking both Id2 and E2A (Boos et al. 2007). However, NK cells do not reach normal numbers in the spleen and lymph nodes of these mice, again highlighting that NK cell development and maturation is influenced by tissue-specific cues.

In the absence of GATA-3, IRF-2 and T-bet, NK cells are generated, but peripheral cells are reduced in number and are phenotypically similar to immature cells. Moreover, they fail to produce normal amounts of IFN-g. IFN-g is a 'Th1' cytokine, thus the defective production of IFN-g in T-bet-deficient NK cells is consistent with a role of this TF in regulating inflammatory cytokine secretion in T and NK cells. On the other hand, GATA-3 is a typical 'Th2' promoting TF in T cells, and it suppresses the production of IFN-g. Thus, it was surprising to find that GATA-3 actually promotes IFN-g production in NK cells (Samson et al. 2003). These results clearly highlight the fundamental differences in regulation of cytokine gene expression in NK and T cells. Along these lines, cytokine genes are constitutively transcribed in NK and NKT cells, whereas de novo gene transcription is required for naïve T cells. Therefore post-transcriptional regulation of cytokine gene expression in lymphocytes of innate immunity is likely to be important in keeping the balance between protective and pathogenic inflammation.

Further down in the differentiation process, another set of TFs regulates functional maturation of NK cells. Thus, mice lacking CEBP-g, MEF and MITF appear to develop normal numbers of NK cells, which show a normal phenotype, but fail to develop normal cytotoxicity or cytokine production. The functional NK cell defect in mice expressing a mutant form of microphtalmia TF (mi-MITF) is due to reduced expression of the perforin gene. MEF, which is a transcriptional activator of the ETS family of TFs, also directly regulates the expression of the perforin gene in NK cells.

Signalling Pathways

What are the cell surface receptors and the signalling pathways that regulate NK cell development? B- and T-cell development strongly relies on signals emanating from their antigen receptors, and the BCR and TCR are also the main driving forces behind B- and T-cell effector functions. Such a *master* receptor is unknown for NK cells, which instead can be activated by a number of receptors. Moreover, none of the signalling pathways triggered by activating receptors is absolutely required for NK cell development. Perhaps the most striking feature of NK cell development in this regard is that the intracellular signalling requirement for development and effector functions can be largely dissociated. The majority of NK cell-activating receptors lack intrinsic signalling domains and enzymatic activity (Lanier 2005; Vivier et al. 2004; Zompi and Colucci 2005). The trans-membrane portion of activating receptors associates with small adapters possessing intracellular signalling domains, which are essential for recruitment and activation of downstream signalling components. NK cell receptors can associate with four known adapters. Three of these adapters, FcεRIγ, CD3ζ and DNAX adapter protein (DAP)-12 contain immunoreceptor tyrosine-based activating motifs (ITAM) that activate signal transduction pathways, including activation of protein tyrosine kinases (PTK) of the SRC family (such as LCK, FYN, YES and LYN), followed by the recruitment and activation of SYK-family PTK (ZAP-70 and SYK). This leads to the activation of downstream regulators such as adaptor molecules (LAT, SLP-76, 3BP2), lipid enzymes including PLC-γ and phosphatidylinositol 3-kinase (PI3K), VAV-family guanine nucleotide exchange factors and RHO-RAC low molecular weight GTP-binding proteins and their effectors. By contrast, DAP-10 does not contain ITAMs, but possesses instead a motif that initiates an alternative signal transduction pathway that resembles the pathway activated by the co-stimulatory CD28 receptor on T cells. This pathway, through PI3K and Grb2, activates critical downstream regulators (VAV, PLC-γ), without the need of a SYK-dependent activation step. A third pathway for NK cell activation is represented by the 2B4-FYN-SAP axis, which has been shown to be an ITAM-independent pathway that activates natural cytotoxicity (Bloch-Queyrat et al. 2005). The intracellular tail of 2B4 contains a signalling motif called immunorecep-tor tyrosine-based switch motif (ITSM). Upon binding to the relevant ligand (CD48), ITSM can recruit the signalling lymphocyte activation molecule-associated protein (SAP), which mediates activation of NK cells. However, the ITSM motif can also recruit SH2-domain containing PTPs, thereby initiating inhibitory pathways.

The fundamental differences of NK cell development, on one hand, and B- and T-cell development, on the other, are perhaps best illustrated by mice lacking both Syk and ZAP-70, in which ITAM-bearing receptors are non-functional. The only lymphocytes generated in these mice are NK cells, demonstrating that ITAM-based signals are not required for NK cell development (Colucci et al. 2002). Similarly, NK cells develop with only minor defects in mice lacking either all three ITAM-bearing adapters FcεRIγ, CD3ζ and DAP12, or in mice lacking both DAP12 and DAP10. Yet some abnormalities are found in the Ly49 repertoire of Syk/ZAP-70-deficient mice, suggesting that these two PTKs are necessary to establish a normal Ly49 repertoire. Similarly, mice lacking CD45, PLC-γ2, or PI3K catalytic isoforms p110γ or p110δ develop Ly49 aberrant repertoires. Mice lacking both p110γ or p110δ have profound defects in NK cell development, suggesting that PI3Ks may control pivotal checkpoints during NK cell differentiation, although the PI3K-dependent receptors and downstream pathways that are required during NK cell development are not defined (Kim et al. 2007; Tassi et al. 2007).

Education

The take-home message from the previous section is that, with the exception of the IL-15 signalling axis, we do not really know the critical intracellular signals required to generate NKPs, immature NK cells, Ly49 (or KIR)-expressing mature NK cells and functionally competent, yet self-tolerant NK cells. Interactions between MHC and receptors on developing NK cells have recently been proposed as central components of a pathway leading to development of functional competence, which has alternatively been referred to as 'licencing' and 'education' (Fernandez et al. 2005; Kim et al. 2005). Regardless of the semantic issue, the evidence supports a model according to which ITIM-competent inhibitory receptors on developing NK cells must engage in functional interactions with cognate, self-MHC in order for the cells to acquire the ability of delivering effector functions at later stages. Thus, both mouse and human NK cells have been shown to be more potent effector cells if they developed in an environment where the self-MHC was a cognate ligand for their inhibitory receptors (Anfossi et al. 2006). Although it was shown that an intact ITIM was necessary, thus attesting the requirement for signalling downstream of the receptor for self-MHC, the mechanisms of this phenomenon have not been elucidated yet. Nevertheless, the proposed model has so far stood the experimental test. Thus, receptor stimulation of 'licensed' cells resulted in more potent effector functions than 'unlicensed' cells. For example, Ly49C+ NK cells in C57BL/6 mice bind cognate MHC-I (H-2^b) and become licensed. Instead, Ly49C– cells do not have a cognate MHC-I in this strain and stay unlicensed. When bulk NK cells of this strain are activated through stimulation of the activating receptor NK1.1, which is expressed on all NK cells, Ly49C+ cells will produce more IFN-g than Ly49C– cells. The hyporeactive NK cells that develop in a MHC-I-deficient mouse, such as β2m–/– mice (Kim et al. 2005), confirm the prediction and illustrate the most extreme case of licensing requirement.

Yet, it is difficult to reconcile the model, however elegant it may be, with three sets of observations. First, stimulation with pro-inflammatory cytokines overcomes the need for licensing, and thus both licensed and unlicensed cells can equally respond to the stimulus. Importantly, exposure to inflammatory cytokines is likely to be biologically relevant in most NK-cell activation scenarios. Secondly, β2m–/– mice can control cytomegalovirus infection, demonstrating biologically relevant functional competence. Finally, effector functions can be acquired in NK cells of some of the gene-targeted mice discussed earlier (mice lacking specific transcription factors, for example), which have more or less severe blocks in NK cell development and fail, in some instances, to generate normal Ly49 repertoire. Therefore, other mechanisms of acquisition of functional competence and NK cell regulation may be discovered in the future.

Acknowledgments Work in our lab is funded by the BBSRC, the MRC and the CR-UK. I would like to thank Marta Roche-Molina for help in preparing the figure, Daniel Hampshire for reading the manuscript and all the present and past members of the Colucci group.

References

Anfossi, N., Andre, P., Guia, S., Falk, C. S., Roetynck, S., Stewart, C. A., Breso, V., Frassati, C., Reviron, D., Middleton, D., et al. (2006). Human NK cell education by inhibitory receptors for MHC class I. *Immunity* 25, 331–342.

Bloch-Queyrat, C., Fondaneche, M. C., Chen, R., Yin, L., Relouzat, F., Veillette, A., Fischer, A., and Latour, S. (2005). Regulation of natural cytotoxicity by the adaptor SAP and the Src-related kinase Fyn. *J Exp Med* 202, 181–192.

Boos, M. D., Yokota, Y., Eberl, G., and Kee, B. L. (2007). Mature natural killer cell and lymphoid tissue-inducing cell development requires Id2-mediated suppression of E protein activity. *J Exp Med* 204, 1119–1130.

Caraux, A., Lu, Q., Fernandez, N., Riou, S., Di Santo, J. P., Raulet, D. H., Lemke, G., and Roth, C. (2006). Natural killer cell differentiation driven by Tyro3 receptor tyrosine kinases. *Nat Immunol* 7, 747–754.

Carlyle, J. R., Jamieson, A. M., Gasser, S., Clingan, C. S., Arase, H., and Raulet, D. H. (2004). Missing self-recognition of Ocil/Clr-b by inhibitory NKR-P1 natural killer cell receptors. *Proc Natl Acad Sci USA* 101, 3527–3532.

Colucci, F., Di Santo, J. P., and Leibson, P. J. (2002). Natural killer cell activation in mice and men: different triggers for similar weapons? *Nat Immunol* 3, 807–813.

Colucci, F., Caligiuri, M. A., and Di Santo, J. P. (2003). What does it take to make a natural killer? *Nat Rev Immunol* 3, 413–425.

Cooper, M. A., Fehniger, T. A., and Caligiuri, M. A. (2001). The biology of human natural killer-cell subsets. *Trends Immunol* 22, 633–640.

Croy, B. A., van den Heuvel, M. J., Borzychowski, A. M., and Tayade, C. (2006). Uterine natural killer cells: a specialized differentiation regulated by ovarian hormones. *Immunol Rev* 214, 161–185.

Di Santo, J. P. (2006). Natural killer cell developmental pathways: a question of balance. *Annu Rev Immunol* 24, 257–286.

Di Santo, J. P., and Vosshenrich, C. A. (2006). Bone marrow versus thymic pathways of natural killer cell development. *Immunol Rev* 214, 35–46.

Fernandez, N. C., Treiner, E., Vance, R. E., Jamieson, A. M., Lemieux, S., and Raulet, D. H. (2005). A subset of natural killer cells achieves self-tolerance without expressing inhibitory receptors specific for self-MHC molecules. *Blood* 105, 4416–4423.

Freud, A. G. and Caligiuri, M. A. (2006). Human natural killer cell development. *Immunol Rev* 214, 56–72.

Freud, A. G., Becknell, B., Roychowdhury, S., Mao, H. C., Ferketich, A. K., Nuovo, G. J., Hughes, T. L., Marburger, T. B., Sung, J., Baiocchi, R. A., et al. (2005). A human CD34(+) subset resides in lymph nodes and differentiates into CD56bright natural killer cells. *Immunity* 22, 295–304.

Haller, O., Kiessling, R., Orn, A., and Wigzell, H. (1977). Generation of natural killer cells: an autonomous function of the bone marrow. *J Exp Med* 145, 1411–1416.

Hayakawa, Y. and Smyth, M. J. (2006). CD27 dissects mature NK cells into two subsets with distinct responsiveness and migratory capacity. *J Immunol* 176, 1517–1524.

Huntington, N. D., Vosshenrich, C. A., and Di Santo, J. P. (2007). Developmental pathways that generate natural-killer-cell diversity in mice and humans. *Nat Rev Immunol* 7, 703–714.

Karre, K., Ljunggren, H. G., Piontek, G., and Kiessling, R. (1986). Selective rejection of H-2-deficient lymphoma variants suggests alternative immune defence strategy. *Nature* 319, 675–678.

Kim, N., Saudemont, A., Webb, L., Camps, M., Ruckle, T., Hirsch, E., Turner, M., and Colucci, F. (2007). The p110delta catalytic isoform of PI3K is a key player in NK-cell development and cytokine secretion. *Blood* 110, 3202–3208.

Kim, S., Poursine-Laurent, J., Truscott, S. M., Lybarger, L., Song, Y. J., Yang, L., French, A. R., Sunwoo, J. B., Lemieux, S., Hansen, T. H., and Yokoyama, W. M. (2005). Licensing of natural killer cells by host major histocompatibility complex class I molecules. *Nature* 436, 709–713.

Koopman, L. A., Kopcow, H. D., Rybalov, B., Boyson, J. E., Orange, J. S., Schatz, F., Masch, R., Lockwood, C. J., Schachter, A. D., Park, P. J., and Strominger, J. L. (2003). Human decidual natural killer cells are a unique NK cell subset with immunomodulatory potential. *J Exp Med* 198, 1201–1212.

Kumar, V. and McNerney, M. E. (2005). A new self: MHC-class-I-independent natural-killer-cell self-tolerance. *Nat Rev Immunol* 5, 363–374.

Lanier, L. L. (2005). NK cell recognition. *Annu Rev Immunol* 23, 225–274.

Lanier, L. L., Le, A. M., Civin, C. I., Loken, M. R., and Phillips, J. H. (1986). The relationship of CD16 (Leu-11) and Leu-19 (NKH-1) antigen expression on human peripheral blood NK cells and cytotoxic T lymphocytes. *J Immunol* 136, 4480–4486.

Lee, K. N., Kang, H. S., Jeon, J. H., Kim, E. M., Yoon, S. R., Song, H., Lyu, C. Y., Piao, Z. H., Kim, S. U., Han, Y. H., et al. (2005). VDUP1 is required for the development of natural killer cells. *Immunity* 22, 195–208.

Lucas, M., Schachterle, W., Oberle, K., Aichele, P., and Diefenbach, A. (2007). Dendritic cells prime natural killer cells by trans-presenting interleukin 15. *Immunity* 26, 503–517.

Moffett, A. and Loke, C. (2006). Immunology of placentation in eutherian mammals. *Nat Rev Immunol* 6, 584–594.

Paffaro, V. A., Jr., Bizinotto, M. C., Joazeiro, P. P., and Yamada, A. T. (2003). Subset classification of mouse uterine natural killer cells by DBA lectin reactivity. *Placenta* 24, 479–488.

Rosmaraki, E. E., Douagi, I., Roth, C., Colucci, F., Cumano, A., and Di Santo, J. P. (2001). Identification of committed NK cell progenitors in adult murine bone marrow. *Eur J Immunol* 31, 1900–1909.

Ruggeri, L., Capanni, M., Urbani, E., Perruccio, K., Shlomchik, W. D., Tosti, A., Posati, S., Rogaia, D., Frassoni, F., Aversa, F., et al. (2002). Effectiveness of donor natural killer cell alloreactivity in mismatched hematopoietic transplants. *Science* 295, 2097–2100.

Salcedo, M., Colucci, F., Dyson, P. J., Cotterill, L. A., Lemonnier, F. A., Kourilsky, P., Di Santo, J. P., Ljunggren, H. G., and Abastado, J. P. (2000). Role of Qa-1(b)-binding receptors in the specificity of developing NK cells. *Eur J Immunol* 30, 1094–1101.

Samson, S. I., Richard, O., Tavian, M., Ranson, T., Vosshenrich, C. A., Colucci, F., Buer, J., Grosveld, F., Godin, I., and Di Santo, J. P. (2003). GATA-3 promotes maturation, IFN-gamma production, and liver-specific homing of NK cells. *Immunity* 19, 701–711.

Seaman, W. E., Gindhart, T. D., Greenspan, J. S., Blackman, M. A., and Talal, N. (1979). Natural killer cells, bone, and the bone marrow: studies in estrogen-treated mice and in congenitally osteopetrotic (mi/mi) mice. *J Immunol* 122, 2541–2547.

Sivori, S., Falco, M., Marcenaro, E., Parolini, S., Biassoni, R., Bottino, C., Moretta, L., and Moretta, A. (2002). Early expression of triggering receptors and regulatory role of 2B4 in human natural killer cell precursors undergoing in vitro differentiation. *Proc Natl Acad Sci USA* 99, 4526–4531.

Tassi, I., Cella, M., Gilfillan, S., Turnbull, I., Diacovo, T. G., Penninger, J. M., and Colonna, M. (2007). p110gamma and p110delta phosphoinositide 3-kinase signaling pathways synergize to control development and functions of murine NK cells. *Immunity* 27, 214–227.

Vivier, E., Nunes, J. A., and Vely, F. (2004). Natural killer cell signaling pathways. *Science* 306, 1517–1519.

Vosshenrich, C. A., Ranson, T., Samson, S. I., Corcuff, E., Colucci, F., Rosmaraki, E. E., and Di Santo, J. P. (2005a). Roles for common cytokine receptor gamma-chain-dependent cytokines in the generation, differentiation, and maturation of NK cell precursors and peripheral NK cells in vivo. *J Immunol* 174, 1213–1221.

Vosshenrich, C. A., Samson-Villeger, S. I., and Di Santo, J. P. (2005b). Distinguishing features of developing natural killer cells. *Curr Opin Immunol* 17, 151–158.

Vosshenrich, C. A., Garcia-Ojeda, M. E., Samson-Villeger, S. I., Pasqualetto, V., Enault, L., Richard-Le Goff, O., Corcuff, E., Guy-Grand, D., Rocha, B., Cumano, A. et al. (2006). A thymic pathway of mouse natural killer cell development characterized by expression of GATA-3 and CD127. *Nat Immunol* 7, 1217–1224.

Yadi, H., Burke, S., Madeja, Z., Hemberger, M., Moffett, A., Colucci, F. (2008). Unique receptor repertoire in mouse uterine NK cells. *J. Immunol* 181:6140–6147.

Yokoyama, W. M., Kim, S., and French, A. R. (2004). The dynamic life of natural killer cells. *Annu Rev Immunol* 22, 405–429.

Zompi, S. and Colucci, F. (2005). Anatomy of a murder – signal transduction pathways leading to activation of natural killer cells. *Immunol Lett* 97, 31–39.

Chapter 10
Leukemic Stem Cells: New Therapeutic Targets?

Dominique Bonnet

Abstract An emerging concept in cancer biology is that a subset of cancer cells among the heterogeneous cell mass that constitutes the tumor may drive the growth of the tumor. This so-called *cancer stem cell* (CSC) shared some main features with normal stem cells. These include self-renewal, differentiation into the cell types of the original cancer and potent tumor formation. Despite the clear importance of CSCs in the genesis and perpetuation of cancers, little is currently known about the biological and molecular properties that make CSCs distinct from normal stem cells, the developmental/cellular origin of CSCs, the mechanisms responsible for their emergence in the course of the disease, and identification of candidate molecular targets for therapeutic intervention. This report will focus more specifically on the blood-related cancer leukemia, which was the first disease where human CSCs, or leukemic stem cells (LSCs), were isolated. In this chapter we will summarize our knowledge of LSCs notably in acute myeloid leukemia (AML) and will discuss different issues that are arising in trying to eradicate these cells.

Concept of Leukemia-Initiating Cells or Leukemic Stem Cells

Although it was previously believed that most or all cancer cells possess the property to self-renew and replenish new cancer cells, it has recently become clear that cancers, as most normal tissues, are organized in a hierarchical fashion, and that only a small fraction of tumor cells have the ability to reconstitute a new tumor (Reya et al. 2001). The existence of such cancer stem cells (CSCs) with self-renewal potential was first documented in leukemias (Bonnet and Dick 1997; Lapidot et al. 1994),

D. Bonnet
Haematopoietic Stem Cell Laboratory, Cancer Research UK,
London Research Institute, 44 Lincoln's Inn Fields,
London, WC2A 3PXUK
e-mail: dominique.bonnet@cancer.org.uk

A. Wickrema and B. Kee (eds.), *Molecular Basis of Hematopoiesis*,   215
DOI: 10.1007/978-0-387-85816-6_1, © Springer Science+Business Media, LLC 2009

but has since then been extended to a wide varieties of solid tumors, such as breast, brain, colon, and prostate cancers raising the possibility that such cells are the apex of all neoplastic systems (Singh et al. 2004; Al Hajj et al. 2003; O'Brien et al. 2007; Ricci-Vitiani et al. 2007; Collins et al. 2005). As these rare CSCs are both required and sufficient to reconstitute a new tumor, they have immediate and important clinical implications. A better identification and characterization of CSCs should provide a better understanding of tumor developmental biology and the genetic events involved in the transformation process. Recent studies have shown that CSCs, as normal somatic stem cells, express high levels of multidrug resistance (MDR) pumps that efficiently efflux cytotoxic drugs (Zhou et al. 2001), making them particularly difficult to target with cytotoxic therapies. Similarly, they appear to be less immunogenic than more differentiated cancer cells in the tumor (Costello et al. 2000), and therefore also more difficult to eradicate with immunotherapy. Thus, it has become evident that CSCs might prove not only to be the most important but also the most difficult cancer cells to eliminate with conventional therapies, and that a specific monitoring and targeting of these elusive CSCs could become an important tool toward identification and characterization of improved cellular and molecular targets for development of cancer therapies.

Based on these recent studies, the paradigm of cancer as a hierarchical disease, sustained by a rare population of CSCs has emerged. Implicit in this model of cancer development is the notion that CSCs are biologically distinct from other cells in the tumor and are able to initiate and sustain tumor growth in vivo whereas the bulk cells are not.

Identification of the Leukemia-Initiating Cell/LSC

Isolation of LSC

The adaptation of xenotransplantation assays to examine the propagation of AML in vivo has allowed the phenotypic identification of the AML-IC. Transplantation of primary AML cells into NOD/SCID mice led to the finding that only rare cells, termed AML-initiating cells (AML-IC), are capable of initiating and sustaining growth of the leukemic clone in vivo, and serial transplantation experiments showed that AML-IC possess high self-renewal capacity, and thus can be considered to be the leukemic stem cells. By LSC we refer to a cell that has self-renewal and differentiation potential and is able to reinitiate the leukemia when transplanted into NOD/SCID. This definition does not preclude the nature of the cells that is being transformed (i.e., normal HSC, progenitors, or mature cells).

Importantly, AML-IC can be prospectively identified and purified as CD34$^+$/CD38$^-$ cells in AML patient samples, regardless of the phenotype of the bulk blast population, and represented the only AML cells capable of self-renewal (Lapidot et al. 1994; Bonnet and Dick 1997). This phenotype has been extended via the use of immunodeficient mice to include an absence of CD90, CD71, HLA-DR, and CD117, but include expression of CD123 (Blair et al. 1997; Blair et al. 1998; Blair and Sutherland 2000; Jordan et al. 2000). We have extended this phenotype further to include expression of CD33 and CD13 on AML-IC from the vast majority of patients (Taussig et al. 2005). However, subsequently, considerable heterogeneity has been revealed. Use of lentiviral gene marking to track the behavior of individual LSCs, following serial transplantation, has revealed heterogeneity in their ability to self-renew, similar to what is seen in the normal HSC compartment (Hope et al. 2004).

Moreover, recent evidence from our laboratory suggests that in some patients (including patients with MLL translocations), AML-ICs could have a progenitor phenotype (CD34$^+$/CD38$^+$). Based on the heterogeneity of AML in terms of karyotype, differentiation stage of the blasts, and clinical outcome; it is not surprising that AML-ICs could be heterogeneous too (Taussig et al. 2008). Thus, the phenotype of AML-ICs is more complex than previously thought and can vary from patient to patient and also probably in the same patient depending on the stage of the disease (Taussig et al. 2008).

Apart from immunophenotype, other methods have been utilized to identify the AML-IC. For instance, we have published data on the aldehyde dehydrogenase activity (ALDH-1) of AML-IC from various AML patients. It seems that in approximately one-third of patients, at least a subset of AML-IC possesses a high ALDH activity that is detectable using a fluorescent ALDH substrate (Pearce et al. 2005).

The Heterogeneity of Leukemia-Initiating Cells

Given the various possible routes to AML from a normal hematopoietic cell, it is not surprising that there is great heterogeneity in AML. Indeed, AML may be thought of as a large collection of different diseases that merely share a similar morphology. Indeed, the most effective risk stratification approach so far has been to examine the genetic abnormalities associated with a particular case of AML and compare with previous experience of AML cases with the same abnormality (Grimwade et al. 1998, 2001). Although cytogenetic analysis allows the definition of the hierarchical groups with favorable, intermediate, and poor prognosis, the intermediate risk group contains patients with variable outcomes. Assessing the prognosis of this large group of patients is currently difficult.

We have recently reported that the ability of a particular AML to engraft in the NOD/SCID model is related to the prognosis of individual AML cases. Specifically, examination of the follow-up of younger patients with intermediate-risk AML revealed a significant difference in overall survival between NOD/SCID-engrafting and nonengrafting cases. We could not detect a difference between engrafting and non-engrafting cases in various engraftment variables including homing ability, AML-IC frequency, immune rejection by the host, or alternative tissue sources. Hence, the ability to engraft NOD/SCID recipients seems to be an inherent property of the cells that is directly related to prognosis.

Alternatively, it may be that certain AMLs require a cytokine or cell–cell interaction that cannot be provided by the NOD/SCID microenvironment. Most cytokines cross react, but some do not, and a lack of a certain cytokine or cellular interaction may be important for the growth of particular AML cases.

Further studies determining the difference between engrafted versus non-engrafted AML should provide valuable information on the mechanisms necessary to obtain engraftment and might potentially shed light on important biological differences that can be distinguished between favorable versus poor prognosis patients.

Evolution of LSCs During the Course of the Disease

In a recent study by Barabe et al., human cord blood cells that were transduced with the MLL-ENL fusion protein developed an acute leukemia when transplanted into NOD/SCID mice (Barabe et al. 2007). The original LSCs in these mice were derived from primitive cells with germline immunoglobulin genes. Nevertheless, during the course of the disease and after expansion of these cells ex vivo, a *second generation* of CSCs was detected. The second generation of LSCs was derived from malignant progenitor cells (the immunoglobulin genes were rearranged in these cells) that had acquired self-renewal ability. With time these second generation LSCs came to dominate. This model suggests that in acute leukemia the LSCs change over time, and thus depending at the stage of the disease two or more types of LSCs may be present. We have observed recently two types of cell that could initiate leukemia in immunodeficient mice, one with a progenitor phenotype ($CD34^+ CD38^+$) and one with a more primitive phenotype ($CD34^+ CD38^-$) from a same AML sample (Taussig et al. 2008). The evolution of LSCs is thought to underlie the progression of CML. It was indeed shown that in acute phase CML, the malignant myeloid progenitors gained self-renewal capacity (as assessed by replating ability in vitro) via activation of β-catenin pathway (Jamieson et al. 2004). The numbers of granulocyte–monocyte progenitors (thought to represent second generation LSCs) increased compared with that in the chronic phase. The original first generation LSCs (i.e., present in chronic phase CML) were detectable at similar frequencies in all phases of the disease.

Thus, this indicates that different cells or acquisition of other mutations in the original LSC might occur during the progression of the leukemia. This represents an even more important challenge in the development of LSC-specific targeting therapy, as LSC, even in the same patient, might represent a moving target.

Notion of Preleukemic Stem Cells

A recent study from Dr. T. Enver's group indicates that the overexpression of TEL/AML1 in umbilical cord blood cells gives rise to phenotypically aberrant preleukemic stem cells that lack the additional DNA hits to allow full leukemic transformation. Study of twins, where one had ALL with the TEL-AML1 fusion oncogene and one was healthy but had a long-lived population of progenitor B cells with the same TEL-AML1 oncogene, demonstrated that TEL-AML1 conferred the ability to self-renew on progenitor B cells (as well as altering the phenotype) (Hong et al. 2008).

Although TEL-AML1 confers phenotypic and functional changes there is no reason to think that all preleukemic stem cells would be phenotypically and/or functionally different from normal HSCs. A number of oncogenes have been introduced into hematopoietic cells without apparent functional effect (Huntly et al. 2004; Schessl et al. 2005). Thus, many preleukemic stem cells may not be readily detectable as that occurred in the study of twins.

This study represents the first direct proof and isolation of these preleukemic stem cells, which have been postulated originally by Dr. Mel Greaves's group. They show indeed the existence of leukemia-associated fusion oncogene sequences in neonatal blood spots taken years before the onset of leukemia in children (Gale et al. 1997; Kim-Rouille et al. 1999; Wiemels et al. 1999, 2002). These sequences were also found in children who did not develop leukemia. Indeed, only a minority of children with leukemia-associated sequences developed leukemia.

The existence of a common (JAK2 wild type) preleukemic stem cell might also potentially explain the presence of wild-type JAK2 in secondary AML patients that arise from patients with JAK2-mutated myeloproliferative disorders (the JAK2 mutation being a later event) (Jelinek et al. 2005).

The evolution of precancerous stem cells to CSCs may explain the very late recurrences seen in some types of leukemia and solid tumors such as breast cancer. Eliminating these precancerous stem cells may be important for true cure.

Implications of the LSC Model for the Design and Evaluation of Anticancer Treatments

According to the LSC model, therapeutic approaches that do not eradicate the LSC compartment are likely to achieve little success; they will indeed produce a temporary regression of the tumor mass by killing the majority of the leukemic blasts and thus induce a temporary remission but will fail to prevent disease relapse.

Most of the current therapies against leukemia have been developed using either cell lines or blast cells as control. The preclinical assessment of potential cancer treatments has generally relied on short-term culture systems that assess the effect of drugs on the proliferation or survival of malignant cells. Many drugs that have activity in preclinical assessments go on to clinical trials. There is, however, a low success rate of anticancer drugs in clinical trials suggesting that the current preclinical validation in vitro is inadequate. This may be explained by the fact that conventional short-term assays will assess malignant progenitors/mature leukemic blasts, but not LSCs. The same problem can apply to the evaluation of efficiency of new drugs during clinical trials where the screening is based on the capacity to induce a dramatic regression of the tumor. This approach again tends to select for treatments that are active on the bulk leukemic blasts but not necessary on the LSC compartment. Thus, it appears that alongside the development of new therapeutic strategies against LSC, new approaches for preclinical and clinical evaluation of the efficacy of these new drugs will need to be devised.

Considerations in Targeting LSC

The Cell of Origin in Leukemia

An important question for a more complete understanding of leukemogenesis is the cellular origin of the leukemic stem cell. Although candidate CSCs have been identified in many hematological as well as solid tumors, their normal cellular origins remain elusive in most cases. The importance of establishing this is not only to better understand how the functional properties (such as self-renewal potential) of the cellular targets of primary and subsequent transforming genetic events (which might occur at the same or distinct developmental and commitment stages) can impact on the biology of the disease, but also for the development of therapeutic strategies.

This question has been addressed directly through the use of retroviral-mediated gene transfer and cell fractionation. Cozzio et al. (2003) have tested the leukemogenic potential of the MLL-ENL fusion protein in purified HSC, CMP, GMP, and MEP cell populations (Cozzio et al. 2003). Transduction of HSC, CMP, and GMP cells by MLL-ENL reportedly produces a similar leukemia as defined by phenotype and leukemogenic latency, whereas transduction of MEP cells did not confer any increase in proliferative or leukemogenic potential. Hence, in the case of MLL-ENL fusion protein, it seems that transformation of stem cells and myeloid progenitors is possible, but transformation of erythroid progenitors is not. Presumably, this is due to the selective expression of a factor that interacts with the MLL-ENL fusion protein in myeloid but not erythroid progenitors. In a similar study by So et al. (2003), almost identical results were generated with the MLL-GAS7 fusion protein. HSC, CMP, and GMP cell subsets were susceptible to MLL-GAS7-mediated transformation, whereas MEP cell populations could not be transformed. Interestingly, biphenotypic leukemia was generated when the MLL-GAS7 fusion protein was transduced in HSCs, whereas transduction of CMP and GMP populations resulted in myeloid leukemia only (So et al. 2003).

These results confirm that the cell type is important for the leukemogenic effect of certain fusion proteins. However, retroviral-mediated transduction of fusion proteins into various stages of hematopoiesis is a very artificial model, which may not represent the leukemogenic event in vivo, and hence the leukemia obtained may not behave in a similar fashion. For instance, the phenotype of leukemias generated from HSC, CMP, and GMP cell populations is very uniform with no apparent hierarchy (Cozzio et al. 2003). In contrast, there is much greater heterogeneity and hierarchy within human AML samples (Bonnet and Dick 1997). Related to this observation, the frequency of AML-IC within the leukemias generated by MLL-ENL fusions is extremely high (Cozzio et al. 2003). Indeed, the transfer of 13–46 transduced cells into irradiated mice resulted in high-level engraftment of all five recipients. This suggests a frequency of AML-IC that is completely different to the $1/10^6$ to $1/10^7$ reported for human AML cell populations (Pearce et al. 2006; Rombouts et al. 2000).

A model that is possibly more analogous to the human AML situation is the Cre-loxP-mediated translocations described by Rabbitts et al. (2001). In this translocator mouse model, de novo translocations are generated in vivo in adult mice by Cre-mediated excision of LoxP sites. The expression of Cre can be regulated by various lineage-specific promoters to direct the excision of LoxP and hence translocation to a specific cell lineage. These Cre-loxP-mediated translocation experiments confirm the ability of the MLL-ENL fusion protein to transform LMO2-expressing HSCs and reveal that the MLL-AF9 fusion protein can also transform HSCs when Cre is under the control of the LMO2 promoter. MLL-ENL translocations may also cause leukemia when the T-cell specific promoter, Lck, controls Cre expression. Interestingly, the MLL-AF9 fusion protein, which usually causes myeloid leukemia, cannot produce any malignancy when Cre is under the control of the Lck promoter (Corral et al. 1996). These data seem to be consistent with the known incidence of either lymphoid or myeloid malignancies with each of these fusion proteins: MLL-ENL is associated with both ALL and AML, whereas MLL-AF9 is associated with myeloid leukemias only. These data provide an example that leukemogenesis depends on the cell type and the mutation as some fusion proteins may cause leukemia regardless of cell type (MLL-ENL), whereas some fusion proteins are dependent on cell type (MLL-AF9).

Whether this is possible in vivo and how AML is generated in humans still remain to be determined.

In human samples, the study of the cellular origin of the leukemia transformation has till now relied in large part on the immunophenotyping of the LSCs. It is clear that the immunophenotype and morphology of the bulk of AML cells are very similar to various stages of hematopoietic progenitor development. However, the LSC reportedly has a phenotype that is very similar to a subset of non-malignant HSCs. This is indirect evidence that relies in the premise that the phenotype of the AML-IC does not change significantly during leukemogenesis. Some reported observations support this hypothesis. For instance, the immunophenotype of primitive hematopoietic cell does not change significantly when transformed with the AML-ETO fusion protein (Delaney and Bernstein 2004). However, other studies have described profound alterations in immunophenotype upon MLL-ENL and MLL-GAS7 transformations of HSCs (Cozzio et al. 2003; So et al. 2003). One consideration, however, is that only bulk immunophenotying has been reported; it may be that the LSC in these MLL-generated leukemias has a different immunophenotype than in the majority of the leukemia.

Furthermore, there are more similarities between LSC and HSC than between LSC and progenitors. Hallmark features of both cell types are the ability to self-renew and proliferate extensively. The cellular properties of HSCs are thus very close to the behavior of LSC, and fewer changes are required to transform an HSC into an LSC.

Another aspect of the multi-hit hypothesis is that the original hit (potential preleukemic stem cell) may occur in one cell type, but the final leukemogenic event may occur in a more differentiated progeny of the original cell. In this situation, the first hit could predispose in which cell the second hit may be leukemogenic.

As mentioned previously, it is extremely interesting to note that although the NOD/SCID model assesses AML independent of the response to chemotherapy, engraftment still correlates with the response to this treatment (prognosis group). A possible explanation is that NOD/SCID engraftment reflects the stem cell nature of each individual AML case. AML cases that engraft in the NOD/SCID assay at 6 weeks may represent diseases driven by potent leukemia-initiating cells with stem cell-like self-renewal and proliferation abilities whereas nonengrafting AML cases may involve less potent leukemia-initiating cells with more restricted progenitor-type self-renewal and proliferation abilities. Indeed, clinically it will make sense that leukemia in patients in which the transformation events occur in the original normal HSC cells will be more difficult to eradicate than in patients where the transformation events occur in myeloid-committed progenitors. Indeed these progenitor cells might reacquire some defined features of stem cell but might be less potent than a bona fide HSC. These cells might thus be more sensitive to conventional drug treatment.

Knowing the nature of cells from where the LSC originated might thus represent an important question on to how to treat leukemia.

Targeting Cell Surface Markers

Cell surface antigens have been used to isolate CSCs in a range of conditions. These surface antigens can be used as potential targets for therapy. CD123 (Jordan et al. 2000) and CD33 (Taussig et al. 2005) are both expressed in AML LSCs, and therapies that target these antigens are undergoing clinical trials. One therapy comprises diphtheria toxin fused to interleukin 3. Interleukin 3 binds to its receptor (CD123), which is expressed in AML LSCs but not in most normal bone marrow HSCs. This compound inhibits growth of AML in vivo while having a limited effect on normal hematopoietic cells (Yalcintepe et al. 2006). Monoclonal antibody against CD33 has been conjugated to a cytotoxic agent to form gemtuzumab ozagamicin (GO). Antibody to CD33 is internalized on binding to the cell surface, and so it brings the cytotoxic agent into the cell. We demonstrated recently that CD33, CD13, and CD123 are indeed present on most AML-LSCs (Taussig et al. 2005). Nevertheless, these markers thought to be myeloid specific are also expressed in the majority of normal HSCs indicating that careful assessment of antigen expression in both normal HSCs and LSCs will be needed when designing new targeted therapies to ensure that they are selective for LSCs. For some diseases the CSCs share a number of features in common with normal HSCs, and thus the task of identifying antigens specific for the LSC might be difficult to achieve.

Targeting LSC Signaling Pathways

Fusion oncoproteins are an attractive target for therapy. The tyrosine kinase produced by the BCR-ABL oncogene is inhibited by a number of relatively selective tyrosine kinase inhibitors (TKIs). These induce complete cytogenetic responses in the majority of patients with CML and result in long-term disease control. The disease recurs, however, if the drug is withdrawn. This suggests that the tyrosine kinase inhibitors do not eliminate CML CSCs (Bhatia et al. 2003). The CML CSCs appear to be resistant to killing by TKIs even when there is no mutation in the BCR-ABL leading to TKI resistance. A number of mechanisms may be involved in resistance, including membrane pumps that remove TKIs from CSCs, expression of high levels of BCR-ABL in CSCs (relative to other CML cells), and quiescence of CSCs (Copland et al. 2006).

Another approach is to target signal transduction pathways that are active in CSCs. Phosphoinositide 3 kinases (PI3Ks) are a family of enzymes involved in signal transduction. PI3Ks generate lipid second messengers and these activate downstream effectors such as AKT, which is important in regulation of cell survival. Components of the PI3K-AKT pathway are a frequent target for mutation in malignant disease. In 50% of AML cases there is constitutive activation of the PI3K-AKT pathway (Tamburini et al. 2007). Activation of this pathway is seen not just in blast cells but also in the CD34$^+$ CD123$^+$ CD38$^-$ fraction that is enriched in CSCs (Bardet et al. 2006). Inhibiting PI3K pathway in vitro enhances chemotherapy-induced cytotoxicity. There are a number of isoforms of PI3K, and it is the p110 δ isoform that is expressed in AML. Mice with disrupted p110 δ have only mild hematopoietic defects, suggesting a relatively limited role in HSC function (Billottet et al. 2006). Targeting p110 δ with specific inhibitors would be predicted to be selective for AML and target the LSCs.

Another pathway that seems to be activated in LSCs but not within normal HSCs is nuclear factor-κB (NF-κB). The NF-κB is constitutively expressed in the CD34$^+$ CD38$^-$ fraction of AML (where most AML LSCs reside). It is not expressed in normal HSCs. An inhibitor of NF-κB, parthenolide, reduces engraftment of AML in NOD/SCID mice while having no significant effect on normal hematopoietic cells (Guzman et al. 2005). Parthenolide derivatives are undergoing clinical trial.

Signaling Pathways Regulating the Self-Renewal of LSCs

When a hematopoietic stem cell has entered the cell cycle, two fates are possible: self-renewal and differentiation with associated proliferation. The control of self-renewal is under the control of various regulators. Deregulated self-renewal would involve an expansion of primitive cells and is probably very important in leukemogenesis. The hedgehog, Wnt, and Notch pathways have been implicated in promoting normal stem cell self-renewal (Blank et al. 2008; Huang et al. 2007; Nemeth et al. 2007).

The molecular mechanisms by which all these factors interact to regulate stem and progenitor cell self-renewal remain to be elucidated.

The question of whether these signaling molecules play an important role in the regulation of LSCs remains mainly unknown. Nevertheless, Bmi-1 has been reported to play a key role in self-renewal determination in both normal and leukemic murine stem cells (Lessard and Sauvageau 2003; Park et al. 2003). Similarly, the role of Wnt pathway for survival, proliferation, and self-renewal of normal HSCs has raised the hypothesis that aberrant Wnt signaling might also contribute to the pathogenesis of AML/CML. Recently, aberrant activation of Wnt signaling and downstream effectors has been demonstrated in AML as well as in CML patients (Jamieson et al. 2004; Mikesch et al. 2007; Muller-Tidow et al. 2004; Zhao et al. 2007). Thus, it seems that the same molecular pathways that govern the self-renewal in normal stem cell are being usurped by LSCs and other CSCs. Thus, it remains to be seen whether any of this pathway might be used for targeting LSCs.

LSC and Their Microenvironment

The hematopoietic microenvironment consists of bone cells, stromal cells, macrophages, fat cells, and extracellular matrix. Stem cells and their immediate progeny interact with the hematopoietic microenvironment. Stem/progenitor cells adhere to stromal cells via adhesion molecules. Cytokines and chemokines are also retained in the microenvironment; they bind to extracellular matrix and some are presented on the surface of stromal cells (Verfaillie 1998). This coadhesion of progenitors and cytokines to the microenvironment results in the colocalization of progenitors at a specific stage of differentiation with the appropriate array of cytokines, in niches (Verfaillie 1998).

Both cellular as well as extracellular matrix components of the stem cell microenvironment or niche are critical in stem cell regulation. By labeling stem cells and reinfusing them, it has been shown that HSCs reside in close proximity to the endosteal bone surface (Haylock et al. 2005; Potocnik et al. 2000). Using a combinatorial Gata-2 and Sca-1 expression system, Suzuki et al. confirmed these results using in vivo imaging (Suzuki et al. 2006).

The microenvironment has been described to influence survival, proliferation, and differentiation (Haylock et al. 2005; Rafii et al. 1997). However, the molecules that regulate stem cell niche interactions and how these may influence the balance between self-renewal and differentiation are just beginning to be revealed. Indeed, Arai et al. (2004) show that the Angiopoietin-1/Tie-2 signaling pathway maintains repopulating quiescent HSC in the niche, presumably via the activation of cell adhesion molecules such as N-cadherin (Arai et al. 2004). Two recent studies demonstrate that another osteoblast-secreted molecule, osteopontin (Opn), also plays a key role in the attraction, retention, and regulation of HSC (Nilsson et al. 2005; Strier et al. 2005).

Although osteoblasts have been shown to have a central role in HSC regulation, other stromal and microenvironmmental cell types and extracellular matrix proteins also contribute to this process (Calvi et al. 2003). Hyaluronic acid as well as the membrane-bound form of stem cell factor is also key components of this niche (Driessen et al. 2003; Nilsson et al. 2003). More recently, it has been shown that calcium sensing receptor (CaR) has a function in retaining HSCs in close physical proximity to the endosteal surface and the regulatory niche components associated with it (Adams et al. 2006).

Overall, it appears that the regulation of hematopoiesis is the result of multiple processes involving cell–cell and cell–extracellular matrix interactions, the action of specific growth factors and others cytokines, as well as intrinsic modulators of haematopoietic development.

Currently, the question of whether leukemic stem cells depend on their niche for self-renewal remains unresolved. Both normal and LSCs depend on SDF-1/CXCR4 axis for homing (Lapidot et al. 2002; Spoo et al. 2007). In an AML study published recently, in vivo administration of anti-CD44 antibody to NOD/SCID mice transplanted with human AML significantly reduced leukemic engraftment. Absence of leukemia in serially transplanted mice demonstrated that AML stem cells have been targeted. The mechanisms underlying this eradication included interference with the homing of these LSCs to their niche and alteration of LSC fate. This report thus suggests that LSCs require interaction with a niche to maintain their stem cell properties (Jin et al. 2006).

In addition, it appears that leukemia cells depend on the bone marrow microenvironment for survival. Leukemia cells from the peripheral blood of patients with AML are transplantable into immunodeficient mice, indicating that LSCs circulate (Pearce et al. 2006). Yet many of these patients have no extra-hematopoietic organ involvement. This suggests that the circulating LSCs cannot seed all organs. One explanation is that LSCs are dependent on survival signals from the bone marrow. Furthermore, after chemotherapy, the clearance of blast cells occurs first from the peripheral blood and then from the bone marrow, and relapse is usually observed first within the bone marrow again supporting the idea of a supportive LSC niche in the bone marrow.

Xenograft experiments provide additional evidence for a supportive LSC niche; few primary leukemias seed organs other than the bone marrow and related hematopoieitc organs (e.g., spleen). In immunodeficient mice, LSCs home to the endosteal region of bone marrow, and leukemia cells spread from there to the rest of the marrow. The LSCs within the endosteal region survive treatment by cytarabine chemotherapy, while cells within the central cavity undergo apoptosis (Ishikawa et al. 2007).

Thus, building evidences suggest an involvement of a niche for LSCs development. It is nevertheless still unclear whether LSCs depend on the same niche as HSCs. This has not been proven yet and remains an area of active investigation.

It has also been well documented especially for solid tumor that aberrations could occur in the functional interactions between cancer cells and cells of the microenvironment, a scenario whereby the normal stem cell niche is replaced by the 'tumor microenvironment' (Clarke et al. 2006). Thus, it is important to understand how CSCs are regulated within the context of their natural tumor microenvironment. In the context of LSC it is unclear to what extent the stem cell niche is or not perturbed by leukemic burden but it is extremely worthwhile to explore this issue in future study.

Generalization of the Concept of CSCs to Solid Tumors

Recent studies in solid tumors indicate that the concept of cancer as a hierarchy that is initiated and maintained by a rare population of stem cells may have broader implications beyond the field of hematopoiesis. In 2003, the group lead by Dr. Clarke has identified CSCs in breast. A subpopulation of human breast cancer cells was shown to express specific surface antigens and to induce tumor growth in a xenotransplantation model. Furthermore, when these cells were injected into mice, the tumors generated contained multiple cell types, similar to the original patient sample. This indicates that injected cells were not only capable of self-renewal but also of generating different progenies, like stem cells (Al Hajj et al. 2003). Since this discovery, a multitude of papers have revealed the evidence that rare CSCs drive the formation of a number of different tumors such as brain, colon, prostate, head, and neck tumors raising the question of whether all cancers originate from and are maintained by CSCs (Singh et al. 2004; O'Brien et al. 2007; Ricci-Vitiani et al. 2007; Collins et al. 2005). The question of whether these CSCs shared similarity needs to be further investigated. It appears nevertheless that cell surface markers such as CD133 and CD44 might be present on a number of different CSCs as is the case for major regulatory pathways such as β-catenin, Notch, and sonic hedgehog. Thus, investigations on LSC might shed light on other CSC types.

Acknowledgments We apologize to authors whose work has not been discussed in an effort to focus on issues that we feel are important. We thank Rachael Swallow for her assistance in the preparation of this manuscript.

References

Adams, G. B., Chabner, K. T., Alley, I. R., Olson, D. P., Szczepiorkowski, Z. M., Poznansky, M. C., Kos, C. H., Pollak, M. R., Brown, E. M. & Scadden, D. T. (2006) Stem cell engraftment at the endosteal niche is specified by the calcium-sensing receptor. Nature, **439**, 599–603.

Al-Hajj, M., Wicha, M. S., Benito-Hernandez, A., Morrison, S. J. & Clarke, M. F. (2003) Prospective identification of tumorigenic breast cancer cells. Proc Natl Acad Sci USA, **100**, 3983–3988.

Arai, F., Hirao, A., Ohmura, M., Sato, H., Matsuoka, S., Takubo, K., Ito, K., Koh, G. Y. & Suda, T. (2004) Tie2/angiopoietin-1 signaling regulates hematopoietic stem cell quiescence in the bone marrow niche. Cell, **118**, 149–161.

Barabe, F., Kennedy, J. A., Hope, K. J. & Dick, J. E. (2007) Modeling the initiation and progression of human acute leukemia in mice. Science, **316**, 600–604.

Bardet, V., Tamburini, J., Ifrah, N., et al. (2006) Single cell analysis of phosphoinositide 3-kinase/Akt and ERK activation in acute myeloid leukemia by flow cytometry. Haematologica, **91**, 757–764.

Bhatia, R., Holtz, M., Niu, N., et al. (2003) Persistence of malignant hematopoietic progenitors in chronic myelogenous leukemia patients in complete cytogenetic remission following imatinib mesylate treatment. Blood, **101**, 4701–4707.

Billottet, C., Grandage, V. L., Gale, R. E., et al. (2006) A selective inhibitor of the p110delta isoform of PI 3- kinase inhibits AML cell proliferation and survival and increases the cytotoxic effects of VP16. Oncogene, **25**, 6648–6659.

Blair, A. & Sutherland, H. J. (2000) Primitive acute myeloid leukemia cells with long-term pro-liferative ability in vitro and in vivo lack surface expression of c-kit (CD117). Exp Hematol, **28**, 660–671.

Blair, A., Hogge, D. E., Ailles, L. E., Lansdorp, P. M. & Sutherland, H. J. (1997) Lack of expres-sion of Thy-1 (CD90) on acute myeloid leukemia cells with long-term proliferative ability in vitro and in vivo. Blood, **89**, 3104–3112.

Blair, A., Hogge, D. E. & Sutherland, H. J. (1998) Most acute myeloid leukemia progenitor cells with long-term proliferative ability in vitro and in vivo have the phenotype CD34(+)/CD71(–)/HLA-DR. Blood, **92**, 4325–4335.

Blank, U., Karlsson, G., & Karlsson, S. (2008) Signaling pathways governing stem-cell fate. Blood, **111**, 492–503.

Bonnet, D. & Dick, J. E. (1997) Human acute myeloid leukemia is organized as a hierarchy that originates from a primitive hematopoietic cell. Nat Med, **3**, 730–737.

Calvi, L. M., Adams, G. B., Weibrecht, K. W., Weber, J. M., Olson, D. P., Knight, M. C., Martin, R. P., Schipani, E., Divieti, P., Bringhurst, F. R., Milner, L. A., Kronenberg, H. M. & Scadden, D. T. (2003) Osteoblastic cells regulate the haematopoietic stem cell niche. Nature, **425**, 841–846.

Clarke, M. F. & Fuller, M. (2006) Stem cells and cancer: two faces of eve. Cell, **124**, 1111–1115.

Collins, A. T., Berry, P. A., Hyde, C., Stower, M. J., & Maitland, N. J. (2005) Prospective isolation of tumorogenic prostate cancer stem cells. Cancer Res, **65**, 10946–10951.

Copland, M., Hamilton, A., Elrick, L. J., et al. (2006) Dasatinib (BMS-354825) targets an earlier progenitor population than imatinib in primary CML but does not eliminate the quiescent frac-tion. Blood, **107**, 4532–4539.

Corral, J., Lavenir, I., Impey, H., Warren, A. J., Forster, A., Larson, T. A., Bell, S., Mckenzie, A. N., King, G. & Rabbitts, T. H. (1996) An Mll-AF9 fusion gene made by homologous recombination causes acute leukemia in chimeric mice: a method to create fusion onco-genes. Cell, **85**, 853–861.

Costello, R. T., Mallet, F., Gaugler, B., Sainty, D., Arnoulet, C., Gastaut, J. A. & Olive, D. (2000) Human acute myeloid leukemia CD34+/CD38– progenitor cells have decreased sensitivity to chemotherapy and Fas-induced apoptosis, reduced immunogenicity, and impaired dendritic cell transformation capacities. Cancer Res, **60**, 4403–4411.

Cozzio, A., Passegue, E., Ayton, P. M., Karsunky, H., Cleary, M. L. & Weissman, I. L. (2003) Similar MLL-associated leukemias arising from self-renewing stem cells and short-lived myeloid progenitors. Genes Dev, **17**, 3029–3035.

Delaney, C. & Bernstein, I. D. (2004) Establishment of a pluripotent preleukaemic stem cell line by expression of the AML1-ETO fusion protein in Notch1-immortalized HSCN1cl10 cells. Br J Haematol, **125**, 353–357.

Driessen, R. L., Johnston, H. M. & Nilsson, S. K. (2003) Membrane-bound stem cell factor is a key regulator in the initial lodgment of stem cells within the endosteal marrow region. Exp Hematol, **31**, 1284–1291.

Gale, K. B., Ford, A. M., Repp, R., et al. (1997) Backtracking leukemia to birth: identification of clonotypic gene fusion sequences in neonatal blood spots. Proc Natl Acad Sci USA, **94**, 13950–13954.

Grimwade, D., Walker, H., Oliver, F., Wheatley, K., Harrison, C., Harrison, G., Rees, J., Hann, I., Stevens, R., Burnett, A. & Goldstone, A. (1998) The importance of diagnostic cytogenetics on outcome in AML: analysis of 1,612 patients entered into the MRC AML 10 trial. The Medical Research Council Adult and Children's Leukaemia Working Parties. Blood, **92**, 2322–2333.

Grimwade, D., Walker, H., Harrison, G., Oliver, F., Chatters, S., Harrison, C. J., Wheatley, K., Burnett, A. K. & Goldstone, A. H. (2001) The predictive value of hierarchical cytogenetic clas-sification in older adults with acute myeloid leukemia (AML): analysis of 1065 patients entered into the United Kingdom Medical Research Council AML11 trial. Blood, **98**, 1312–1320.

Guzman, M. L., Rossi, R. M., Karnischky, L., et al. (2005) The sesquiterpene lactone parthenolide induces apoptosis of human acute myelogenous leukemia stem and progenitor cells. Blood, **105**, 4163–4169.

Haylock, D. N. & Nilsson, S. K. (2005) Stem cell regulation by the hematopoietic stem cell niche. Cell Cycle, **4**, 1353–1355.

Hope, K. J., Jin, L. & Dick, J. E. (2004) Acute myeloid leukemia originates from a hierarchy of leukemic stem cell classes that differ in self-renewal capacity. Nat Immunol, **5**, 738–743.

Hong, D., Gupta, R., Ancliff, P., et al. (2008) Initiating and cancer-propagating cells in TEL-AML1-associated childhood leukemia. Science, **319**, 336–339.

Huang, X., Cho, S., & Spangrude, G. J. (2007) Hematopoietic stem cells: generation and self-renewal. Cell Death Differ, **14**, 1851–1859.

Huntly, B. J., Shigematsu, H., Deguchi, K., et al. (2004) MOZ-TIF2, but not BCR-ABL, confers properties of leukemic stem cells to committed murine hematopoietic progenitors. Cancer Cell, **6**, 587–596.

Ishikawa, F., Yoshida, S., Saito, Y., et al. (2007) Chemotherapy-resistant human AML stem cells home to and engraft within the bone-marrow endosteal region. Nat Biotechnol, **25**, 1315–1321.

Jamieson, C. H., Ailles, L. E., Dylla, S. J., et al. (2004) Granulocyte-macrophage progenitors as candidate leukemic stem cells in blast-crisis CML. N Engl J Med, **351**, 657–667.

Jelinek, J., Oki, Y., Gharibyan, V., et al. (2005) JAK2 mutation 1849G > T is rare in acute leuke-mias but can be found in CMML, Philadelphia chromosome-negative CML, and megakaryo-cytic leukemia. Blood, **106**, 3370–3373.

Jin, L., Hope, K. J., Zhai, Q., Smadja-Joffe, F., & Dick, J. E. (2006) Targeting of CD44 eradicates human acute myeloid leukemic stem cells. Nat Med, **12**, 1167–1174.

Jordan, C. T., Upchurch, D., Szilvassy, S. J., Guzman, M. L., Howard, D. S., Pettigrew, A. L., Meyerrose, T., Rossi, R., Grimes, B., Rizzieri, D. A., Luger, S. M. & Phillips, G. L. (2000) The interleukin-3 receptor alpha chain is a unique marker for human acute myelogenous leukemia stem cells. Leukemia, **14**, 1777–1784.

Kim-Rouille, M. H., MacGregor, A., Wiedemann, L. M., Greaves, M. F. & Navarrete, C. (1999) MLL-AF4 gene fusions in normal newborns. Blood, **93**, 1107–1108.

Lapidot, T. & Kollet, O. (2002) The essential roles of the chemokine SDF-1 and its receptor CXCR4 in human stem cell homing and repopulation of transplanted immune-deficient NOD/SCID and NOD/SCID/B2m(null) mice. Leukemia, **16**, 1992–2003.

Lapidot, T., Sirard, C., Vormoor, J., Murdoch, B., Hoang, T., Caceres-Cortes, J., Minden, M., Paterson, B., Caligiuri, M. A. & Dick, J. E. (1994) A cell initiating human acute myeloid leukaemia after transplantation into SCID mice. Nature, **367**, 645–648.

Lessard, J. & Sauvageau, G. (2003) Bmi-1 determines the proliferative capacity of normal and leukaemic stem cells. Nature, **423**, 255–260.

Mikesch, J. H., Steffen, B., Berdel, W. E., Serve, H. & Muller-Tidow, C. (2007) The emerging role of Wnt signaling in the pathogenesis of acute myeloid leukemia. Leukemia, **21**, 1638–1647.

Muller-Tidow, C., Steffen, B., Cauvet T, et al. (2004) Translocation products in acute myeloid leuke-mia activate the Wnt signaling pathway in hematopoietic cells. Mol Cell Biol, 24, 2890–2904.

Nemeth, M. J. & Bodine, D. M. (2007) Regulation of hematopoiesis and the hematopoietic stem cell niche by Wnt signaling pathways. Cell Res, **17**, 746–758.

Nilsson, S. K., Haylock, D. N., Johnston, H. M., Occhiodoro, T., Brown, T. J. & Simmons, P. J. (2003) Hyaluronan is synthesized by primitive hemopoietic cells, participates in their lodg-ment at the endosteum following transplantation, and is involved in the regulation of their proliferation and differentiation in vitro. Blood, **101**, 856–862.

Nilsson, S. K., Johnston, H. M., Whitty, G. A., Williams, B., Webb, R. J., Denhardt, D. T., Bertoncello, I., Bendall, L. J., Simmons, P. J. & Haylock, D. N. (2005) Osteopontin, a key component of the hematopoietic stem cell niche and regulator of primitive hematopoietic progenitor cells. Blood, **106**, 1232–1239.

O'Brien, C.A., Pollett, A., Gallinger, S. & Dick, J. E. (2007) A human colon cancer capable of initiating tumor growth in ummunodeficient mice. Nature, **445**, 106–110.

Park, I. K., Qian, D., Kiel, M., Becker, M. W., Pihalja, M., Weissman, I. L., Morrison, S. J. & Clarke, M. F. (2003) Bmi-1 is required for maintenance of adult self-renewing haematopoietic stem cells. Nature, **423**, 302–305.

Pearce, D. J., Taussig, D., Simpson, C., Allen, K., Rohatiner, A. Z., Lister, T. A. & Bonnet, D. (2005) Characterization of cells with a high aldehyde dehydrogenase activity from cord blood and acute myeloid leukemia samples. Stem Cells, **23**, 752–760.

Pearce, D. J., Taussig, D., Zibara, K., Smith, L. L., Ridler, C. M., Preudhomme, C., Young, B. D., Rohatiner, A. Z., Lister, T. A. & Bonnet, D. (2006) AML engraftment in the NOD/SCID assay reflects the outcome of AML: implications for our understanding of the heterogeneity of AML. Blood, **107**, 1166–1173.

Potocnik, A. J., Brakebusch, C. & Fassler, R. (2000) Fetal and adult hematopoietic stem cells require beta1 integrin function for colonizing fetal liver, spleen, and bone marrow. Immunity, **12**, 653–663.

Rabbitts, T. H., Appert, A., Chung, G., Collins, E. C., Drynan, L., Forster, A., Lobato, M. N., Mccormack, M. P., Pannell, R., Spandidos, A., Stocks, M. R., Tanaka, T. & Tse, E. (2001) Mouse models of human chromosomal translocations and approaches to cancer therapy. Blood Cells Mol Dis, **27**, 249–259.

Rafii, S., Mohle, R., Shapiro, F., Frey, B. M. & Moore, M. A. (1997) Regulation of hematopoiesis by microvascular endothelium. Leuk Lymphoma, **27**, 375–386.

Reya, T., Morrison, S. J., Clarke, M. F. & Weissman, I. L. (2001) Stem cells, cancer, and cancer stem cells. Nature, **414**, 105–111.

Ricci-Vitiani, L., Lombardi, D. G., Pilozzi, E., Biffoni, M., Todaro, M., Peschle, C. & De Maria, R. (2007) Identification and expansion of human colon-cancer-initiating cells. Nature, **445**, 111–115.

Rombouts, W. J., Martens, A. C. & Ploemacher, R. E. (2000) Identification of variables determining the engraftment potential of human acute myeloid leukemia in the immunodeficient NOD/SCID human chimera model. Leukemia, **14**, 889–897.

Schessl, C., Rawat, V. P., Cusan, M., Deshpande, A., Kohl, T. M., Rosten, P. M., Spiekermann, K., Humphries, R. K., Schnittger, S., Kern, W., Hiddemann, W., Quintanilla-Martinez, L., Bohlander, S. K., Feuring-Buske, M. & Buske, C. (2005) The AML1-ETO fusion gene and the FLT3 length mutation collaborate in inducing acute leukemia in mice. J Clin Invest, **115**, 2159–2168.

Singh, S. K., Hawkins, C., Clarke, I. D., Squire, J. A., Bayani, J., Hide, T., Henkelman, R. M., Cusimano, M. D. & Dirks, P. B. (2004) Identification of human brain tumour initiating cells. Nature, **432**, 396–401.

So, C. W., Karsunky, H., Passegue, E., Cozzio, A., Weissman, I. L. & Cleary, M. L. (2003) MLL-GAS7 transforms multipotent hematopoietic progenitors and induces mixed lineage leukemias in mice. Cancer Cell, **3**, 161–171.

Spoo, A. C., Lubbert, M., Wierda, W. G. & Burger, J. A. (2007) CXCR4 is a prognostic marker in acute myelogenous leukemia. Blood, **109**, 786–791.

Stier, S., Ko, Y., Forkert, R., Lutz, C., Neuhaus, T., Grunewald, E., Cheng, T., Dombkowski, D., Calvi, L. M., Rittling, S. R. & Scadden, D. T. (2005) Osteopontin is a hematopoietic stem cell niche component that negatively regulates stem cell pool size. J Exp Med, **201**, 1781–1791.

Suzuki, N., Ohneda, O., Minegishi, N., Nishikawa, M., Ohta, T., Takahashi, S., Engel, J. D. & Yamamoto, M. (2006) Combinatorial Gata2 and Sca1 expression defines hematopoietic stem cells in the bone marrow niche. Proc Natl Acad Sci USA, **103**, 2202–2207.

Tamburini, J., Elie, C., Bardet, V., et al. (2007) Constitutive phosphoinositide-3kinase/AKT activation represents a favourable prognostic factor in de novo AML patients. Blood, **110**, 1025–1028.

Taussig, D. C., Pearce, D. J., Simpson, C., Rohatiner, A. Z., Lister, T. A., Kelly, G., Luongo, J. L., Danet-Desnoyers, G. A. & Bonnet, D. (2005) Hematopoietic stem cells express multiple myeloid markers: implications for the origin and targeted therapy of acute myeloid leukemia. Blood, **106**, 4086–4092.

Taussig, D. C., Miraki-Moud, F., Anjos-Afonso, F., Pearce D.J., et al. (2008). Anti-CD38 antibody mediated clearancee of human repopulating cells masks the heterogeneity of leukemia initiating cells. Blood, In Press.

Verfaillie, C. M. (1998)Adhesion receptors as regulators of the hematopoietic process. Blood, **92**, 2609–2612.

Wiemels, J. L., Ford, A. M., Van Wering, E. R., Postma, A., & Greaves, M. (1999) Protracted and variable latency of acute lymphoblastic leukemia after TEL-AML1 gene fusion in utero. Blood **94**, 1057–1062.

Wiemels, J. L., Xiao, Z., Buffler P. A., et al. (2002) In utero origin of t(8;21) AML1-ETO translocations in childhood acute myeloid leukemia. Blood **99**, 3801–3805.

Yalcintepe, L., Frankel, A. E., & Hogge, D. E. (2006) Expression of interleukin-3 receptor subunits on defined subpopulations of acute myeloid leukemia blasts predicts the cytotoxicity of diphtheria toxin interleukin-3 fusion protein against malignant progenitors that engraft in immunodeficient mice. Blood. **108**, 3530–3537.

Zhao, C., Blum, J., Chen, A., et al. (2007) Loss of beta-catenin impairs the renewal of normal and CML stem cells in vivo. Cancer Cell **12**, 528–541.

Zhou, S., Schuetz, J. D., Bunting, K. D., Colapietro, A. M., Sampath, J., Morris, J. J., Lagutina, I., Grosveld, G. C., Osawa, M., Nakauchi, H. & Sorrentino, B. P. (2001) The ABC transporter Bcrp1/ABCG2 is expressed in a wide variety of stem cells and is a molecular determinant of the side-population phenotype. Nat Med, **7**, 1028–1034.

Chapter 11
Targeting Signal Transduction Pathways in Hematopoietic Disorders

Li Zhou and Amit Verma

Abstract Hematopoiesis is the process during which pluripotent stem cells give rise to lineage committed progenitors that differentiate into mature erythroid, myeloid, and lymphoid cells. This is a finely regulated process that is required for optimal maintenance of peripheral blood counts. Alterations in hematopoiesis can lead to a variety of bone marrow failure syndromes and hematological malignancies. Treatment of these diseases requires a detailed study of pathogenic mechanisms that lead to hematopoietic failure.

Cytokines Play Important Roles in the Regulation of Normal Hematopoiesis

A fine balance between the actions of hematopoietic growth factors and myelosuppressive factors is required for optimal production of cells of different hematopoietic lineages. Growth factors such as erythropoietin (EPO), granulocyte-colony stimulating factor (G-CSF), granulocyte macrophage-colony stimulating factor (GM-CSF), stem cell factor (SCF), and IL-3 stimulate the growth and differentiation of CD34+ cells into erythroid and myeloid cells. On the other hand, inhibitory cytokines such as TNFα, TGFβ, IL-1β, and interferons inhibit and regulate this process. All cytokines attach to cell surface receptors and trigger distinct signaling cascades that lead to eventual changes in gene transcription. Since the regulation of hematopoiesis is a very complex and specific process, there is a lot of crosstalk between these signaling pathways. Thus, the discovery of common central signaling

L. Zhou(✉) and A. Verma
Albert Einstein College of Medicine,
1300 Morris Park Ave, Chanin 302B,
Bronx, NY10461, USA

A. Wickrema and B. Kee (eds.), *Molecular Basis of Hematopoiesis*,
DOI: 10.1007/978-0-387-85816-6_1, © Springer Science+Business Media, LLC 2009

pathways activated by these different cytokines can have important functional implications. One such pathway identified by us is the p38 MAP kinase signaling pathway. Mitogen activated protein (MAP) family of kinases are evolutionary conserved enzymes, which include the p38, extracellular signal regulated kinase (erk), c-Jun N-terminal kinase (jnk), and erk-5 kinase (Uddin et al. 2000; Romerio and Zella 2002). p38 MAPK is a serine-threonine kinase, originally discovered as a stress-activated kinase that has now been shown to be involved in cell cycle control and apoptosis, with its effects being cell and context specific (Johnson and Lapadat 2002; Kumar et al. 2003; Platanias 2003). It has recently been shown that activation of p38 is required for transcriptional activation of IFN-sensitive genes in hematopoiesis, and this effect appears to be unrelated to the effects on DNA binding of Stat complexes or serine phosphorylation of STATs apparently involving a Stat-independent nuclear mechanism (Uddin et al. 2000). Although, TGFβ and TNFα act through different signaling cascades, they have also been shown to strongly activate the p38 MAPK in primary human hematopoietic cells (Verma et al. 2002a, b). Since the p38 MAPK pathway is a common signaling pathway activated by inhibitory cytokines in hematopoietic progenitor cells, studies were done to clarify the functional role of this pathway in hematopoiesis. We have previously shown that IFNs α and β, TGFβ, and TNFα treatment leads to dose-dependant inhibition of both myeloid and erythroid colonies in methylcellulose colony forming assays performed with normal human hematopoietic progenitors. Concomitant treatment of hematopoietic cells with pharmacologic inhibitors of p38 MAPK (SB 203580 and SB 202190) leads to reversal of the growth inhibitory effects of these cytokines. On the other hand, the inactive structure analogue SB202474 (control) or inhibitors of the MEK/Erk pathway (PD 98059) do not reverse the growth inhibition by these four cytokines (Verma et al. 2002a, b). SiRNA-mediated downregulation of p38α also makes hematopoietic progenitors resistant to the inhibitory effects of these cytokines (Mohindru et al. 2004). Further mechanistic studies have shown that p38 activation regulates hematopoietic inhibition by mediating interferon and TGFβ-induced cell cycle arrest and TNFα-induced apoptosis in progenitors (Verma et al. 2002a, b) (Fig. 11.1). Thus, it appears that the p38 MAP kinase pathway is a central cytokine-stimulated inhibitory pathway in hematopoeisis.

These findings were translated to cytokine-mediated bone marrow diseases such as aplastic anemia. Aplastic anemia is characterized by a quantitative and qualitative reduction in hematopoietic progenitors in the bone marrow. This hematopoietic failure leads to severe reduction in blood counts. Overproduction of TNFα and IFNγ has been implicated in the generation of the myelosuppressive state (Welsh et al. 2004; Dufour et al. 2003) and has been shown to be associated with stem cell suppression in this disease. Studies showed that p38 MAPK inhibition can stimulate hematopoietic colony formation in this disease by interrupting myelosuppressive cytokine signaling (Uddin et al. 2000; Verma et al. 2002a). Thus, based on these studies, abnormal activation of the p38 MAPK pathway may be playing a role in the pathogenesis of aplastic anemia. These studies have prompted the examination of p38 and other signaling pathways in other cytokine-mediated hematologic diseases such as myelodysplastic syndromes (MDS).

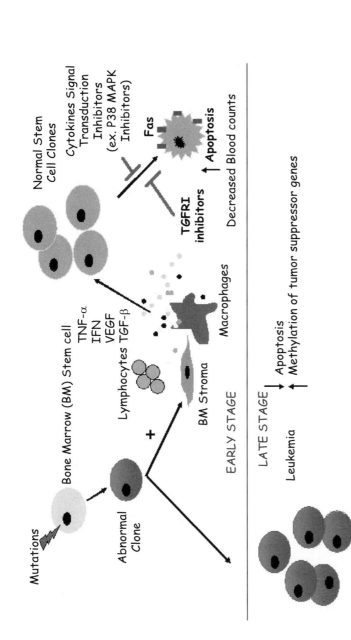

Fig. 11.1 Pathogenesis of MDS and the role of cytokine signal transduction inhibitors in rescuing normal hematopoiesis. Myelodysplastic syndromes (MDS) arise as a result of genetic or epigenetic alterations in a hematopoietic stem cell compartment. The abnormal clones interact with the bone marrow microenvironment and result in a proinflammatory state in the marrow. It is hypothesized that interrupting inhibitory cytokine signaling may stimulate hematopoiesis in MDS and may result in potential clinical benefit to these patients. In the late stages of the disease, MDS may transform to leukemia after the abnormal clones acquire secondary genetic abnormalities

Various Inhibitory Cytokines Have Been Implicated in the Pathophysiology of Hematopoietic Failure in Myelodysplasia

MDS are a heterogeneous group of diseases characterized by ineffective medullary hematopoiesis that leads to refractory cytopenias. These are the commonest hematologic malignancies in the elderly and are a common cause of cytopenias. These bone marrow failure syndromes can transform into leukemias and can be divided into low- and high-grade subtypes based on risk of malignant transformation (Greenberg et al. 1997). Early/low grade cases (low and intermediate-1 IPSS scores) are characterized by increased rates of apoptosis seen in progenitors (Greenberg 1998; Westwood and Mufti 2003), which leads to the paradox of hypercellular appearing bone marrows with peripheral cytopenias as most of the hematopoietic cells are unable to mature and escape into the circulation. Advanced/higher grade cases (intermediate-2 and high IPSS score) are closer to leukemic transformation, have lesser apoptotic cells, and have higher percentage of myeloblasts in the marrow. Even though it is known that cytogenetic alterations are associated with MDS, the exact molecular pathways involved in the pathogenesis of ineffective hematopoiesis are not very well studied. There is considerable evidence though that dysregulation of cytokine levels are involved in these processes.

Since hematopoietic failure is the cause of low blood counts in MDS, investigators have implicated cytokine dysregulation as one of the etiologies of this syndrome. TNFα is the classic apoptotic cytokine that has been implicated in the increased stem cell apoptosis seen in MDS (Allampallam et al. 2002; Claessens et al. 2005). This is based on the presence of high levels of TNFα in the peripheral blood (Zorat et al. 2001) and bone marrow plasma (Deeg et al. 2000) of MDS patients. A higher expression of TNF receptors and TNF mRNA has also been reported in MDS bone marrows (Allampallam et al. 1999; Mundle et al. 1999). Bone marrow macrophages from MDS patients have also been found to overproduce TNFα (Kitagawa et al. 1997).

TGFβ is another myelosuppressive cytokine that has been found to be raised in MDS bone marrow plasma (Zorat et al. 2001). Higher amounts of bound TGFβ on the surface of progenitors have also been observed by immunohistochemical examination of MDS bone marrows (Allampallam et al. 2002). Interferon γ has been implicated strongly in aplastic anemia, and studies have shown that MDS bone marrow mononuclear cells have higher IFNγ mRNA transcripts than healthy controls (Kitagawa et al. 1997). Macrophages and CD4+ lymphocytes from MDS patients have been also found to secrete large amount of IFNγ (Kitagawa et al. 1997; Selleri et al. 2002).

Vascular endothelial growth factor (VEGF) is an angiogenic cytokine that also regulates the generation of inflammatory cytokines (Bellamy et al. 2001). Neutralization of VEGF has been shown to suppress TNFα generation by bone marrow mononuclear and stromal cells. VEGF has also been shown to support the self renewal of cytogenetically abnormal clones in the bone marrow (Bellamy et al. 2001). Increased VEGF levels and higher expression of high-affinity VEGFR-1 receptor have been observed in MDS patients (Aguayo et al. 2000).

IL-1β is a proinflammatory cytokine that has variable regulatory effects on hematopoiesis (Dinarello 1996). At physiological concentrations, IL-1β acts as a hematopoietic growth factor that induces other colony stimulating factors (CSF), such as granulocyte-macrophage CSF (GM-CSF) and IL-3 (Bagby 1989). At higher concentrations, as in chronic inflammatory bone marrow states, IL-1β leads to the suppression of hematopoiesis through the induction of TNFα and PGE$_2$, a potent suppressor of myeloid stem cell proliferation (Dinarello 1996). Increased levels of IL-1β mRNA have been observed in MDS bone marrow mononuclear cells (Allampallam et al. 1999), and the spontaneous production of IL-1β in AML blast cells has been implicated in the pathogenesis of leukemia transformation (Kurzrock et al. 1993; Griffin et al. 1987).

In addition to these cytokines, high levels of fibroblast growth factor (FGF) and hepatocyte growth factor (HGH) have also been seen in patients with MDS (Aguayo et al. 2000). Thus, it seems that multiple cytokine cascades are activated in MDS and future proteomic and genomic analysis should shed light on the exact roles of these pathways in the pathogenesis of this disease.

The Bone Marrow Microenvironment Participates in the Overproduction of Cytokines Seen in MDS

Bone marrow stromal cells and infiltrating mononuclear cells have both been implicated in the production of pathogenic cytokines in human bone marrow failure syndromes. Stromal cells are potent producers of cytokines and play a role in the pathogenesis of multiple myeloma, myelofibrosis, and many other hematologic diseases. It is still unclear if stromal cells in MDS have a primary defect (Flores-Figueroa et al. 2005; Narendran et al. 2004; Tauro et al. 2001; Aizawa et al. 1999) or are just innocent bystanders (Deeg et al. 2000; 2002; Aizawa et al. 2000). Most studies have not found karyotypic alterations in MDS BM stromal cells and have suggested that contact with MDS hematopoietic clones may be necessary for triggering excessive cytokine generation by these cells (Narendran et al. 2004; Navas et al. 2004). Analysis of stromal cell culture supernatants by cytokine antibody arrays shows that bone marrow stromal cells can produce IL-6, VEGF, TGFβ, and FGF, but not TNF or IFNγ (Navas et al. 2004). Studies have also shown that conditioned media from MDS stromal cells alone cannot inhibit hematopoiesis, but media from a mixture of stromal and MDS cells (Narendran et al. 2004; Navas et al. 2004) can suppress normal hematopoiesis. In separate studies, coculture with pure population of MDS stromal cells has been shown to suppress hematopoiesis from normal BM CD34+ cells implying that these cells have a primary defect (Aizawa et al. 1999; Aizawa et al. 2000). All of these results, while uniformly implicating stroma in the pathogenesis of MDS, do point to the heterogeneous nature of stromal abnormalities in different cases of this disease (Fig. 11.1).

Macrophages and lymphocytes are potent producers of TNFα and IFNγ, cytokines implicated in the increased apoptosis seen in MDS hematopoietic progenitors

(Allampallam et al. 2002; Young and Maciejewski 1997). In fact, both lymphocytes and macrophages derived from MDS patients have been found to secrete higher amounts of these cytokines when compared those from normal controls (Kitagawa et al. 1997; Selleri et al. 2002). Studies have shown that lymphocytes are clonally expanded in some cases of MDS, which provides further evidence that host immune cells may play a role in the pathogenesis of this disease (Selleri et al. 2002, Sloand, Kook et al. 2001; Sloand et al. 2004). In fact, recent findings have shown that clonally expanded CD8+ lymphocytes in MDS cases with trisomy of chromosome 8 react against WT-1, a protein found to be overexpressed in this MDS subtype. These clonal lymphocytes have been shown to directly suppress hematopoiesis by progenitors containing trisomy 8 chromosomal alteration, providing evidence of involvement of immune mechanisms in the pathogenesis of ineffective hematopoiesis (Sloand et al. 2004; Sloand et al. 2005).

These studies suggest that both stromal cells and infiltrating mononuclear cells may be interacting with the MDS hematopoietic clones and creating a cytokine milieu promoting ineffective hematopoiesis.

Targeting Hematopoietic Cytokines in Bone Marrow Failure

Stimulatory Hematopoietic Growth Factors as Treatment for MDS

Bone marrow erythroid progenitors derived from bone marrows of MDS patients demonstrate poor growth in vitro and lead to formation of quantitatively and qualitatively deficient erythroid colonies (Navas et al. 2006). This poor erythroid growth can be augmented in vitro by the addition of growth factors such as EPO, granulocyte-colony stimulating factor (G-CSF), and stem cell factor (SCF). Studies have shown that EPO can promote erythroid differentiation and prevent progenitors from undergoing apoptosis (Kelley et al. 1994). This is accomplished by the activation of STAT5 transcription factor and upregulation of genes responsible for promoting erythroid differentiation (GATA1) and inhibiting apoptosis (*bcl-xl*) (Kelley et al. 1994). These results led to numerous clinical trials of EPO in MDS in the 1990s. A meta-analysis of these trials showed that 16% of patients responded with either increase in hemoglobin or decrease in transfusion requirements (Hellstrom-Lindberg 1995). Pretreatment serum EPO levels <200 U/L, non-RARS (refractory anemia with ringed sideroblasts) subtype, and no previous history of transfusions were factors associated with good response to EPO therapy. Only 8% of patients with RARS subtype responded to EPO. In another randomized double blinded placebo controlled trial an overall benefit of EPO over placebo ($p < .007$) was shown for patients with <10% blasts in the marrow (Italian Cooperative Study Group 1998). Unfortunately, only one-third of MDS patients are able to sustain the initial response or elicit a response at reintroduction of EPO following relapse of their anemia. This is probably due to the relative refractoriness of MDS progenitors to EPO when compared to normal controls. The refractoriness is probably due to the extrinsic effects of inhibitory cytokines and intrinsic cellular characteristics of these stem cells. To increase erythropoiesis in refractory patients, other growth factors have been used in combination with EPO. GCSF is a classical myeloid growth factor that signals through pathways different from the EPO receptor and also leads to stimulatory effects on erythroid development (Millot et al. 2001). Thus, GCSF may be able to stimulate erythroid growth in MDS cells where STAT5 signaling may be disrupted. Observation of potential synergy between these two cytokines in vitro has led to clinical trials of a combination of EPO and low-dose GCSF. Results of these studies have demonstrated in vivo synergy, leading to higher response rates than EPO alone (Hellstrom-Lindberg et al. 1998; Miller et al. 2004). Moreover, the use of this combination resulted in high response rates in RARS subtype and in patients with moderate transfusion need. This combination has also resulted in longer duration of responses with a median duration of 24 months. A recent phase III ECOG trial determined that this combination did not lead to increased risk of leukemic transformation and was effective in patients nonresponsive to EPO alone. In this trial patients who developed an erythroid response lived longer than nonresponders, demonstrating the need to growth factor therapies in these patients (Miller et al. 2004).

Targeting Inhibitory Cytokine Signaling in MDS

Increased progenitor apoptosis seen in MDS leads to reduced stem cell numbers, which leads to a reduction in red cell and white cell counts in patients. For early/low grade cases of MDS, the rate of transformation to leukemia is generally insidious and most of the morbidity experienced by these patients is due to chronically low blood counts (Greenberg 1998). Thus, interventions to preserve normal stem cells and increase peripheral blood counts would be advantageous to these patients. Various treatment strategies are in use and in development based on these concepts (Table 11.1).

Immunomodulatory (Imid) Drugs Target Myelosuppressive Cytokine Signaling

Thalidomide and the newer, more efficacious analogue, Lenalidomide (Revlimid©, Celgene, Warren, NJ) are immunomodulatory and antiangiogenic agents that can alter the cytokine milieu in MDS. In addition to their well-known antiangiogenic effects, these Imids alter the Th1/Th2 immune balance and also change cytokine secretion patterns of immune cells (Dredge et al. 2002; Corral et al. 1999). These agents have been shown to inhibit TNF production by downregulating the expression

Table 11.1 Treatments targeting cytokine signaling in hematopoietic failure in MDS

Treatment	Biological effects	Efficacy
Augmentation of hematopoietic growth factors		
Erythropoietin (EPO)	Inhibits erythroid progenitor apoptosis	16–23% Erythroid responses (Hellstrom-Lindberg 1995; Italian Cooperative Study Group 1998)
Epo + G-CSF	Synergy between two distinct stimulatory signaling pathways	Responses in Epo refractory patients (Hellstrom-Lindberg et al. 1998; Miller et al. 2004)
Targeting Inhibitory cytokines		
Thalidomide	Degradation of TNFα mRNA, downregulation of TNF receptors	16/83 (19%) Erythroid responses (Raza et al. 2001)
Lenalidomide	Anti-TNF, alters Th1/Th2 balance	24/36 (56%) Erythroid responses, 83% in patients with 5q-, 50% CR rate (List et al. 2005)
Infliximab	Anti-TNF monoclonal antibody	5/28 (19%) Erythroid responses (Raza et al.)
Ethanercept	Soluble TNF receptor	3/12 (25%) erythroid responses (Deeg et al. 2002)
Anti-thymocyte globulin	Targets immune effector cells	11/25 (44%) Transfusion independence (Molldrem et al. 2002)
PTK-787	VEGF receptor tyrosine kinase inhibitor	5/17 Remissions in AML (Roboz et al. 2006)
SCIO-469	p38 MAPK inhibitor	11/47 (23%) Erythroid responses (Sokol et al. 2006)

of TNF receptors (Marriott et al. 2002). Thalidomide also increases the degradation of TNF-a mRNA by reducing the half life of the molecule from 30 to 17 min, thereby suppressing proinflammatory cytokine release (Melchert et al. 2007). Thalidomide was shown to inhibit TNF-a production in response to LPS-induced stimulation of the PBMC (Moreira et al. 1993).

Thalidomide has shown modest efficacy in low-grade MDS and has resulted in erythroid responses in 18% of patients unresponsive to EPO, after a median duration of 16 weeks (Raza et al. 2001). The newer, more potent Imid, Lenalidomide, has shown considerably more promise in MDS patients without the neurotoxicity and teratogenicity seen with thalidomide. In the recent pilot trial, 24 (56%) of a total of 36 patients with refractory anemia treated with Lenalodimide™ experienced erythroid responses. Twenty of these responders experienced sustained transfusion independence. Interestingly, a high response rate of 83% was observed in patients with 5q31.1 karyotypic deletion. Most importantly, 50% of the patients with this cytogenetic abnormality were able to achieve a complete cytogenetic remission (List et al. 2005). It is known that Lenalidomide leads to alterations in functions of immune and NK cells and can alter the Th1/Th2 cytokine balance (Tai et al. 2005; Anderson 2005; Verma and List 2005).

The exact mechanism of action of Lenalidomide™ in enhancing erythropoiesis in deletion 5q- cases of MDS in not known. Preliminary studies support stimulation of normal hematopoietic progenitors (Verhelle et al. 2007) as well as selective cytotoxicity of del 5q clones. Pellagati et al. studied the effect of Lenalidomide™ on isolated differentiating erythroblasts from del 5q MDS patients and healthy controls. The addition of Lenalidomide™ significantly inhibited the in vitro proliferation of erythroblasts harboring del 5q while the proliferation of cells from normal controls and cells without 5q deletion was not affected. Gene expression profiling was performed at day 7 when a median of 97% cells in culture from MDS patients still harbored the 5q deletion and thus was representative of the in vivo disease state. There was altered gene expression in many genes, but a set of four genes was consistently upregulated (VSIG4, PPIC, TPBG, and SPARC) by more than twofold in all samples. The upregulation of SPARC (secreted protein acidic and rich in cysteine) after treatment with Lenalidomide™ is particularly interesting given its location at 5q 31-32 genomic location and its role as a tumor suppressor with its antiproliferative, antiadhesion, antiangiogeneic properties. The modulation of SPARC expression by Lenalidomide™ in del 5q MDS patients suggests that it may play a role in the pathogenic role in MDS (Pellagatti et al. 2007).

Another study recently published studied the gene expression profiles of CD34+ stem cells derived from MDS patients with del 5q and compared them between healthy controls and MDS patients with normal karyotypes using the Affymetrix platform. Approximately 40% of the probe sets showing reduced expression levels localized to the del 5q region, also known as common deleted segment (CDR). CDR region is thought to comprise approximately 40 genes and is hypothesized to have a tumor-suppressive role given the observation that deletion of the 5q region leads to clonal proliferation of myelodysplastic clone. Majority of the genes associated with CDR showed lower expression due to haploinsufficiency, but several candidate genes showed marked downregulation (RBM22, CSNK1A1, SPARC, and RPS14) (Boultwood et al. 2007). RBM22 is a highly conserved ribosomal protein, and the effects of downregulation may include deregulated apoptosis by its action on ALG-2 (apoptosis-linked gene). CSNK1A1 has recently been shown to be important in hedgehog signaling that governs cell growth and has been implicated in carcinogenesis. Downregulation of CSNK1A1 may contribute to MDS by altering the Hh signaling. Ribosomal protein S14 (RPS14) is involved in the synthesis of 40S subunit of the ribosome and is downregulated in CD34+ cells from MDS patients with del 5q (Boultwood et al. 2007). Another recent study has postulated that RPS-14 is the candidate causative gene in this 5q del MDS. A large shRNA screening strategy was used to knockdown each of the 40 genes in the CDR in normal CD34+ stem cells. These cells were differentiated into erythroid and megakaryotic lineages in vitro. Since patients with 5q- del MDS have anemia but have high platelet counts, an in vitro decrease in erythroid progenitors coupled with an increase in megakaryocytic progenitors was used to identify candidate genes in the shRNA screen. RPS-14 knockdown was able to recapitulate this phenotype and was also then seen to be downregulated in primary bone marrow cells derived from 5q- del MDS patients (Ebert et al. 2008). This important finding implicates a ribosomal synthesis gene in erythropoietic disease and highlights the important role of increased protein synthesis seen in late stages of erythropoiesis. This finding also links 5q- del MDS with Diamond Blackfan anemia (DBA) (Flygare et al. 2007; Liu and Ellis 2006). DBA is a congenital anemia seen in children and has been associated with mutations and haploinsuficiency in ribosomal protein synthesis gene, *RPS-19*.

Targeting TNF-α in Bone Marrow Failure

Since TNFα is the prototypical proapoptotic cytokine implicated in ineffective hematopoiesis, various strategies have targeted this cytokine. Etanercept™ (Enbrel©, Amgen, Thousand Oaks, CA) is a soluble TNF receptor that acts as a competitive inhibitor of TNF and prevents its binding to cellular receptors. A phase I trial evaluated twice weekly subcutaneous use of Etanercept™ in low-grade MDS (Deeg et al. 2002). In this trial, 16 weeks of treatment resulted in erythroid responses in 3/12 patients and platelet and neutrophil responses in 2/12 patients. All patients had increased TNFα levels in marrow plasma when compared with normal controls. Interestingly, responses did not correlate with drop in TNF levels in the marrow, suggesting that the regulation and production of cytokines in MDS may be regulated at many different cellular levels. Another small Phase II study did not reveal any benefits of this approach in MDS (Rosenfeld and Bedell 2002). Infliximab™ (Remicade©, Centocor, Malvern, PA) is a chimeric anti-TNFα mono-clonal antibody that binds to both soluble and membrane-bound TNFα. Case reports of efficacy in MDS (Stasi and Amadori 2002) prompted a Phase II study of two i.v. dosage regimens in low-risk disease (Raza et al. 2004). In this trial, 5/28 patients had a hematologic response with minimal side effects. Thus, it appears that these agents only have modest activity as single agents and may have a role in combination with other agents.

Targeting the P38 MAP Kinase in Hematopoietic Failure

Since multiple cytokines are involved in causing abnormal hematopoietic development in MDS, targeting one single cytokine may not yield appreciable clinical benefit. Targeting common pathways that would be regulated by many cytokines may have better therapeutic potential in this disease. Having previously shown that p38 MAPK activation is needed for cytokine-mediated inhibition of hematopoiesis, we have determined the role of this kinase in MDS. Our recent results show that p38 MAPK pathway is constitutively activated in MDS bone marrows (Navas et al. 2006). This activation is uniformly seen in a variety of different subtypes of early/low grade MDS and is associated with the increased apoptosis seen in MDS hematopoietic progenitors. p38 MAPK activation appears to play an important role in the pathogenesis of ineffective hematopoiesis as siRNA-mediated downregulation of p38α can stimulate hematopoiesis in MDS in vitro. Most importantly, pharmacological inhibition of p38 by novel clinically relevant small molecule inhibitor SCIO-469 results in dose-dependant increase in hematopoietic colony formation from primary MDS CD34+ hematopoietic progenitors. Based on these results a multicenter clinical trial of SCIO-469 is presently accruing in low-grade MDS. Early clinical results have shown modest efficacy in this disease (Sokol et al. 2006).

P38 MAP kinase is also involved in production of cytokines by the bone marrow microenvironment by being involved in signal crosstalk. For example, TNFα has also been shown to induce IL-1β through the activation of NFκB (Werner et al. 2005), and TNFα-induced NFκB activation, in turn, has been shown to be regulated by p38 MAPK (Brinkman et al. 1999). In addition to regulating transcription, p38 MAPK has also been shown to regulate posttranscriptional modification of TNFα and IL-1β, through message stabilization involving MapKapk-2 (Kotlyarov et al. 1999). Thus, targeting p38 MAPK can impact various steps in hematopoietic failure.

Targeting TGF-β Signaling in Hematopoietic Failure

TGF-β Signaling in Normal Hematopoiesis

Several members of the TGF-β superfamily, including TGF-β, bone morphogenic proteins (BMPs), and activin, are known to regulate the fate of hematopoietic progenitor and stem cells. TGF-β has been shown to selectively inhibit the growth and differentiation of early hematopoietic progenitor cells that are largely quiescent, but not of more mature progenitors. In an early murine study, it inhibited the growth and colony formation of IL-3-dependent bone marrow cells but had no effect on the proliferation and differentiation induced by later acting hematopoietic growth factors, such as G-CSF, GM-CSF, and EPO (Keller et al. 1988). Similar studies in human hematopoietic cells demonstrated that TGF-β selectively inhibited colony formation from early CD34+ CD33− progenitors and did not affect more differentiated G-CSF-responsive cells. In some of these studies the same effects were observed with both TGF-β1 and TGF-β2 while in others, hematopoietic progenitor cells were shown to be much more sensitive to inhibition of colony formation and antiproliferative effects of TGF-β1 than TGF-β2 (Ohta et al. 1987; Sargiacomo et al. 1991). Since the aforementioned studies used freshly aspirated unseparated bone marrow cells, it was possible that the action of TGF-β on these hematopoietic progenitors was indirectly due to its effects on other subpopulations. A later study demonstrated that TGF-β can inhibit the growth of single cell suspensions of primitive Thy+ Lin−** murine progenitors. Additionally, the study showed that TGF-β1 can also inhibit the growth of high proliferative potential colony forming cells (HPP-CFC), the earliest hematopoietic progenitors measured in vitro (Keller et al. 1990). The effects of disrupting TGF-β signaling at a single cell level were also studied using a dominant negatively acting mutant of TGF-β type II receptor. Adenoviral vector-mediated gene delivery was used to transiently block TGF-β signaling, and this was shown to enhance the survival, proliferation, and growth kinetics of human hematopoietic stem cells (Fan et al. 2002). Further evidence was obtained from a highly enriched population of Lin⁻, c-kit⁺, Sca1⁺, CD34⁻ quiescent stem cells isolated from TβRI knockout mice that were cultured in serum-free medium with low stimulation (SCF alone). These TβRI null HSCs had a twofold increase in the number of proliferating clones compared with control HSCs when stimulated by active autocrine TGF-β1. These findings were consistent with an inhibitory effect of TGF-β on HSC proliferation in vitro (Larsson et al. 2003).

There are conflicting data on whether TGF-β is a committed inducer of apoptosis or whether its inhibitory effects on hematopoietic progenitor cells are reversible. It has been shown that TGF-β inhibits survival of Lin⁻, Sca1⁺ hematopoietic progenitors dormant in the G_0-G_1 phase of the cell cycle, through induction of apoptosis as detected by an in situ TdT assay (Jacobsen et al. 1995). Conversely, another in vitro murine study suggested that TGF-induced inhibition of long-term repopulating hematopoietic stem cells (LTR-HSC) via sustained downregulation of bone marrow cell surface cytokine receptors was reversible and not accompanied by programmed cell death (Sitnicka et al. 1996). Other studies using human cells have confirmed this by showing that TGF-β1 is not apoptotic and can reversibly maintain primitive human hematopoietic progenitor cells in quiescence. In these studies TGF-β1 has also been shown to upregulate expression of the CD34 membrane antigen, which is a marker of immature hematopoietic stem cells. These findings are important as they point to a potential clinical utility of TGF-β1 in maintaining pools of quiescent hematopoietic stem cells.

The role of TGF-β in hematopoiesis has been studied in vivo through the use of murine knockout models (Shull et al. 1992). Homozygous deletion of TGF-β resulted in embroyonic lethality with defective yolk sac vasculogenesis secondary to abnormal endothelial differentiation, and defective hematopoiesis with a reduced yolk sac erythroid cell number. Although in vitro neutralization of TGF-β resulted in increased hematopoietic cell growth and proliferation, in vivo, loss of TGF-β1 function resulted in defective erythroid differentiation (Dickson et al. 1995). Homozygous TGF-β receptor II (TβRII) deletion causes an indistinguishable phenotype from that of TGF-β1 null mice, with defective vasculogenesis and hematopoiesis (Oshima et al. 1996). Mice lacking TβRI die due to defects in vascular development of the yolk sac as well. In vitro, TBRI–/– endothelial cells have impaired migration and fibronectin production, which would explain the malformed vascular structures in vivo.

The role of TGF-β in the differentiation of committed hematopoietic progenitors has been clarified by many recent observations. TGF-β1 also been shown to stimulate erythroid differentiation by triggering the conversion of early BFU-E into later BFU-E or CFU-E, and by inducing hemoglobin synthesis in mature cells (Krystal et al. 1994). A recent report has shown that a nuclear protein transcriptional intermediary factor 1-γ (TIF1γ) mediates the differentiating effects of TGF-β in human hematopoiesis. TIF-γ competes with Smad4 for Smad2/3 complex binding after TGF-β stimulation and while Smad2/3–Smad4 complexes mediate antiproliferative responses, Smad2/3–TIF1γ complexes mediate erythroid differentiation in response to TGF-β (He et al. 2006).

Several of these growth, proliferation, and differentiation pathways affected by TGF-β in either an inhibitory or stimulatory way have physiologic relevance in the pathogenesis of leukemias and myelodysplasia.

Role of TGF-β in Leukemia

Acute myeloid leukemia is characterized by malignant proliferation of leukemic cells ultimately resulting in a suppression of the normal hematopoietic system. Several factors have been implicated in contributing to the growth advantage of the malignant cells over their normal counterparts. Increased responsiveness to stimulatory growth factors or insensitivity to inhibitory growth factors can lead to transformation of cells and render them resistant to normal homeostatic mechanisms (Wierenga et al. 2002). In support of these hypotheses, leukemic cell lines generally are less sensitive to TGF-β in vitro. Many leukemia specific mechanisms have been described to explain this loss of sensitivity to TGF-β.

As Smad family proteins are known critical components in TGF-β signaling pathway, mutations in the *Smad4* gene have been associated with the pathogenesis of acute myelogenous leukemia (Imai et al. 2001). A missense mutation (P102L) in the Smad4 MH1 domain results in disruption with its DNA binding activity and disrupts TGF-mediated transcriptional activity. A second frame shift mutation resulting in termination in the MH2 domain (Delta 483–552) prevents nuclear localization of Smad4 and thus disrupts TGF signaling. Both these mutants also exert dominant negative effects and can result in insensitivity of leukemic cells to TGF-β.

Smad3, a receptor-stimulated Smad has also been implicated in the pathogenesis of T-cell leukemias. A relative reduction of Smad3 levels has been shown to impair the tumor suppressor ability of the TGF-β pathway and can lead to oncogenic transformation in the presence of additional mutations. Loss of the p27 tumor suppressor gene in combination with one Smad3 allele results in spontaneous T-cell leukemias in mice. Strikingly, nearly absent levels of Smad3 protein were seen in a series of pediatric T-cell leukemias, when compared with normal levels seen in B-cell leukemias and normal T cells. These low protein levels were seen in the presence of normal Smad3 mRNA levels and in the absence of any gene mutations (Wolfraim et al. 2004). Thus, the level of Smad3 was found to be a determinant of the T-cell response to TGF-β, and a reduction in Smad3 works synergistically with other known oncogenic pathways to promote T-cell leukemogenesis.

Evi-1 is a zinc-finger nuclear protein that is highly expressed in human myeloid leukemias and MDS. Evi-1 is also expressed at a very low level in a limited stage of normal myeloid differentiation. Aberrant overexpression of Evi-1 as a fusion transcript with AML-1 leads to blastic transformation in patients with CML. It has been shown that evi-1 may disrupt TGF signaling by direct interactions with Smad3 protein (Sood et al. 1999). Additionally, Evi-1 has also been shown to repress Smad-induced transcription by recruiting C-terminal binding protein (CtBP) as a corepressor. These mechanisms have been postulated to contribute to evi-1-mediated leukemogenesis (Izutsu et al. 2001).

The 8:21 chromosomal translocation leads to acute myelogenous leukemia by leading to the formation of fusion oncoprotein AML1/ETO. AML/ETO is a chimeric transcription factor that represses the expression of many genes involved in myeloid cell development and leads to a maturation arrest. In addition to direct effects on transcription, the AML/ETO protein may also disrupt normal TGF signaling. Both AML1 and AML/ETO have been shown to physically associate with Smads and participate in their nuclear localization. Even though wild-type AML1 stimulates TGF-mediated gene transcription, AML1/ETO interrupts it. Thus, insensitivity to TGF-mediated growth regulation may partly explain the leukogenic potential of AML/ETO (Jakubowiak et al. 2000).

The PML/RARα fusion oncoprotein is formed by the reciprocal chromosomal 15:17 translocation and leads to the pathogenesis of acute promyelocytic leukemia. This transcriptional repressor oncoprotein represses retinoic acid-mediated transcription of genes necessary for myeloid development and causes a maturation arrest. Recent evidence suggests that this fusion protein disrupts normal PML function in a dominant negative manner. The cytoplasmic form of pml (c-pml) is an essential modulator of TGF-β signaling and has been shown to directly bind to Smads and facilitate their association with Smad anchor for receptor activation (SARA) protein for their accumulation in the endosome. Consequently, cells devoid of PML are resistant to growth arrest and apoptosis induced by TGF-β (Lin et al. 2004). Disrupted nuclear transport and phosphorylation of Smads2/3 are seen in PML null cells and results in impaired induction of the TGF-β target genes. These phenomena contribute to the pathogenesis of APL (Lin et al. 2004).

The human T-cell lymphotropic virus type 1 (HTLV-1) infection leads to an aggressive T-cell leukemia and has been linked to the overexpression of the Tax oncoprotein. Tax has been shown to interact with Smads 2, 3, and 4 and interrupt their association with each other and their target DNA sequences. Tax also competes for Smad binding with transcriptional coactivator p300 and thus decreases the transcription of TGF-mediated genes such as plasminogen activator (Mori et al. 2001; Lee et al. 2002). The decrease in TGF-β signaling likely provides the optimal conditions for development of aggressive, resistant T-cell leukemias.

Role of TGF-β in Myelodysplasia

Since TGF has been shown to cause a G0/G1 cell cycle arrest in primary human hematopoietic progenitors it has been strongly implicated in the pathogenesis of MDS. The plasma levels of TGF-β have been reported to be elevated in some (Allampallam et al. 2002; Zorat et al. 2001; Allampallam et al. 1999; Powers et al. 2007; Akiyama et al. 2005) but not all studies (Aguayo et al. 2000, Taketazu et al. 1992; Gyulai et al. 2005; Yoon et al. 2000) and are supported by greater TGF-β immunohistochemical staining in selected studies. In addition to direct myelosuppressive effects, TGF-β has also been implicated in the autocrine production of other myelosuppressive cytokines (TNF, IL-6, and IFN γ) in MDS (Verma and List 2005). Conflicting data may arise from technical limitations of bone marrow immunohisto-chemical analyses of a secreted protein as well as the biological heterogeneity of the disease itself. Additionally, plasma levels of TGF-β may not be an accurate reflection of the biological effects of this cytokine in the MDS bone marrow microenvironment. Thus, we investigated the role of TGF-β in MDS by direct examination of receptor signal activation to conclusively determine its role in the pathogenesis of ineffective hematopoiesis in MDS. We determined that the Smad2 protein is heavily phospho-rylated in MDS bone marrow progenitors, thereby demonstrating sustained TGF-β signal activation in this disease (Zhou et al. 2007). Ongoing studies are evaluating the potential of TGFβ receptor kinase inhibitors and other signaling inhibitors in MDS and similar cytokine-driven hematopoietic diseases.

Immunosuppressive Strategies in Human Bone Marrow Failure Syndromes

Preventing cytokine production by targeting immune cells has also been attempted in MDS. Further rationale for immunosuppressive treatments in MDS is provided by similarities of some cases to aplastic anemia. Antithymocyte globulin (ATG) has been accepted as the standard of care in aplastic anemia and results in long-term remissions and increase in blood counts in patients (Young 2002). These arguments prompted a trial of ATG in red cell transfusion-dependant cases of low-grade MDS. In this trial, 11 out of a total of 25 patients treated with ATG achieved transfusion independence (Molldrem et al. 2002). These responses were found to correlate with the loss of T-cell clones that could suppress hematopoiesis in vitro. Follow-up studies in 20 patients demonstrated that HLA-DR15, younger age, and shorter duration of RBC transfusions correlated significantly with response (Saunthararajah et al. 2003). Thus, patients with these characteristics and hypoplastic marrow phenotype can derive potential benefit from immunosuppressive treatment strategies. These studies are supported by observations of clonal T cells in the bone marrows of MDS patients and expression of autoantigens on some MDS hematopoietic progenitors (Sloand et al. 2005; Epperson et al. 2001; Risitano et al. 2002).

Antiangiogenic Strategies in Bone Marrow Failure

Since VEGF is a cytokine that is elevated in some cases of MDS, attempts have been made to target it in this disease. In addition to inhibiting angiogenesis, anti-VEGF strategies can also inhibit the autocrine and paracrine cytokine loops that play a role in the pathogenesis of ineffective hematopoiesis. Initial trials with VEGF receptor tyrosine kinase inhibitor SU5416 resulted in minimal responses in high-risk cases of MDS (Giles et al. 2003). A recent trial of PTK 787 (Novartis) showed that a more specific inhibitor of VEGF receptor tyrosine kinase led to clinical responses in a few cases of AML and MDS but was associated with significant toxicities, such as headache, dizziness, and other CNS side effects (Roboz et al. 2006). The results of other antiangiogenic agents in bone marrow failure are awaited.

Conclusions and Future Directions

There is considerable evidence to suggest that signal transduction cascades stimulated by various inhibitory cytokines are involved in the pathogenesis of ineffective hematopoiesis seen in MDS. It is also clear that interactions between the MDS clones, bone stromal cells, and infiltrating immune cells are necessary for the secretion and propagation of these cytokines. It appears likely that expression of certain antigens by the MDS clone may be triggering these reactions from the bone marrow microenvironment. Gene array analysis led to the discovery of WT-1 as one such offending antigen in MDS cases with trisomy of chromosome 8. Future genomic studies may answer similar questions on other subsets of MDS. It is also possible that the MDS clones may themselves produce cytokines that may stimulate autocrine and paracrine loops from the surrounding cells that may perpetuate the disease. Advances in proteomic screening of patient samples may shed more light on these questions and also reveal newer cytokines that may play a role in the pathogenesis of this disease.

Recent studies have also shown that cytokine signal transduction pathways may be potential therapeutic targets in MDS. Based on these principles, various kinase inhibitors are being tested in MDS. Because of the heterogeneity of this disease, no single therapy will work in all patients with this disease. It is very likely that there may be different pathways activated in different subsets of this disease, and thus custom tailor-made therapies may be needed. It is our hope that technological developments, increased knowledge of signaling pathways, and cooperative tissue repositories may help answer these questions in the near future.

References

Aguayo, A., Kantarjian, H., Manshouri, T., Gidel, C., Estey, E., Thomas, D., Koller, C., Estrov, Z., O'Brien, S., Keating, M., et al. 2000. Angiogenesis in acute and chronic leukemias and myelodysplastic syndromes. *Blood* 96:2240–2245.

Aizawa, S., Nakano, M., Iwase, O., Yaguchi, M., Hiramoto, M., Hoshi, H., Nabeshima, R., Shima, D., Handa, H., and Toyama, K. 1999. Bone marrow stroma from refractory anemia of myelodysplastic syndrome is defective in its ability to support normal CD34-positive cell proliferation and differentiation in vitro. *Leuk Res* 23:239–246.

Aizawa, S., Hiramoto, M., Hoshi, H., Toyama, K., Shima, D., and Handa, H. 2000. Establishment of stromal cell line from an MDS RA patient which induced an apoptotic change in hematopoietic and leukemic cells in vitro. *Exp Hematol* 28:148–155.

Akiyama, T., Matsunaga, T., Terui, T., Miyanishi, K., Tanaka, I., Sato, T., Kuroda, H., Takimoto, R., Takayama, T., Kato, J., et al. 2005. Involvement of transforming growth factor-beta and thrombopoietin in the pathogenesis of myelodysplastic syndrome with myelofibrosis. *Leukemia* 19:1558–1566.

Allampallam, K., Shetty, V., Hussaini, S., Mazzoran, L., Zorat, F., Huang, R., and Raza, A. 1999. Measurement of mRNA expression for a variety of cytokines and its receptors in bone marrows of patients with myelodysplastic syndromes. *Anticancer Res* 19:5323–5328.

Allampallam, K., Shetty, V., Mundle, S., Dutt, D., Kravitz, H., Reddy, P.L., Alvi, S., Galili, N., Saberwal, G.S., Anthwal, S., et al. 2002. Biological significance of proliferation, apoptosis, cytokines, and monocyte/macrophage cells in bone marrow biopsies of 145 patients with myelodysplastic syndrome. *Int J Hematol* 75:289–297.

Anderson, K.C. 2005. Lenalidomide and thalidomide: mechanisms of action – similarities and differences. *Semin Hematol* 42:S3–S8.

Bagby, G.C., Jr. 1989. Interleukin-1 and hematopoiesis. *Blood Rev* 3:152–161.

Bellamy, W.T., Richter, L., Sirjani, D., Roxas, C., Glinsmann-Gibson, B., Frutiger, Y., Grogan, T.M., and List, A.F. 2001. Vascular endothelial cell growth factor is an autocrine promoter of abnormal localized immature myeloid precursors and leukemia progenitor formation in myelodysplastic syndromes. *Blood* 97:1427–1434.

Boultwood, J., Pellagatti, A., Cattan, H., Lawrie, C.H., Giagounidis, A., Malcovati, L., Della Porta, M.G., Jadersten, M., Killick, S., Fidler, C., et al. 2007. Gene expression profiling of CD34+ cells in patients with the 5q- syndrome. *Br J Haematol* 139:578–589.

Brinkman, B.M., Telliez, J.B., Schievella, A.R., Lin, L.L., and Goldfeld, A.E. 1999. Engagement of tumor necrosis factor (TNF) receptor 1 leads to ATF-2- and p38 mitogen-activated protein kinase-dependent *TNF-α* gene expression. *J Biol Chem* 274:30882–30886.

Claessens, Y.E., Park, S., Dubart-Kupperschmitt, A., Mariot, V., Garrido, C., Chretien, S., Dreyfus, F., Lacombe, C., Mayeux, P., and Fontenay, M. 2005. Rescue of early stage myelodysplastic syndrome-deriving erythroid precursors by the ectopic expression of a dominant negative form of FADD. *Blood*105:4035–4042.

Corral, L.G., Haslett, P.A., Muller, G.W., Chen, R., Wong, L.M., Ocampo, C.J., Patterson, R.T., Stirling, D.I., and Kaplan, G. 1999. Differential cytokine modulation and T cell activation by two distinct classes of thalidomide analogues that are potent inhibitors of TNF-α. *J Immunol* 163:380–386.

Deeg, H.J. 2002. Marrow stroma in MDS: culprit or bystander? *Leuk Res* 26:687–688.

Deeg, H.J., Beckham, C., Loken, M.R., Bryant, E., Lesnikova, M., Shulman, H.M., and Gooley, T. 2000. Negative regulators of hemopoiesis and stroma function in patients with myelodysplastic syndrome. *Leuk Lymphoma* 37:405–414.

Deeg, H.J., Gotlib, J., Beckham, C., Dugan, K., Holmberg, L., Schubert, M., Appelbaum, F., and Greenberg, P. 2002. Soluble TNF receptor fusion protein (etanercept) for the treatment of myelodysplastic syndrome: a pilot study. *Leukemia* 16:162–164.

Dickson, M.C., Martin, J.S., Cousins, F.M., Kulkarni, A.B., Karlsson, S., and Akhurst, R.J. 1995. Defective haematopoiesis and vasculogenesis in transforming growth factor-beta 1 knock out mice. *Development* 121:1845–1854.

Dinarello, C.A. 1996. Biologic basis for interleukin-1 in disease. *Blood* 87:2095–2147.

Dredge, K., Marriott, J.B., and Dalgleish, A.G. 2002. Immunological effects of thalidomide and its chemical and functional analogs. *Crit Rev Immunol* 22:425–437.

Dufour, C., Corcione, A., Svahn, J., Haupt, R., Poggi, V., Beka'ssy, A.N., Scime, R., Pistorio, A., and Pistoia, V. 2003. TNF-α and IFN-γ are overexpressed in the bone marrow of Fanconi anemia patients and TNF-α suppresses erythropoiesis in vitro. *Blood* 102:2053–2059.

Ebert, B.L., Pretz, J., Bosco, J., Chang, C.Y., Tamayo, P., Galili, N., Raza, A., Root, D.E., Attar, E., Ellis, S.R., et al. 2008. Identification of *RPS14* as a 5q- syndrome gene by RNA interference screen. *Nature* 451:335–339.

Epperson, D.E., Nakamura, R., Saunthararajah, Y., Melenhorst, J., and Barrett, A.J. 2001. Oligoclonal T cell expansion in myelodysplastic syndrome: evidence for an autoimmune process. *Leuk Res* 25:1075–1083.

Fan, X., Valdimarsdottir, G., Larsson, J., Brun, A., Magnusson, M., Jacobsen, S.E., ten Dijke, P., and Karlsson, S. 2002. Transient disruption of autocrine TGF-β signaling leads to enhanced survival and proliferation potential in single primitive human hemopoietic progenitor cells. *J Immunol* 168:755–762.

Flores-Figueroa, E., Arana-Trejo, R.M., Gutierrez-Espindola, G., Perez-Cabrera, A., and Mayani, H. 2005. Mesenchymal stem cells in myelodysplastic syndromes: phenotypic and cytogenetic characterization. *Leuk Res* 29:215–224.

Flygare, J., Aspesi, A., Bailey, J.C., Miyake, K., Caffrey, J.M., Karlsson, S., and Ellis, S.R. 2007. Human *RPS19*, the gene mutated in Diamond-Blackfan anemia, encodes a ribosomal protein required for the maturation of 40S ribosomal subunits. *Blood* 109:980–986.

Giles, F.J., Stopeck, A.T., Silverman, L.R., Lancet, J.E., Cooper, M.A., Hannah, A.L., Cherrington, J.M., O'Farrell, A.M., Yuen, H.A., Louie, S.G., et al. 2003. SU5416, a small molecule tyrosine kinase receptor inhibitor, has biologic activity in patients with refractory acute myeloid leukemia or myelodysplastic syndromes. *Blood* 102:795–801.

Greenberg, P., Cox, C., LeBeau, M.M., Fenaux, P., Morel, P., Sanz, G., Sanz, M., Vallespi, T., Hamblin, T., Oscier, D., et al. 1997. International scoring system for evaluating prognosis in myelodysplastic syndromes. *Blood* 89:2079–2088.

Greenberg, P.L. 1998. Apoptosis and its role in the myelodysplastic syndromes: implications for disease natural history and treatment. *Leuk Res* 22:1123–1136.

Griffin, J.D., Rambaldi, A., Vellenga, E., Young, D.C., Ostapovicz, D., and Cannistra, S.A. 1987. Secretion of interleukin-1 by acute myeloblastic leukemia cells in vitro induces endothelial cells to secrete colony stimulating factors. *Blood* 70:1218–1221.

Gyulai, Z., Balog, A., Borbenyi, Z., and Mandi, Y. 2005. Genetic polymorphisms in patients with myelodysplastic syndrome. *Acta Microbiol Immunol Hung* 52:463–475.

He, W., Dorn, D.C., Erdjument-Bromage, H., Tempst, P., Moore, M.A., and Massague, J. 2006. Hematopoiesis controlled by distinct TIF1γ and Smad4 branches of the TGFβ pathway. *Cell* 125:929–941.

Hellstrom-Lindberg, E. 1995. Efficacy of erythropoietin in the myelodysplastic syndromes: a meta-analysis of 205 patients from 17 studies. *Br J Haematol* 89:67–71.

Hellstrom-Lindberg, E., Ahlgren, T., Beguin, Y., Carlsson, M., Carneskog, J., Dahl, I.M., Dybedal, I., Grimfors, G., Kanter-Lewensohn, L., Linder, O., et al. 1998. Treatment of anemia in myelodysplastic syndromes with granulocyte colony-stimulating factor plus erythropoietin: results from a randomized phase II study and long-term follow-up of 71 patients. *Blood* 92:68–75.

Imai, Y., Kurokawa, M., Izutsu, K., Hangaishi, A., Maki, K., Ogawa, S., Chiba, S., Mitani, K., and Hirai, H. 2001. Mutations of the *Smad4* gene in acute myelogeneous leukemia and their functional implications in leukemogenesis. *Oncogene* 20:88–96.

Italian Cooperative Study Group for rHuEpo in Myelodysplastic Syndromes. 1998. A randomized double-blind placebo-controlled study with subcutaneous recombinant human erythropoietin in patients with low-risk myelodysplastic syndromes. *Br J Haematol* 103:1070–1074.

Izutsu, K., Kurokawa, M., Imai, Y., Maki, K., Mitani, K., and Hirai, H. 2001. The corepressor CtBP interacts with Evi-1 to repress transforming growth factor beta signaling. *Blood* 97:2815–2822.

Jacobsen, F.W., Stokke, T., and Jacobsen, S.E. 1995. Transforming growth factor-beta potently inhibits the viability-promoting activity of stem cell factor and other cytokines and induces apoptosis of primitive murine hematopoietic progenitor cells. *Blood* 86:2957–2966.

Jakubowiak, A., Pouponnot, C., Berguido, F., Frank, R., Mao, S., Massague, J., and Nimer, S.D. 2000. Inhibition of the transforming growth factor beta 1 signaling pathway by the AML1/ETO leukemia-associated fusion protein. *J Biol Chem* 275:40282–40287.

Johnson, G.L. and Lapadat, R. 2002. Mitogen-activated protein kinase pathways mediated by ERK, JNK, and p38 protein kinases. *Science* 298:1911–1912.

Keller, J.R., Mantel, C., Sing, G.K., Ellingsworth, L.R., Ruscetti, S.K., and Ruscetti, F.W. 1988. Transforming growth factor beta 1 selectively regulates early murine hematopoietic progenitors and inhibits the growth of IL-3-dependent myeloid leukemia cell lines. *J Exp Med* 168:737–750.

Keller, J.R., McNiece, I.K., Sill, K.T., Ellingsworth, L.R., Quesenberry, P.J., Sing, G.K., and Ruscetti, F.W. 1990. Transforming growth factor beta directly regulates primitive murine hematopoietic cell proliferation. *Blood* 75:596–602.

Kelley, L.L., Green, W.F., Hicks, G.G., Bondurant, M.C., Koury, M.J., and Ruley, H.E. 1994. Apoptosis in erythroid progenitors deprived of erythropoietin occurs during the G1 and S phases of the cell cycle without growth arrest or stabilization of wild-type p53. *Mol Cell Biol* 14:4183–4192.

Kitagawa, M., Saito, I., Kuwata, T., Yoshida, S., Yamaguchi, S., Takahashi, M., Tanizawa, T., Kamiyama, R., and Hirokawa, K. 1997. Overexpression of tumor necrosis factor (TNF)-α and interferon (IFN)-γ by bone marrow cells from patients with myelodysplastic syndromes. *Leukemia* 11:2049–2054.

Kook, H., Zeng, W., Guibin, C., Kirby, M., Young, N.S., and Maciejewski, J.P. 2001. Increased cytotoxic T cells with effector phenotype in aplastic anemia and myelodysplasia. *Exp Hematol* 29:1270–1277.

Kotlyarov, A., Neininger, A., Schubert, C., Eckert, R., Birchmeier, C., Volk, H.D., and Gaestel, M. 1999. MAPKAP kinase 2 is essential for LPS-induced TNF-α biosynthesis. *Nat Cell Biol* 1:94–97.

Krystal, G., Lam, V., Dragowska, W., Takahashi, C., Appel, J., Gontier, A., Jenkins, A., Lam, H., Quon, L., and Lansdorp, P. 1994. Transforming growth factor beta 1 is an inducer of erythroid differentiation. *J Exp Med* 180:851–860.

Kumar, S., Boehm, J., and Lee, J.C. 2003. p38 MAP kinases: key signalling molecules as therapeutic targets for inflammatory diseases. *Nat Rev Drug Discov* 2:717–726.

Kurzrock, R., Kantarjian, H., Wetzler, M., Estrov, Z., Estey, E., Troutman-Worden, K., Gutterman, J.U., and Talpaz, M. 1993. Ubiquitous expression of cytokines in diverse leukemias of lymphoid and myeloid lineage. *Exp Hematol* 21:80–85.

Larsson, J., Blank, U., Helgadottir, H., Bjornsson, J.M., Ehinger, M., Goumans, M.J., Fan, X., Leveen, P., and Karlsson, S. 2003. TGF-β signaling-deficient hematopoietic stem cells have normal self-renewal and regenerative ability in vivo despite increased proliferative capacity in vitro. *Blood* 102:3129–3135.

Lee, D.K., Kim, B.C., Brady, J.N., Jeang, K.T., and Kim, S.J. 2002. Human T-cell lymphotropic virus type 1 tax inhibits transforming growth factor-beta signaling by blocking the association of Smad proteins with Smad-binding element. *J Biol Chem* 277:33766–33775.

Lin, H.K., Bergmann, S., and Pandolfi, P.P. 2004. Cytoplasmic PML function in TGF-β signalling. *Nature* 431:205–211.

List, A., Kurtin, S., Roe, D.J., Buresh, A., Mahadevan, D., Fuchs, D., Rimsza, L., Heaton, R., Knight, R., and Zeldis, J.B. 2005. Efficacy of lenalidomide in myelodysplastic syndromes. *N Engl J Med* 352:549–557.

Liu, J.M. and Ellis, S.R. 2006. Ribosomes and marrow failure: coincidental association or molecular paradigm? *Blood* 107:4583–4588.

Marriott, J.B., Clarke, I.A., Dredge, K., Muller, G., Stirling, D., and Dalgleish, A.G. 2002. Thalidomide and its analogues have distinct and opposing effects on TNF-α and TNFR2 during co-stimulation of both CD4(+) and CD8(+) T cells. *Clin Exp Immunol* 130:75–84.

Melchert, M., Kale, V., and List, A. 2007. The role of lenalidomide in the treatment of patients with chromosome 5q deletion and other myelodysplastic syndromes. *Curr Opin Hematol* 14:123–129.

Miller, K., Kim, H.T., Greenberg, P., Van der Jagt, R., Bennett, J., Tallman, M.S., Paietta, E., Dewald, G., Houston, J.G., Thomas, M., et al. 2004. Phase III prospective randomized trial of EPO with or without G-CSF versus supportive therapy alone in the treatment of myelodysplastic syndromes (MDS): results of the ECOG-CLSG trial(E1996). *Blood* 104:70.

Millot, G.A., Svinarchuk, F., Lacout, C., Vainchenker, W., and Dumenil, D. 2001. The granulocyte colony-stimulating factor receptor supports erythroid differentiation in the absence of the erythropoietin receptor or Stat5. *Br J Haematol* 112:449–458.

Mohindru, M., Pahanish, P., Katsoulidis, E., Collins, R., Rogers, T., Navas, T., Higgins, L.S., Platanias, L.C., and Verma, A. 2004. Novel P38 MAP kinase inhibitor and anti-P38 RNA interference as potential therapeutic approaches in myelodysplastic syndromes. *Blood* 104:470.

Molldrem, J.J., Leifer, E., Bahceci, E., Saunthararajah, Y., Rivera, M., Dunbar, C., Liu, J., Nakamura, R., Young, N.S., and Barrett, A.J. 2002. Antithymocyte globulin for treatment of the bone marrow failure associated with myelodysplastic syndromes. *Ann Intern Med* 137:156–163.

Moreira, A.L., Sampaio, E.P., Zmuidzinas, A., Frindt, P., Smith, K.A., and Kaplan, G. 1993. Thalidomide exerts its inhibitory action on tumor necrosis factor alpha by enhancing mRNA degradation. *J Exp Med* 177:1675–1680.

Mori, N., Morishita, M., Tsukazaki, T., Giam, C.Z., Kumatori, A., Tanaka, Y., and Yamamoto, N. 2001. Human T-cell leukemia virus type I oncoprotein Tax represses Smad-dependent transforming growth factor beta signaling through interaction with CREB-binding protein/p300. *Blood* 97:2137–2144.

Mundle, S.D., Reza, S., Ali, A., Mativi, Y., Shetty, V., Venugopal, P., Gregory, S.A., and Raza, A. 1999. Correlation of tumor necrosis factor alpha (TNF α) with high Caspase 3-like activity in myelodysplastic syndromes. *Cancer Lett* 140:201–207.

Narendran, A., Hawkins, L.M., Ganjavi, H., Vanek, W., Gee, M.F., Barlow, J.W., Johnson, G., Malkin, D., and Freedman, M.H. 2004. Characterization of bone marrow stromal abnormalities in a patient with constitutional trisomy 8 mosaicism and myelodysplastic syndrome. *Pediatr Hematol Oncol* 21:209–221.

Navas, T., Nguyen, A.N., Ma, J., Stebbins, E.G., Haghnazari, E., Heaton, R., List, A., and Higgins, L.S. 2004. Inhibition of p38 MAPK by SCIO-469 suppresses TNF generation and promotes CD34+ cell survival in an in vitro MDS cell culture model. *Blood* 104:3424.

Navas, T.A., Mohindru, M., Estes, M., Ma, J.Y., Sokol, L., Pahanish, P., Parmar, S., Haghnazari, E., Zhou, L., Collins, R., et al. 2006. Inhibition of overactivated p38 MAPK can restore hematopoiesis in myelodysplastic syndrome progenitors. *Blood* 108:4170–4177.

Ohta, M., Greenberger, J.S., Anklesaria, P., Bassols, A., and Massague, J. 1987. Two forms of transforming growth factor-beta distinguished by multipotential haematopoietic progenitor cells. *Nature* 329:539–541.

Oshima, M., Oshima, H., and Taketo, M.M. 1996. TGF-β receptor type II deficiency results in defects of yolk sac hematopoiesis and vasculogenesis. *Dev Biol* 179:297–302.

Pellagatti, A., Jadersten, M., Forsblom, A.M., Cattan, H., Christensson, B., Emanuelsson, E.K., Merup, M., Nilsson, L., Samuelsson, J., Sander, B., et al. 2007. Lenalidomide inhibits the malignant clone and up-regulates the *SPARC* gene mapping to the commonly deleted region in 5q- syndrome patients. *Proc Natl Acad Sci USA* 104:11406–11411.

Platanias, L.C. 2003. Map kinase signaling pathways and hematologic malignancies. *Blood* 101:4667–4679.

Powers, M.P., Nishino, H., Luo, Y., Raza, A., Vanguri, A., Rice, L., Zu, Y., and Chang, C.C. 2007. Polymorphisms in TGFβ and TNFα are associated with the myelodysplastic syndrome phenotype. *Arch Pathol Lab Med* 131:1789–1793.

Raza, A., Meyer, P., Dutt, D., Zorat, F., Lisak, L., Nascimben, F., du Randt, M., Kaspar, C., Goldberg, C., Loew, J., et al. 2001. Thalidomide produces transfusion independence in longstanding refractory anemias of patients with myelodysplastic syndromes. *Blood* 98:958–965.

Raza, A., Candoni, A., Khan, U., Lisak, L., Tahir, S., Silvestri, F., Billmeier, J., Alvi, M.I., Mumtaz, M., Gezer, S., et al. 2004. Remicade as TNF suppressor in patients with myelodysplastic syndromes. *Leuk Lymphoma* 45:2099–2104.

Risitano, A.M., Kook, H., Zeng, W., Chen, G., Young, N.S., and Maciejewski, J.P. 2002. Oligoclonal and polyclonal CD4 and CD8 lymphocytes in aplastic anemia and paroxysmal nocturnal hemoglobinuria measured by V beta CDR3 spectratyping and flow cytometry. *Blood* 100:178–183.

Roboz, G.J., Giles, F.J., List, A.F., Cortes, J.E., Carlin, R., Kowalski, M., Bilic, S., Masson, E., Rosamilia, M., Schuster, M.W., et al. 2006. Phase 1 study of PTK787/ZK 222584, a small molecule tyrosine kinase receptor inhibitor, for the treatment of acute myeloid leukemia and myelodysplastic syndrome. *Leukemia* 20:952–957.

Romerio, F. and Zella, D. 2002. MEK and ERK inhibitors enhance the anti-proliferative effect of interferon-α2b. *FASEB J* 16:1680–1682.

Rosenfeld, C. and Bedell, C. 2002. Pilot study of recombinant human soluble tumor necrosis factor receptor (TNFR:Fc) in patients with low risk myelodysplastic syndrome. *Leuk Res* 26:721–724.

Sargiacomo, M., Valtieri, M., Gabbianelli, M., Pelosi, E., Testa, U., Camagna, A., and Peschle, C. 1991. Pure human hematopoietic progenitors: direct inhibitory effect of transforming growth factors-beta 1 and -beta 2. *Ann N Y Acad Sci* 628:84–91.

Saunthararajah, Y., Nakamura, R., Wesley, R., Wang, Q.J., and Barrett, A.J. 2003. A simple method to predict response to immunosuppressive therapy in patients with myelodysplastic syndrome. *Blood* 102:3025–3027.

Selleri, C., Maciejewski, J.P., Catalano, L., Ricci, P., Andretta, C., Luciano, L., and Rotoli, B. 2002. Effects of cyclosporine on hematopoietic and immune functions in patients with hypoplastic myelodysplasia: in vitro and in vivo studies. *Cancer* 95:1911–1922.

Shull, M.M., Ormsby, I., Kier, A.B., Pawlowski, S., Diebold, R.J., Yin, M., Allen, R., Sidman, C., Proetzel, G., Calvin, D., et al. 1992. Targeted disruption of the mouse transforming growth factor-beta 1 gene results in multifocal inflammatory disease. *Nature* 359:693–699.

Sitnicka, E., Ruscetti, F.W., Priestley, G.V., Wolf, N.S., and Bartelmez, S.H. 1996. Transforming growth factor beta 1 directly and reversibly inhibits the initial cell divisions of long-term repopulating hematopoietic stem cells. *Blood* 88:82–88.

Sloand, E.M., Rezvani, K., Barrett, J., Mainwaring, L., Kurlander, R., Gostick, E., Ramkissoon, S., Tang, Y., Douek, D., Price, D., et al. 2004. Myelodysplasia with Trisomy 8 is associated with a cytotoxic CD8 T cell immune response to Wilms tumor-1 protein (WT1). *Blood* 104:474.

Sloand, E.M., Mainwaring, L., Fuhrer, M., Ramkissoon, S., Risitano, A.M., Keyvanafar, K., Lu, J., Basu, A., Barrett, A.J., and Young, N.S. 2005. Preferential suppression of trisomy 8 versus normal hematopoietic cell growth by autologous lymphocytes in patients with trisomy 8 myelodysplastic syndrome. *Blood* 106:841–851.

Sokol, L., Cripe, L., Kantarjian, H., Sekeres, M., Parmar, S., Greenberg, P., Goldberg, S., Bhushan, V., Shammo, J., Hohl, R., et al. 2006. Phase I/II, Randomized, multicenter multicenter, dose, dose-ascension study of the p38 MAPK inhibitor ascension study of the p38 MAPK inhibitor Scio Scio-469 in patients with myelodysplastic syndromes (MDS). *Am Soc Hematol* 108:2657.

Sood, R., Talwar-Trikha, A., Chakrabarti, S.R., and Nucifora, G. 1999. MDS1/EVI1 enhances TGF-β1 signaling and strengthens its growth-inhibitory effect but the leukemia-associated fusion protein AML1/MDS1/EVI1, product of the t(3;21), abrogates growth-inhibition in response to TGF-β1. *Leukemia* 13:348–357.

Stasi, R. and Amadori, S. 2002. Infliximab chimaeric anti-tumour necrosis factor alpha monoclonal antibody treatment for patients with myelodysplastic syndromes. *Br J Haematol* 116:334–337.

Tai, Y.T., Li, X.F., Catley, L., Coffey, R., Breitkreutz, I., Bae, J., Song, W., Podar, K., Hideshima, T., Chauhan, D., et al. 2005. Immunomodulatory drug lenalidomide (CC-5013, IMiD3) augments anti-CD40 SGN-40-induced cytotoxicity in human multiple myeloma: clinical implications. *Cancer Res* 65:11712–11720.

Taketazu, F., Miyagawa, K., Ichijo, H., Oshimi, K., Mizoguchi, H., Hirai, H., Miyazono, K., and Takaku, F. 1992. Decreased level of transforming growth factor-beta in blood lymphocytes of patients with aplastic anemia. *Growth Factors* 6:85–90.

Tauro, S., Hepburn, M.D., Bowen, D.T., and Pippard, M.J. 2001. Assessment of stromal function, and its potential contribution to deregulation of hematopoiesis in the myelodysplastic syndromes. *Haematologica* 86:1038–1045.

Uddin, S., Lekmine, F., Sharma, N., Majchrzak, B., Mayer, I., Young, P.R., Bokoch, G.M., Fish, E.N., and Platanias, L.C. 2000. The Rac1/p38 mitogen-activated protein kinase pathway is required for interferon alpha-dependent transcriptional activation but not serine phosphorylation of Stat proteins. *J Biol Chem* 275:27634–27640.

Verhelle, D., Corral, L.G., Wong, K., Mueller, J.H., Moutouh-de Parseval, L., Jensen-Pergakes, K., Schafer, P.H., Chen, R., Glezer, E., Ferguson, G.D., et al. 2007. Lenalidomide and CC-4047 inhibit the proliferation of malignant B cells while expanding normal CD34+ progenitor cells. *Cancer Res* 67:746–755.

Verma, A. and List, A.F. 2005. Cytokine targets in the treatment of myelodysplastic syndromes. *Curr Hematol Rep* 4:429–435.

Verma, A., Deb, D.K., Sassano, A., Kambhampati, S., Wickrema, A., Uddin, S., Mohindru, M., Van Besien, K., and Platanias, L.C. 2002a. Cutting edge: activation of the p38 mitogen-activated protein kinase signaling pathway mediates cytokine-induced hemopoietic suppression in aplastic anemia. *J Immunol* 168:5984–5988.

Verma, A., Deb, D.K., Sassano, A., Uddin, S., Varga, J., Wickrema, A., and Platanias, L.C. 2002b. Activation of the p38 mitogen-activated protein kinase mediates the suppressive effects of type I interferons and transforming growth factor-beta on normal hematopoiesis. *J Biol Chem* 277:7726–7735.

Welsh, J.P., Rutherford, T.R., Flynn, J., Foukaneli, T., Gordon-Smith, E.C., and Gibson, F.M. 2004. In vitro effects of interferon-gamma and tumor necrosis factor-alpha on CD34+ bone marrow progenitor cells from aplastic anemia patients and normal donors. *Hematol J* 5:39–46.

Werner, S.L., Barken, D., and Hoffmann, A. 2005. Stimulus specificity of gene expression programs determined by temporal control of IKK activity. *Science* 309:1857–1861.

Westwood, N.B. and Mufti, G.J. 2003. Apoptosis in the myelodysplastic syndromes. *Curr Hematol Rep* 2:186–192.

Wierenga, A.T., Eggen, B.J., Kruijer, W., and Vellenga, E. 2002. Proteolytic degradation of Smad4 in extracts of AML blasts. *Leuk Res* 26:1105–1111.

Wolfraim, L.A., Fernandez, T.M., Mamura, M., Fuller, W.L., Kumar, R., Cole, D.E., Byfield, S., Felici, A., Flanders, K.C., Walz, T.M., et al. 2004. Loss of Smad3 in acute T-cell lymphoblastic leukemia. *N Engl J Med* 351:552–559.

Yoon, S.Y., Li, C.Y., Lloyd, R.V., and Tefferi, A. 2000. Bone marrow histochemical studies of fibrogenic cytokines and their receptors in myelodysplastic syndrome with myelofibrosis and related disorders. *Int J Hematol* 72:337–342.

Young, N.S. 2002. Acquired aplastic anemia. *Ann Intern Med* 136:534–546.

Young, N.S. and Maciejewski, J. 1997. The pathophysiology of acquired aplastic anemia. *N Engl J Med* 336:1365–1372.

Zhou, L.N., Pahanish, A., Hayman, P., Gundabolu, J.K., Chubak, A., Parmar, S., Garry, D., Wickrema, A., Navas, T., Higgins, L., Friedman, E., List, A., Bitzer, M., Verma, A. 2007. Inhibition of the TGF- receptor I can stimulate hematopoiesis in primary myelodysplastic syndrome progenitors as well as in TGF-driven transgenic mouse model of bone marrow failure. *Blood* 110.

Zorat, F., Shetty, V., Dutt, D., Lisak, L., Nascimben, F., Allampallam, K., Dar, S., York, A., Gezer, S., Venugopal, P., et al. 2001. The clinical and biological effects of thalidomide in patients with myelodysplastic syndromes. *Br J Haematol* 115:881–894.

Index